The Science of Sadness

The Science of Sadness

A New Understanding of Emotion

David Huron

The MIT Press
Cambridge, Massachusetts
London, England

The MIT Press would like to thank the anonymous peer reviewers who provided comments on drafts of this book. The generous work of academic experts is essential for establishing the authority and quality of our publications. We acknowledge with gratitude the contributions of these otherwise uncredited readers.

This book was set in Stone Serif and Stone Sans by Westchester Publishing Services. Printed and bound in the United States of America.

Library of Congress Cataloging-in-Publication Data

Names: Huron, David Brian, author.
Title: The science of sadness : a new understanding of emotion / David Huron.
Description: Cambridge, Massachusetts : The MIT Press, [2024] |
   Includes bibliographical references and index.
Identifiers: LCCN 2023027997 (print) | LCCN 2023027998 (ebook) |
   ISBN 9780262547772 (paperback) | ISBN 9780262378307 (epub) |
   ISBN 9780262378314 (pdf)
Subjects: LCSH: Sadness. | Emotions.
Classification: LCC BF575.S23 H87 2024 (print) | LCC BF575.S23 (ebook) |
   DDC 152.4—dc23/eng/20231128
LC record available at https://lccn.loc.gov/2023027997
LC ebook record available at https://lccn.loc.gov/2023027998

10 9 8 7 6 5 4 3 2 1

Contents

Preface

Emotions are quite possibly the most important stuff of life. Without the ability to feel pleasure or pain, heaven and hell would be indistinguishable. A person incapable of feeling emotion would seem soulless—an empty shell. A life lived without emotions would seem meaningless.

The value of pleasure seems patently obvious. Pleasure encourages us to eat sustaining foods, engage in procreative acts of love, nurture our offspring, build supportive friendships, and participate in dozens of other life-promoting behaviors. All of these make us feel good, and feeling good is, well, good.

So what good is there in feeling bad? What good is irritation, embarrassment, jealousy, or boredom? In particular, what good is there in feeling sad? Sadness is widely regarded as the very antithesis of happiness, so why would anyone ever welcome sad feelings?

Sadness comes in various forms, such as melancholy, grief, nostalgia, and depression. One of these (depression) is a genuine pathology that appears to serve no useful purpose.[1] But psychological research in recent decades has chronicled the notable benefits of the other forms of sadness. As we will see over the course of this book, apart from depression, the utility of sadness is evident in improved individual well-being as well as in valuable social interaction. Moreover, the benefits are also evident when viewed from a purely survival perspective: in appropriate contexts, feeling sad typically increases biological fitness, which is why the capacity to experience sadness evolved in the first place.

In this book, I offer a novel theory of sadness.[2] The book distinguishes three forms of sadness: the quiet reflective state (*melancholy*), the more energetic state characterized by weeping (*grief*), and the bittersweet state of reminiscing (*nostalgia*). The theory is ambitious in scope and endeavors to account for a wide range of phenomena related to sadness, including physiological, biochemical, behavioral, developmental, psychological, and sociological aspects. Grief, melancholy, and nostalgia also find different expressions in disparate cultures. We will see that the theory provides a fruitful framework for better understanding many of the rituals, practices, and cultural interpretations associated with sadness.

Perhaps surprisingly, a central character in our story is the immune system. The research suggests that stress-related emotions such as sadness are intimately connected to immune mechanisms. For example, we will learn that the characteristic features of weeping (watery eyes, nasal congestion, tightness in the throat, and staccato breathing) are elaborations of an underlying allergic response. We will also discover that the various forms of sadness are biological modifications of another distinctive feeling associated with stress—the unique experience of *feeling sick*.

The book explicitly links the emotion research to broader issues in animal behavior (ethology) and evolution. In particular, it offers an overarching account of the evolution of melancholy and grief. Evolutionary storytelling is an alluring though hazardous enterprise. The perils of evolutionary speculation notwithstanding, we will see that there is a remarkably coherent story to tell.

It is important to understand at the outset that this book focuses on typical or normal experiences of melancholy, grief, and nostalgia. To be sure, pathological forms of sadness exist that are anything but helpful. In the case of melancholy, pathologies can lead to major depressive disorders. In the case of grief, pathological forms include pseudobulbar affect (PBA), a condition that leads to uncontrollable weeping. Although from time to time our discussion will draw on research pertaining to depression, PBA, and allergies, the focus here is on normal rather than pathological behaviors. Over the course of this book, we will review evidence of how the various forms of sadness contribute directly to human well-being. However, readers should not assume that these benefits are also present in the pathological forms. In particular, readers will discover that the benefits of melancholy described in this book do not apply to depression.

It is also important to understand that the aim of this book is explanatory rather than therapeutic. These pages offer no advice about how to deal with depression, mitigate grief, or escape from a protracted bout of nostalgia.

There is no shortage of self-help books dealing with these and related topics. Instead, my aim is simply to explain sadness in its various forms. We will learn why it makes sense for a beauty pageant winner to cry and why (paradoxically) beauty pageant losers avoid crying. We will discover why nostalgia necessarily involves a bittersweet mixture of positive and negative feelings. We will answer questions about behaviors that are so familiar, few of us ever pause to consider why they occur. For example, we will learn why extended bouts of weeping make the muscles at the back of your throat sore and why crying involves fluid oozing out of your eyes rather than, say, fluid drooling out of your mouth. We will see why philosophers like John Stuart Mill, Søren Kierkegaard, and Friedrich Nietzsche were right to regard melancholy as beneficial to deep thought. There is indeed considerable merit to the phrase "sadder but wiser." We will also learn how listening to sad music might be enjoyable for many listeners. Most important, we will learn how the various puzzle pieces fit together. Sadness does not merely involve a bunch of disparate physiological, neurochemical, affective, cognitive, social, and cultural behaviors. There is an integrated story that accounts for a huge range of sundry observations and simultaneously explains the relationship between melancholy, grief, nostalgia, and (coincidentally) feeling sick.

This book does not suggest that the key to well-being is life, liberty, and the pursuit of sadness. But we will see that grief, melancholy, and nostalgia play essential roles in healthy lives. At the same time, we will review research that casts substantial doubt on the idea of "a good cry." Although popular, the idea of "catharsis" has little scientific support. What the research suggests is a far more fascinating basket of benefits that don't have anything to do with "purging negative emotions."

The value of sadness is not limited to handling the slings and arrows life throws our way. Sadness also plays a prominent role in the arts and entertainment. The most affecting moments in literature, poetry, drama, film, and music are often those that portray stressful or tragic situations that may move us to tears—tears that are commonly enjoyable. While we may properly celebrate happiness, much of the poetry of life—both figurative and literal—originates in the deepest of human sorrows.

I became interested in sadness as part of a three-decades-long effort in my lab to understand music-induced crying, melancholy, and nostalgia. How is it that people can enjoy listening to music that brings tears to their eyes, take pleasure in the glum gloom of melancholic music, or relish the bittersweet feelings of music-induced nostalgia? It should come as no surprise that a proper account of sadness in the arts must ultimately build on a general theory of sadness. This book only briefly addresses the musical phenomenon (the larger story will be left for another occasion). Instead, this volume focuses on the general theory. Although parts of this theory have been published in more technical venues, this book is intended to provide both a more comprehensive account and a more accessible presentation.[3] I have aimed to make the research readily understandable by the general reader without oversimplifying its main points.

In offering an evolutionary account, feelings of sadness will necessarily be linked to biological fitness. That is, I propose to show how melancholy, grief, and nostalgia can ultimately enhance survival and reproductive success—the hallmarks of evolutionary adaptation.[4] In making the evolutionary case, it's impossible not to be struck by the juxtaposition of the prosaic and the profound. Few human experiences match the profundity of weeping. When compared to the intensity of our subjective feelings, any story of the presumed evolutionary origins will necessarily seem shallow and sterile. An account that emphasizes the biological, cognitive, and social utility of sadness is apt to leave some readers feeling disappointed. Yet there is also magic in this story. We will see that the science of sadness highlights the depth of human social connection and the extraordinary cooperation that characterizes our species. We will also see that an evolutionarily grounded theory offers a coherent and insightful vantage point from which to appreciate the various manifestations of sadness and its sublime subtleties.

Charles Darwin recognized that there is grandeur in an evolutionary perspective of life. With each insight, it is a perspective that offers renewed feelings of wonder. In the end, it is nothing short of magical that the competing chemical patterns that lead to life ultimately provide the foundation for some of our most profound and cherished feelings. The real magic is that such deep feelings can arise from such humble origins.

Two Books

This book is really two books in one. Ostensibly, this book is about sadness. However, in endeavoring to assemble a comprehensive account of sadness, our research took us back to reconsider the most fundamental aspects of how we understand emotions in general.

As any emotion researcher can attest, emotion research has been in considerable turmoil for more than a century. Different theories have been proposed and various schools of thought have emerged. Over the past decade, the problems have become more focused, yet little consensus has emerged. Mostly, researchers have become more aware of the various flaws associated with each perspective.

Over the course of this book, we will address several classic problems that have plagued theories of emotion. One such problem is that most facial displays can arise from several different emotions. For example, people weep when experiencing grief, but tears are also common when experiencing joy, adoration, and extreme laughter. Similarly, people commonly smile when happy, but they also smile when feeling stressed, embarrassed, shy, or polite. Laughter or giggling are common in play behaviors, but either one can also arise when people are nervous or anxious, and laughter is sometimes used as a form of hostile mocking—as when we laugh *at* someone. So what is the relationship between emotions and their associated displays?

In working through the phenomenon of sadness, our research ultimately led to unanticipated insights into how to better understand emotions in general. Accordingly, in presenting our work on sadness, readers will come to learn a very different way of understanding how emotions are displayed and recognized. Said another way, instead of viewing this book as a book about sadness, this volume may be regarded as presenting a novel theory of emotional displays with sadness as an extended case study. Readers will find that the theory advocated in this book offers a fresh perspective that provides plausible solutions to problems that have long confounded emotion research.

Elements of the general theory will emerge throughout the early chapters as we explore the phenomenon of sadness. Instead of presenting the theory at the outset, it helps to build a detailed understanding of sadness so that the reader can relate specific theoretical points to concrete examples. For readers interested in emotion in general, the detailed account of sadness should help to better illustrate the strengths of the overarching theory.

Organization

This book is organized into twenty chapters. Chapter 1 offers detailed descriptions distinguishing the two most basic forms of sadness—melancholy and grief. The descriptions involve several different perspectives, including biological, affective, cognitive, developmental, behavioral, social, and cultural.

The distinction between melancholy and grief begs the question: Why do humans have two main sadness-related emotions rather than just one? Chapter 2 addresses this puzzle by reviewing a classic distinction in animal behavior between *signals* and *cues*. Following up on the signal/cue distinction, chapter 3 notes that grief has all the hallmarks of an ethological signal, whereas melancholy conforms to an ethological cue.

Chapter 4 considers the function of weeping. Weeping commonly induces feelings of pity or compassion in spectators, and in many cases (though not all), we feel compelled to help the weeping individual. The chapter identifies six biological (ultimate) and six psychological (proximal) motivations that encourage altruistic or prosocial behaviors.

Chapter 5 addresses the function of melancholy. Melancholy (as opposed to depression) confers a wealth of cognitive benefits. Compared with a happy or neutral mood, melancholy promotes more detail-oriented thinking, reduced stereotyping, less judgment bias, greater memory accuracy, reduced gullibility, more patience, greater task perseverance, more social attentiveness and politeness, more accurate assessments of the emotional states of others, and improved reasoning related to social risks. In stressful situations, melancholy provides important mental tools for realistic strategizing (a phenomenon we call "melancholic realism"). We will also review research distinguishing (normal) melancholy from (pathological) depression and see that depression fails to deliver the cognitive goodies associated with melancholy.

Chapter 6 reviews the evidence suggesting that weeping behavior warrants the rare designation "innate." It also introduces some of the cognitive mechanisms through which we endeavor to control this apparently instinctive behavior.

As we have noted, tears are shed in many circumstances apart from sadness—such as tears of joy, pain, adoration, loyalty, and laughter. Chapter 7 identifies the commonality that underlies all these different situations. We will see that ethological signaling theory provides a strikingly parsimonious account of these otherwise disparate and sometimes seemingly contradictory weeping functions.

Chapter 8 addresses the general question of how we infer the emotional states of others. Can we distinguish tears of grief from tears of joy?

Chapter 9 draws attention to the important role of the immune system. The chapter begins with a discussion of "feeling sick" and the benefits conferred by fever, headache, muscular ache, and lethargy. We then consider some of the neurochemistry of melancholy and grief, highlighting in particular the role of a class of compounds known as proinflammatory cytokines.

Chapters 10 and 11 present a comprehensive story tracing the evolution of melancholy and grief. The chapters offer a detailed causal chronology for how melancholy emerged from feeling sick and how grief emerged from melancholy. All evolutionary accounts are necessarily speculative. However, we will see that our evolutionary narrative explains a remarkable range of disparate observations. We will see that the neurotransmitter histamine is implicated as playing a prominent role in our evolutionary story.

Chapter 12 addresses infant crying. Why do infants and children cry so much? How can a broken crayon evoke an expression of despair resembling the death of a spouse? The chapter provides an explanation for why children have a comparatively low threshold for the advent of crying. We will distinguish two forms of infant crying (bawling and weeping) and see that the two forms serve different social and biological purposes.

Chapter 13 considers how pleasant past memories might lead to a distinctive form of sadness—nostalgia. Nostalgia supplements the negative feelings of grief or melancholy with a distinctly positive feeling—accounting for its mixed "bittersweet" character. The chapter proposes an evolutionary account identifying the function and value of nostalgia. That is, I describe how, far from being unhelpful self-indulgent wallowing, bouts of nostalgia commonly lead to objective improvements in people's lives.

Moving beyond the individual experience of sadness, chapters 14 and 15 consider sadness as a cultural phenomenon. Chapter 14 explores some of the rich and varied forms of crying found in different cultures. For example, we consider common cultural expressions such as wearing distinctive dress when in mourning and mourning-related acts of self-injury. In addition, we also consider more unique cultural expressions such as the anthropological curiosity known as the *Welcome of Tears*. The ensuing chapter—chapter 15—focuses on broader issues of how cultural symbol systems recruit and repurpose innate signals and cues.

Chapter 16 expands our discussion beyond sadness to emotions in general. Based on lessons learned from our work on sadness, chapter 16 offers a summary of the overarching theory—a theory I call the *signal theory of emotion-associated displays* (STEAD). A more detailed exposition of the theory is provided in the book's appendix.

Chapter 17 formally presents my triadic theory of immunological responses to stress (TTIRS, pronounced "tears")—reviewing and summarizing the main ideas concerning sadness. There are many forms of stress, including physical stresses (such as injury or infection), cognitive stresses (such as confusion or boredom), and social stresses (such as loneliness or shame). In response to these stressors, individuals can draw on three classes of resources: *corporeal* resources (for example, where feeling sick enhances immune effectiveness, including tissue repair), *cognitive* resources (where melancholy leads to strategic reflection that benefits from more realistic thought), and *social* resources (where grief-induced weeping solicits the compassionate assistance of others). The chapter also recaps and summarizes the elements of my speculative evolutionary story, suggesting how melancholy and grief arose as elaborations of existing immune responses to stress.

Two ensuing chapters, chapters 18 and 19, address the experience of sadness in the arts and entertainment. How is it that spectators can enjoy portrayals of ostensibly dire or woeful circumstances? How can sad instrumental music move listeners to tears—tears that are paradoxically often enjoyable?

The final chapter, chapter 20, returns to consider the role of culture. Most human emotions existed long before we had language, so emotions were not experiences for which we had "labels" or functional narratives. Mental labels or descriptions of our feelings aren't necessary in order for the emotions to motivate behaviors. Nevertheless, the stories we tell about emotions—such as sadness—shape our phenomenal experiences. At the same time, we struggle to describe experiences whose prelinguistic evolutionary history renders them truly ineffable.

A final postscript addresses why, historically, emotion theorizing has been so vexing. To help readers with technical terminology, a glossary is provided. But first, we offer a brief digression that addresses the question, What is an emotion?

Preamble: What Is an Emotion?

What is an emotion? American psychologist William James posed that question more than a century ago. Proposed answers to James's question remain contentious. Forty years ago, Anne and Paul Kleinginna at Georgia Southern College assembled a list of definitions of "emotion," which they had gleaned from the published research literature.[1] Their list included no less than ninety-two different definitions. Four decades later, the number of definitions has only proliferated.

Emotion researchers don't merely disagree about the fine points of a definition; they hold fundamentally different conceptions regarding emotions. Major emotion theorists such as Paul Ekman, Lisa Feldman Barrett, and Alan Fridlund hold such fundamentally different views that it's hard to believe that they are all intended to explain the same phenomena. In his 2010 review article on the state of emotion research, Stanford professor James Gross began by warning readers, "I do not recommend reading this . . . if you're faint of heart."[2]

There exist an awful lot of theories of emotion. Even efforts to arrange or group the various theories into broad categories present a formidable challenge. In her recent survey of the field, Flemish scholar Agnes Moors characterized emotion theories according to ten dimensions and identified eleven psychological and philosophical theory families.[3] We won't make any effort here to provide a comprehensive review of the various emotion theories. It's nevertheless worthwhile highlighting some of the problems associated with different approaches—even though our descriptions will be little more than caricatures.

Among psychologists, a traditional taxonomy groups emotion theories into four broad schools: evolutionary theories, basic emotions theories, cognitive theories, and social constructivist theories. These perspectives are not necessarily mutually exclusive and most emotion researchers hold a nuanced mix of views. Broadly speaking, each school defines an emotion in ways that reinforce what researchers think is most important.

Evolutionary emotion theorists are right that many feeling states are adaptive traits that have plausible evolutionary origins. States such as fear, aggression, hunger, and itchiness are readily observed in virtually all nonhuman animals—suggesting a shared biological history. However, evolutionary theories have trouble with states like awe and hope, for which there seems to be no compelling evolutionary story.

Partly in light of this difficulty, basic emotions theories propose that there exists a handful of foundational emotions from which all other emotions are derivatives or mixtures. Basic emotions theories suffer from a number of problems,[4] one of which is the proverbial embarrassment of riches. As can be seen in table P.1, there is no agreement on the number or identity of basic emotions.

Cognitive emotion theorists propose that, in any given situation, mental assessment precedes and dictates the induced feeling. Cognitive emotion theorists are right to draw attention to the role of cognitive appraisal and

Table P.1
Lists of "basic emotions" proposed by various emotion researchers

Proposed basic emotions	Authors
anger, disgust, fear, joy, sadness, surprise	Ekman, Friesen, and Ellsworth
desire, happiness, interest, surprise, wonder, sorrow	Frijda
rage/terror, anxiety, joy	Gray
anger, contempt, disgust, distress, fear, guilt, interest, joy, shame, surprise	Izard
fear, grief, love, rage	James
anger, disgust, elation, fear, subjection, tender-emotion, wonder	McDougall
pain, pleasure	Mowrer
anger, disgust, anxiety, happiness, sadness	Oatley and Johnson-Laird
expectancy, fear, rage, panic	Panksepp
acceptance, anger, anticipation, disgust, joy, fear, sadness, surprise	Plutchik

Source: Modified from Ortony and Turner (1990).

labeling. Some emotions can be induced by thought alone. (*Did you remember your mother's recent birthday?* The very thought can, for some people, produce a moment of panic.) However, the emphasis on appraisal necessarily forces cognitive emotion theorists to claim that feeling states with minimal or no cognitive involvement—like smell-induced disgust, thirst, surprise, or pain—cannot be emotions.

Social constructivists posit that emotions do not have an independent existence outside of a shared language community. Social constructivists are right that some feeling states are socially defined. Examples might include the Portuguese concept of *saudade* or the Tahitian concept of *haumani* (more later). Moreover, historians of emotion have chronicled how various feeling states have been interpreted differently over the course of history—even within a single culture. (We'll see an example of this in chapter 20 when we discuss how Western conceptions of "sadness" have changed since the sixteenth century.) However, social constructivists have trouble explaining why so many seemingly identical feeling states can be observed in completely isolated cultures. For example, in every known culture, one can find instances of fear, feeling cold, grief, and maternal love.

It helps, I propose, to focus on the consequences of emotions. A notable observation is that feeling states often lead to particular behaviors—either immediately, or after some period of time. When a person feels "anger," the individual is more likely to strike out or attack; when a person feels "fear," hiding or fleeing is common. Dutch psychologist Nico Frijda usefully characterized emotions as *action tendencies*, while the American psychologist Silvan Tomkins characterized emotions as *motivational amplifiers*. The simple point is that certain behaviors are more likely to occur in the presence of a given emotion than in its absence.

If one accepts a functional conception of emotions, then the list of candidate emotions becomes very large. It includes all the usual suspects such as anger, fear, joy, disgust, affection, grief, and embarrassment. But it also includes feelings of hunger, thirst, satiation, pain, boredom, disappointment, feeling sick, orgasm, craving salt, and the urgent need to urinate.

For many readers, this list will seem far too eclectic. We wonder what the common denominator is for feelings like itchiness, boredom, love, and feeling cold. For those emotions that have an evolutionary origin, it's important to remember that evolution does not follow some master design plan. Useful traits are cobbled together using whatever resources are at hand. Some feeling states (like feeling cold) involve their own special-purpose system of sensory neurons. Others, like satiation, are known to be linked to specific hormones. An emotion like nostalgia is intimately linked to memory. Feelings like fear are known to be associated with particular brain circuits. Emotions like embarrassment or jealousy can be evoked only through social interaction. And an emotion like *saudade* requires that you be raised speaking Portuguese.

What these feeling states share in common is their capacity to provoke functionally useful behaviors accompanied by vivid phenomenological experiences. Of course, some emotions are likely to share resources, and some emotions might piggyback on other emotions. But we should not be surprised that the category "emotion" turns out to be a menagerie of diverse states.

Expanding the Basket of Emotions

When we think about emotions, we tend to focus on major "whiz-bang" feeling states like fear, anger, disgust, or grief. However, dozens of much more subdued feeling states shape our behaviors in ways that typically escape our attention.[5] It's helpful to consider a couple of examples.

Consider the feeling of itchiness. Astronauts commonly wear helmets for extended periods of time—a situation that prevents them from touching their faces. Imagine the potential agony of being unable to scratch an itchy nose for four or five hours. Astronauts' helmets have a stick at their base with a small piece of Velcro at the end, allowing the astronaut to reach the source of a tormenting itch. Incidentally, although there are pathological forms of compulsive scratching, scratching is an important adaptive behavior. Scratching is an effective way of dislodging the nits, lice, and fleas that commonly infest skin (afflictions that are happily less common for modern humans). One shouldn't doubt that the feeling of itchiness can be phenomenologically compelling. Nor should one doubt that scratching is a useful functional behavior.

Although quite rare, some people are born with no ability to experience pain. At first, you might suppose this would be a fantastic advantage. Imagine being able to go through life without having to suffer any of the stings of various bodily insults. Alas, people born without pain sensors are typically dead by early adulthood.[6] One might suppose that death comes from accidents caused by excessive risk-taking or the dangerous inability to feel, for example, the pain of an inflamed appendix. Instead, death more commonly is caused by a much more subtle phenomenon. When sitting or standing, we periodically change our posture: shifting some of our weight from one leg to the other, leaning forward or back. The motivation for these postural changes comes from slight feelings of discomfort from pinched nerves or squeezed blood vessels. Without these feelings, we might simply sit or stand in a single static position for hours without varying our posture. People born without pain sensors commonly die as a consequence of joint failure and major circulatory problems arising because they fail to experience the appropriate feelings of irritation that encourage periodic changes of posture.[7]

Itchiness and postural discomfort offer complementary lessons regarding feeling states. Failing to scratch an itch would seem to have little consequence—after all, ignoring an itch is not going to kill you. Nevertheless, not being able to scratch an itch can easily lead to feelings that some might describe as a form of torture.[8] Conversely, the feeling of postural discomfort may seem inconsequential. Yet, despite the barely noticeable feelings, failing to attend to postural discomfort can indeed ultimately lead to your untimely death.

Feeling of itchiness or postural irritation represent just two of many seemingly trivial feeling states. Other examples include the annoyance we experience when traffic noise interferes with our ability to hear our conversation partner or the irritation we feel when our view is obstructed. Nor are these subtle feelings limited to sensation or perception. There are similar subtle feelings that are purely cognitive. For example, we experience a form of mental irritation when we have difficulty retrieving a word from memory, and we also experience a moment of mental pleasure or relief when we successfully recall that forgotten word.

In my main field of research (musical emotions), listeners commonly report that the main attraction of music listening is that they like how music makes them feel. Whether warranted or not, music has often been poetically described as "the language of emotions." When listening to music, it's possible to experience feelings of fear, anger, disgust, humor, grief, loneliness, pride, and so on. But these whiz-bang feelings aren't common. Music can evoke a rich emotional experience, but in most listening experiences, our emotional responses are far more subtle. The scientific research on music-evoked emotions suggests that our listening experiences are dominated by what might be called micro-emotions.[9] These feelings can be richly reward-ing, even though we don't have names for them and are oblivious to their existence.

It is my research experience with music that makes me wary of much mainstream emotion theorizing. First, focusing on a handful of major emo-tions is unduly restrictive. Second, many of our emotions (feeling states, if you wish) are opaque to us. We experience them—our behaviors are shaped by them—but we are rarely aware of their existence. While I appreciate many insights offered by existing theories, in my view, none of the established per-spectives offer more than a very partial account of a remarkably rich domain of human experience.

In recent decades, cognitive-appraisal theories have garnered a majority following among emotion researchers.[10] As we will see in this book, cog-nition does indeed interact with feeling states in complex and important ways. Many emotions—such as pride and hope—are wholly cognitive in origin. Moreover, as we will see, the overarching purpose of some feeling states is to enhance cognitive performance. However, not all feeling states have a cognitive component. As already noted, smell-induced disgust can arise without the slightest involvement of thought or cognitive assessment. An evocative chord progression in music can lead listeners to experience a powerful emotional response without any thoughtful engagement.[11]

The effect of many definitions of emotion is primarily one of excluding from discussion the full range of human feeling states. In the case of cogni-tive theories, for example, it is problematic to claim that only feelings involv-ing cognitive assessment are proper "emotions" and then suggest that an essential feature of emotions is that they depend on some form of cogni-tive assessment. It's prudent, I propose, to be wary of definitions of emotion, since many definitions seem to exist as a way of circling the wagons around a pet idea.[12]

My own view is that *all* of our feeling states are deserving of sustained research that endeavors to explain how they function and why they exist—whether those feelings have innate or cultural origins. Accordingly, this book assumes a very broad view of emotions. In attempting to understand human experience, I would suggest that it's helpful to begin by casting a wide net. Our ultimate aim is to understand feeling states in their full richness and variety.

The purpose of this preamble is simply to warn readers that the view of emotions presented in this book will be considered unorthodox by many of my professional colleagues. Most emotion researchers will regard my viewpoint as excessively broad. Some will propose that I'm really talking about "drives" or "valenced sensations" rather than emotions. Since readers will have their own preconceptions of what constitutes a valid "emotion," I will frequently resort to the word "affect" and the phrase "feeling state." Unfortunately, for many researchers, terms like *emotion*, *feeling state*, and *affect* have specific technical meanings. This makes it difficult to talk broadly without using words in ways that violate one or another existing technical convention. In using these different terms, my goal here is not to imply some sort of taxonomy of feelings. Instead, my aim is simply to avoid shocking readers who may otherwise resist the idea, for example, that *feeling sick* might be usefully regarded as an emotion.

1 Sadness Observed

The English word "sadness" is commonly used in an ambiguous way. On one hand, the word "sadness" connotes an unhappy, dejected, depressed, glum, blue, forlorn, or despondent state. On the other hand, "sadness" is frequently used more broadly to include an emotional state characterized by grief, crying, weeping, sniffling, sobbing, or wailing. In his classic 1872 book, *The Expression of Emotions in Man and Animals*, Charles Darwin had already described how these two states differ. Darwin used the word "sorrow" to refer to the "languid" and "resigned" state, whereas grief is "frantic" and "energetic." The difference between these two states is readily observed among children: when unhappy, a child may engage in sustained crying or quiet morose sadness. In much modern emotion research, Darwin's distinction between sorrow and grief is regrettably ignored, and "sadness" is often treated as a single basic emotion.[1]

The word "grief" can be similarly ambiguous. Several emotion scholars reserve "grief" to refer solely to the experience of bereavement in response to the death of a loved one. In this usage, one cannot grieve over a failed relationship or a home destroyed by fire. Moreover, since bereavement and mourning can extend over months or even years, grief has sometimes been defined as a process entailing many emotions that unfold over time. Consequently, some emotion researchers regard grief as not itself an emotion.[2]

In light of the potential for confusion, it's useful to define in advance some terminology. In this book, we will use the word "grief" to refer to the more aroused distressed state associated with weeping, whereas the word "melancholy" will be used exclusively to refer to the low-arousal state associated with blue or forlorn feelings. At this point, we can offer a simple rule of thumb to distinguish between these two states—namely, whether the facial and neck muscles are tense or relaxed.[3] If the facial muscles are flexed (e.g., furrowed brow, downturned mouth, or choked-up feeling in the throat), the state is likely symptomatic of *grief*; if the facial muscles are relaxed, it's more likely one of *melancholy*. It is also helpful to have an inclusive or umbrella term, so throughout the remainder of this book, we will reserve the word "sadness" to refer to the general category that includes both melancholy and grief. Later, we'll add nostalgia.

An appropriate place to begin is with detailed descriptions of melancholy and grief. When describing any emotional state, an integrated description may include at least eight different perspectives. We might start by identifying the causes that typically trigger or induce the emotion (what scientists call the *etiology*). A *physiological* description focuses on neurological, endocrine, and general biological aspects. A *developmental* perspective addresses the emergence and possible changes in how the emotion is manifested over a person's lifetime. *Behavioral* description stresses the characteristic actions, postures, vocalizations, or facial displays associated with the emotion. *Cognitive* description draws attention to patterns of thought that are typically associated with a given emotional state. A *phenomenological* description aims to characterize the subjective feelings that a person experiences: What do melancholy and grief *feel* like? *Social* description concentrates on how the emotion is communicated and interpreted by other individuals; a social description will also describe the subjective feelings and ensuing behaviors commonly induced in a social observer. Finally, an *anthropological* description addresses the cultural manifestations of the emotion, including culturally unique expressions, inhibitions, embellishments, and ritual customs.

In this chapter, we will go into considerable detail describing melancholy and grief. In fact, many readers may feel the level of detail to be excessive. As the author, I'm aware that it's a bad idea to begin with an opening chapter that is so dry. However, we will later discover that certain details amount to "smoking guns" that offer revealing clues regarding the origin and function of melancholy and grief. For those readers who can tolerate all the detail, the reward will be an explanatory theory that makes sense of a wide range of eclectic observations. Ultimately, we will introduce an explanatory theory that provides a plausible narrative at each of our eight descriptive levels.

Melancholy

Recall that melancholy is the more passive sadness state associated with feeling glum, blue, or forlorn. So what causes someone to experience melancholy?

Melancholy might be expected in response to failures to achieve life goals, such as romantic, parental, or occupational goals. Melancholy might be caused by the departure of a loved one, the breakup of a romantic relationship, parenting problems, loneliness, unemployment, financial difficulties, poor health, loss of social status, frustration in the pursuit of goals, or the inability to help others. In less affluent environments, melancholy can also arise from more basic vulnerabilities, such as from hunger, thirst, cold, injury, illness, insecurity, or chronic fear. In general, melancholy is associated with failure or powerlessness. In affluent societies, melancholy is most likely to arise from social stressors—giving the misleading impression that melancholy must be exclusively social in origin.[4]

The main physiological symptom of melancholy is low arousal or *anergia*. When we experience a bout of melancholy, our heart rate decreases, and our respiration slows and becomes shallower. Melancholy is linked to several biochemical changes. Foremost among these changes are reduced levels of four important neurotransmitters: epinephrine, norepinephrine, acetylcholine, and serotonin. Each of these neurotransmitters produces a constellation of physiological effects. *Epinephrine* is an energizing hormone related to motor movement. Low levels are associated with reduced activity. Low *acetylcholine* is associated with poor muscle tone and slow muscle reactivity; this makes us feel slow, weak, lethargic, and sleepy.[5] *Norepinephrine* is linked to attention. Low norepinephrine is associated with poor engagement with the world; we tend to daydream more and pay less attention to our immediate environment.[6] *Serotonin* is linked to self-esteem. When serotonin levels are low, we tend not to stick up for ourselves: we are less likely to challenge unreasonable demands and more likely to accept our fate.[7] In addition to reduced levels of epinephrine, acetylcholine, norepinephrine, and serotonin, melancholy is associated with increased levels of cortisol and prolactin—hormones that are commonly released when we experience stress.

The pathological version of melancholy is *depression*. In treating depression through pharmacological means, two approaches have proved to be moderately helpful. One approach aims to increase levels of norepinephrine—with the effect of increasing our engagement with the world. We become less withdrawn and more attentive to the events and people around us. A second approach aims to increase levels of serotonin—with the effect of raising our self-esteem.[8] The result is that we become more optimistic and proactive.

Behaviorally, melancholy commonly leads to reduced activity, slow movement, slumped posture, infrequent speech, weak voice, disrupted sleep,

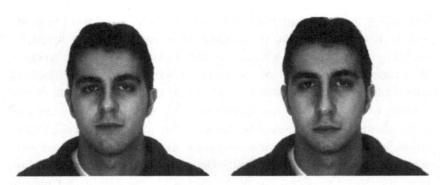

Figure 1.1
Comparison of two photographs in which the distance between the eyebrows and lower lip has been stretched in the righthand image. The "long" face is judged by viewers to be sadder in appearance. Melancholy is linked to low physiological arousal; the resulting muscle relaxation accounts for the lowered chin and flattened cheeks. By permission of the *Journal of Vision*.

changes in appetite, and social withdrawal.[9] When in a state of melancholy, people prefer an event that is characterized as passive or inactive as opposed to the same event when it is portrayed as active or energetic.[10] Further behavioral characteristics include some distinctive facial features. The relaxation of the facial musculature is associated with a lowered head and drooping eyelids. Relaxing the jaw causes the chin to drop downward. Relaxed zygomatic major muscles (which are flexed when smiling) cause the cheeks to flatten—reducing the physical width of the face. Together, the lowered chin and flattened cheeks contribute to the "long face" appearance—a description synonymous in several languages with being sad. This effect is illustrated in the photographs shown in figure 1.1 from the work of Donald Neth and Aleix Martinez.[11] The distance between the eyebrows and mouth has been stretched in the righthand photo. Neth and Martinez showed that viewers judge the stretched faces to be "sadder" in appearance. Melancholy results in a more relaxed expression ("blank face").

Apart from facial features, melancholy is also associated with characteristic changes in speech patterns. People who are melancholic or depressed speak with a quieter voice, more slowly, and with a lower overall pitch; the voice is more monotone (that is, pitch movements are more constrained); and the speech tends to exhibit more mumbled articulation. In addition, melancholic speech tends to be breathier and exhibits a darker timbre or tone color.[12] Overall, you might think of melancholic vocalizing as "lazy speech" or "low-energy speech." However, the single most important observation regarding melancholic speech is that melancholic individuals tend to remain *mute*. When we are sad, we simply speak less.[13]

(At this point, we might note that all of the foregoing observations related to the characteristic facial and vocal features of melancholy will prove to be surprisingly informative in later discussions.)

Cognitively, the principal mental activity when experiencing melancholy is *reflection*. Melancholic individuals tend to become lost in thought, contemplating or brooding on their situation.[14] Such reflections can include replaying events leading to one's current unhappiness and thoughts related to possible future courses of action. We'll have much more to say about the nature of melancholic though in chapter 5, where we will note important cognitive differences that distinguish ordinary melancholy from its pathological form, depression.

Phenomenologically, the principal subjective experience associated with melancholy is *malaise*: a glum, forlorn, despondent, gloomy, morose, or cheerless feeling. The world seems to lose its charm and instead appears drab, wearisome, or soulless. Melancholy is accompanied by a mild form of psychic pain.[15] However, it's noteworthy that the pain of melancholy (as opposed to depression) has none of the discomfort one feels from, say, stubbing your toe or the ache of a problematic tooth. More important, melancholy has a significant impact on other experiences that would normally be pleasurable. Many otherwise enjoyable activities lose their allure, including food, sex, play, and socializing. Consequently, another symptom of melancholy is *anhedonia*— the loss of pleasure. Melancholy is less *unpleasant* than it is *unpleasurable*.

Socially, melancholic sadness is associated with social withdrawal.[16] When experiencing melancholy, people tend to avoid eye contact with others, engage in much less conversation, and seek privacy. Both physical and conversational play lose their appeal, and melancholic individuals tend to resist efforts to cheer them up.

Unlike grief and mourning, melancholic sadness has attracted relatively little attention by cultural anthropologists. Something akin to melancholy has received anthropological description in only a handful of cultures. For example, among the indigenous Nahua of southeastern Mexico, *amantli* is a state that commonly ensues after being separated from a loved one or when experiencing other forms of distress such as bullying or domestic violence.

Amantli is associated with "preoccupation" of thought and linked to a decline of appetite and a loss of energy. The reduced motivation is captured in the local poetic characterization, "My heart can't do what it wants."[17] Despite the similarities, the attributed source or origin of melancholic-like symptoms can differ significantly. For example, in a study in Burundi, South Sudan, and the Democratic Republic of the Congo, sources of mental states resembling melancholy or depression are variously attributed to natural, psychosocial, or supernatural causes.[18] Cultural differences are often echoed in language. Some languages provide emotion terms for which English offers no equivalent, and in other cases, English distinguishes emotions that are not distinguished in other languages. In the Tahitian language, for example, no distinction is made between "fatigued" and "sad" (melancholy or depression)—both are indicated using the same word, *haumani*.[19] In marked contrast to grief (which displays a rich variety of cultural manifestations), melancholy is associated with few cultural expressions—consistent with the disposition toward social withdrawal.

Grief

As in our discussion of melancholy, grief can be described from several different perspectives, including etiological, physiological, developmental, behavioral, cognitive, phenomenological, social, and anthropological.

The etiologies or causes of grief are similar to those for melancholy. Like melancholy, grief accompanies failures to achieve life goals, including romantic, social, or occupational goals. As with melancholy, grief may arise due to extreme hunger, cold, poor health, insecurity, and so on. Compared with melancholy, grief more commonly accompanies the loss of already existing resources, including the reversal of current fortunes or highly anticipated outcomes failing to transpire. Interestingly, grief is more likely than melancholy to occur in response to feelings of guilt or shame.[20] However, in cultures around the globe, grief is most reliably induced by the departure or death of a loved one.[21] Both grief and melancholy are symptoms of adversity, failure, vulnerability, or powerlessness. Although grief and melancholy are caused by similar circumstances, they differ principally in the magnitude of the loss or failure: grief is more likely to be associated with especially onerous failures, losses, or stresses.

From a physiological perspective, grief (in contrast to melancholy) is associated with an increase (rather than decrease) in arousal. Heart rate and blood pressure increase and breathing becomes deeper and more erratic. However, the most characteristic symptom of grief is *weeping*. In its full-blown expression, weeping entails a flushed face, nasal congestion, constriction of the pharynx, staccato exhaling, vocalized wailing, and the shedding of tears (*lacrimation*). The constriction or tightening of the pharynx is technically referred to as the *globus sensation* but more simply described as either "a lump in one's throat" or feeling "choked up." Grief is highly stressful, and this is reflected in elevated levels of cortisol, prolactin, and adrenocorticotropic hormone.[22]

As noted earlier, grief is associated with flexed facial muscles.[23] The muscles of the forehead are flexed in a way that produces a ruffled effect (furrowed brow). At the same time, the inner corners of the eyebrows tend to rise upward, producing one or more vertical creases between the eyebrows. The cheeks are flexed, producing a sort of squinting effect around the eyes. Finally, the corners of the mouth are drawn sharply back and downward.[24]

Of all the visible aspects of weeping, however, the archetypal feature is the shedding of tears. Physiologists distinguish three types of tears: basal, reflex, and psychic.[25] *Basal tears* serve to lubricate the eyeballs and are constantly being secreted. *Reflex tears* are generated in response to irritation, such as when chopping onions. *Psychic tears* are associated with strong emotional states.[26] Apart from grief, psychic tears can also be shed in association with other emotional experiences, as in tears of joy.[27] We will have much more to say about nongrief tears in chapter 7.

Regarding the development of grief, crying undergoes a notable shift between infancy and adulthood. Among adults, prototypical *weeping* involves staccato breathing or vocalized punctuated exhaling. These vocalized "puffs" ("ah-ah-ah-ah . . .") are something weeping shares with laughter (usually transcribed in English as "ha-ha-ha-ha . . ."). In the case of weeping, we will refer to these cough-like convulsions as *sobbing*. Among babies, by contrast, crying consists of a series of long vocalized exhales: *waaaah, waaaah, waaaah . . .* , what we will refer to as *bawling*. For infants, each cry coincides with a full exhale with no vocalization made during inhalation. Notably absent from infant crying is the multiple vocalized "puffs" per exhale evident in the weeping of older children and adults. By twelve months of age, infant crying appears to shift from predominantly *bawling* to the staccato, vocalized puffs characteristic of *weeping*. For adults, bawling virtually disappears as a behavior, leaving only the weeping form. It is possible that the development of punctuated exhaling is delayed due to immature pulmonary (lung) development. For example, Robert Provine has suggested that infant lungs may not yet be able to handle the sophisticated coordination involved in staccato exhaling.[28]

Perhaps the most important developmental aspect of crying is the dramatic reduction in the amount of crying that occurs with age. Babies cry a lot, children less so, and adults rarely. The amount of weeping appears to be related to how much a person depends on others. As an individual becomes more autonomous—able to fend for themselves—the frequency of weeping is reduced. Crying tends to increase among the elderly, but here too, it may be related to feelings of dependency or vulnerability.[29]

Related to the general reduction of crying with age is the raised stress threshold for inducing grief. Children have a comparatively low threshold for the advent of crying: the simplest obstacle can evoke an outburst of utter despair. We'll consider infant and child crying in more detail in chapter 12.

In terms of observable behavior, weeping is easy to recognize. Foremost features include the prototypical appearance of tears and the hallmark facial characteristics just described. However, there are additional visual traits. When a person weeps for an extended period of time, the face tends to become red and puffy with inflammation common around the eyes, including redness of the eyes themselves.[30] Aside from the face, grief-related behaviors may include outstretched arms, covering the face with one's hands, or other gestures.[31]

Apart from the visible behaviors, weeping is also associated with characteristic sounds. We have already noted the distinctive staccato "ah-ah-ah-ah" sound. In addition, weeping can lead to sustained high-pitched wailing. The common experience of weeping-induced nasal congestion also affects the sound of the voice, leading to a muffled or veiled timbre. In addition, a dripping nose encourages rapid inhalation—producing a telltale sniffling sound. In general, grief-induced vocalizing can vary from no sound at all to quiet whimpering, moaning, moderate crying, loud wailing, or heartrending shrieks.

As already mentioned, weeping involves a general tendency to contract the muscles of the face and neck.[32] These contractions lead to a host of acoustical changes. Slight tension in the vocal cords (formally vocal folds) can cause small, rapid pitch fluctuations that are heard as a wavering or trembling voice. Muscle contractions draw the arytenoid cartilages together, which can result in what linguists call laryngealization or vocal fry—a sound best described as "creaky voice." When the edges of the vocal folds themselves are tensed, the result is the high-pitched "falsetto" voice. Strong contractions in the region of the vocal folds also account for the instability between normal and falsetto phonation that is responsible for a cracking or "breaking" voice.[33] (It is the same instability that causes the breaking voice experienced by pubescent boys—and it is the resemblance to crying that is the main reason why boys find the experience so acutely embarrassing.) Constriction of the pharynx leads to a tense- or strained-sounding resonance in the voice, an effect linguists refer to as "pharyngealization." In addition, constriction of the pharynx narrows the air passage so that when inhaling, the weeping person produces a gasping sound. Moreover, it's common for the gasping to be accompanied by inhaled vocalization, a phenomenon known as "ingressive phonation." When weeping is sustained over time, the intense muscle contractions commonly lead to a feeling of soreness at the back of the throat. In summary, muscle contractions in the face and neck produce a number of distinctive vocal changes, including trembling voice, creaky voice, falsetto voice, pharyngealization, gasping, ingressive phonation, and breaking or cracking voice.

It is possible to experience degrees of grief, as indexed by the number and severity of weeping symptoms. For example, a person might simply feel a tightening in the throat ("choked up") without any further symptoms. One might experience "incipient tears"—where there is a feeling that tears are pending. The eyes may appear moist without any teardrops evident, or tears might begin to well up along the lower eyelids without any actually dropping onto the cheeks. The corners of the mouth might turn slightly downward without the mouth opening. In particular, the grief-stricken individual may produce no sound.

These more subtle physiological changes are commonly not evident to observers. For the purposes of this book, we will distinguish between the conspicuous and inconspicuous manifestations of grief. The conspicuous "wailing and waterworks" display we will refer to as *weeping*. The inconspicuous phenomenon we will refer to as *choking up*. Both are symptomatic of grief, although they are also common in other emotions, often characterized as "feeling touched" or "being moved" (discussed in chapter 7). Both weeping and choking up entail the same etiological, biochemical, cognitive, developmental, and affective features. Both involve a rise of physiological arousal, and both are associated with the flexing of facial and neck muscles. Moreover, *choking up* easily morphs into full-fledged *weeping*. Indeed, most weeping begins with the features of choking up: the feeling of a constricted pharynx (globus sensation) or the feeling of imminent tears.

Cognitively, grief contrasts notably with melancholy. Where melancholic individuals tend to become lost in thought, grief-stricken individuals commonly report a partial or complete cessation of thinking. Instead of contemplating, pondering, or reflecting, the weeper's mental state is dominated by feelings of distress or anguish. These negative feelings seem to overwhelm one's consciousness with the result that thought appears to be suspended. In this regard, the cognitive experience of grief resembles the suspension of thought that accompanies intense physical pain.

Phenomenologically, grief is characterized by strongly negative feelings. Grief vies with physical pain as the most negative affective state. Indeed, people who experience chronic intense pain often collapse into grief as well, especially when accompanied by a sense of helplessness—such as when experiencing pain without the promise of remission. Although grief can be agonizing and miserable, the experience of grief has also led to some of the most poignant and beautiful of human expressions: many of the most celebrated achievements in poetry, literature, visual art, and music are linked to expressions or portrayals of grief.

Socially, grief-induced weeping has a marked psychological impact on observers. People are notably sensitive and vigilant for signs that someone might be weeping, and are also notably disposed to come to the assistance of a person in the throes of grief—either by providing direct help, engaging in consoling actions such as physical touching or hugging, or the offering of comforting words.

Grief can be private and individual, but it can also be public and communal and therefore cultural. All over the world, cultures have devised distinctive rituals for grieving. Especially in that most universal of shared experiences—death—all cultures offer funerary rites that shape both public and private contexts for grieving. Anthropologists have chronicled many ways in which grief or grieflike states are expressed differently in different cultures. For example, grief-related behaviors may include occasional acts of self-injury such as slapping one's face, pulling one's hair, or beating one's chest.[34] In many cultures, grief expressions are linked to culturally normative forms of dress (such as wearing black) or other markers (such as applying charcoal to one's face or shaving one's head). In some cultures, loud wailing is culturally acceptable and encouraged, whereas in other cultures, it is considered excessive and so discouraged. The anthropological literature describing such cultural expressions is voluminous.[35] In chapter 14, we'll address in more detail various sociocultural facets of grief and crying.

Many scholars have suggested that grief is an evolved behavior, and a few scholars have proposed phylogenetic precursors for human crying.[36] Jaak Panksepp and Günter Bernatzky, for example, have suggested that human weeping might have biological origins in the phenomenon of *separation distress*—in many species, the distress experienced by offspring when separated from a caregiver.[37]

Over the decades, various anecdotal reports have appeared suggesting the existence of crying in cats, dogs, or other animals such as elephants. Although these reports are constantly recirculated, systematic observation has failed to confirm weeping in nonhuman animals.[38] All mammals produce irritant tears, and tears can also appear in response to various infections. Although something similar to grief may be experienced by many animals, weeping appears to be uniquely human. Tellingly, watery eyes in response to psychic loss have not been observed in our closest relatives—chimpanzees, bonobos, or gorillas.[39]

Melancholy and grief commonly alternate with one another. The death of a beloved pet, for example, may lead to periods of active weeping alternating with periods of quiescent melancholy. In his 1917 book, *Mourning and Melancholia*, Freud proposed that melancholy is a variant of grief. This view has tended to predominate in the psychoanalytic tradition. For example, the British psychiatrist John Bowlby proposed that the active (grief) and passive (melancholy) responses represent stages or phases when mourning. The idea is that grief ultimately gives way to melancholy, which subsequently gives way to recovery.[40] However, research does not support the idea that the two responses represent stages.[41] Instead, when mourning some loss, it is common for active weeping and passive melancholy to alternate back and forth many times.[42] For convenience, we will refer to this oscillation as the *mourning cycle*.[43]

Reprise

As we have seen, melancholy and grief are quite different emotions. Among other differences, they are readily distinguished by their physiology and behavior. Melancholy is a low-arousal emotion, whereas grief involves moderate to high arousal. Melancholy is associated with relaxed vocalizations and slack facial muscles, whereas grief is associated with tense vocalizations and constricted facial muscles. In particular, grief can lead to overt weeping, characterized by a range of distinctive behaviors, including the production of tears and strained cries. However, grief is sometimes less easily observed, as in the case of being choked up.

Despite their differences, melancholy and grief are oddly similar in their etiology: what causes us to feel melancholic is very similar to what causes us to experience grief. Moreover, melancholy and grief commonly appear in tandem, alternating from one to the other, producing a mourning cycle.

This raises a question that might otherwise be overlooked: Why are there two distinct sadness states—melancholy and grief—rather than just one? The answer to this question hinges on an important distinction made in the field of animal communication—namely, the distinction between *signals* and *cues*.

2 Signals and Cues

In the previous chapter, we described melancholy and grief from eight conventional perspectives—such as the biological, cognitive, and social standpoints. In this chapter, we introduce a ninth perspective—*ethology*—the study of animal behavior. An ethological approach ultimately offers an especially compelling account of the different functions of melancholy and grief. Moreover, in subsequent chapters, we will see that ethology opens the door to a wealth of insights regarding emotional displays in general.

Inspired by the writings of American philosopher Charles Sanders Peirce, Austrian zoologist Konrad Lorenz made an important distinction between *signals* and *cues*.[1] This distinction has become a central pillar in modern studies of animal communication.[2]

A *signal* is a functional communicative act involving innate behavioral and physiological mechanisms. A good example of a signal is a rattlesnake's rattle. Rattlesnakes are carnivorous hunters; they never use their rattles when hunting the small mice and other animals that form the bulk of their diet. The rattle is used only as a warning—such as when the snake encounters another animal that could cause it harm. The rattle effective says, "I'm here: don't harm me because I could also harm you." The aim is to avoid unnecessary conflict or injury. Notice that the signal benefits both the rattlesnake and the other animal.[3]

By contrast, a *cue* is a behavior that incidentally conveys information. An example of a cue is the sound of a buzzing mosquito. In a darkened room, the buzzing sound is apt to alert you to the likelihood of being bitten (something you have learned through past experience). Like the sound of the rattlesnake's rattle, the buzzing of the mosquito conveys information—alerting us to the possibility of being attacked. However, the source of the information differs. In the case of the snake's rattle, the communication is a functional behavior on the part of the snake: the snake's interest is best served when the signal is perceived and recognized by the observer. In the case of the mosquito's buzzing, the communication is accidental—an unintended by-product of the need for the mosquito to flap its wings. Unlike the snake's rattle, the information conveyed by the buzzing sound is actually detrimental to the mosquito—but beneficial to the observer.

A personal story might help clarify the concept of an ethological signal. At a park near where I live, people are allowed to run their dogs off the leash. Walking through the park one day, I noticed a cute little white toy poodle wandering aimlessly. At the opposite end of the field, I watched as a dog owner released a Great Dane from its leash. The Great Dane immediately bounded across the field straight toward the little toy poodle. The Great Dane was snarling and growling viciously, sounds that made everyone in the park stop and watch. We were all alarmed by the thought that we were about to witness a petite poodle being torn to bits by a dog many times its size. When the Great Dane got within a couple of meters, the little poodle rolled over, exposing its belly and whimpering. The Great Dane just stopped dead. It was all over: no horrific scene of violence. Moments earlier, the Great Dane had been in a state of vicious attack; it was now wandering around as if nothing at all had happened. By rolling over on its back, the toy poodle had communicated a classic canine signal of appeasement or submission. In effect, the poodle declared to the Great Dane, "You're top dog; I'm not going to challenge you."[4]

It is not always easy to distinguish cues from signals. Biology is perpetually more complicated than you think, so it's nearly impossible to establish a set of fixed principles. Nevertheless, some common attributes help ethologists to distinguish cues from signals. First, signals are much less common than cues. Most information-bearing behaviors are inadvertent rather than overt acts of communication. Second, signals tend to be *conspicuous* rather than subtle or concealed. Since the effectiveness of a signal depends on communicating to an observer, there is no benefit to timidity. Good signals should be loud and clear. By contrast, cues are simply by-products of other behaviors. Cues convey information, but this information is incidental to the behavior. Although cues can also be conspicuous, most are vague or discreet.

Related to the property of conspicuousness is another attribute: signals are more likely to be *multimodal*—that is, signals have a tendency to involve more than just sound or just visual elements.[5] For example, in the case of the rattlesnake, the snake lifts its tail in the air and shakes it. Even if you can't hear the rattle, you might still see the distinctive wagging tail; similarly, if you can't see the tail, you might still hear the rattling sound. Cues, by contrast, are not intended to be communicative, so they are less likely to be multimodal.

Another attribute related to conspicuousness is persistence or *redundancy*. Since the adaptive value of the signal depends on its successful communication, signals are more likely to be repeated or sustained over time.[6]

Of course, cues can also be conspicuous. They may be perceptually obvious, involve more than one sensory modality, and be repeated or sustained over time. But the conspicuousness of a cue does not arise because of selection pressures. One might say that a conspicuous cue was not "designed" to be conspicuous.

Another common feature of signals is that they often employ *purpose-specific anatomy and/or behavior*. In the case of the rattlesnake, a specially evolved anatomical organ—the rattle—does not appear to serve any other purpose. In addition, the shaking of the tail is a distinctive behavior that appears to be used only for this one purpose.

The most important characteristic of an ethological signal is that it changes the behavior of the observer to the mutual benefit of both signaler and observer (*mutually beneficial*). For example, in the case of the rattlesnake, the purpose of the rattle is to prevent another animal from inadvertently harming the snake—either intentionally or unintentionally (such as accidentally stepping on it). If the rattlesnake was capable of eating you, it wouldn't shake its rattle: it would simply attack. At the same time, the rattle benefits the observer, warning the observer to keep clear and so avoid danger. Notice that if a signal wasn't beneficial to the signaler, there would be no reason to make the signal.

In some circumstances, responding to a signal can incur a potential cost or risk for the observer. Consequently, the observer may need to be vigilant for signs that the signal is dishonest or deceptive. One way to ensure that a signal is honest is to make it costly for the signaler. Hence, signals sometimes (though not always) entail a notable cost for the signaler that is readily apparent to the observer.

An example of an honest signal is evident in the mating behaviors of pelicans. In advance of breeding season, male pelicans grow fleshy bumps between their eyes. Females appear to preferentially mate with those males sporting such growths. The bump makes it difficult for the pelican to see in the region around the tip of its bill and so interferes with its ability to catch fish. The effect is like a hunter agreeing to go hunting with one arm tied behind their back. The presence of the bump in an otherwise healthy animal demonstrates that the pelican is a good hunter, despite the impediment. Incidentally, when a brood of chicks appears, the bump shrinks, allowing the male pelican to hunt more effectively and so better care for the brood.[7] The simple fact that the male remains alive and healthy despite the presence of the bump suggests that he is an accomplished hunter. In short, the cost imposed by the bump testifies to the honesty of the signal and so reassures the female that the signal is a reliable indicator of the fitness of the male when choosing a mate.

This idea of ensuring honesty by making a signal costly has long been observed in the field of sociology. In biology, the idea of ensuring honesty by making a signal costly is referred to as the handicap principle.[8] However, the principle was recognized in the field of sociology a century earlier. In his classic 1899 book, *The Theory of the Leisure Class*, sociologist Thorstein Veblen coined the phrase "conspicuous consumption." The person who is able to be cavalier with money must have a lot of it. (The cliché example is the person who lights a cigar using a $100 bill.) The ostentatious consumption of goods therefore provides a relatively trustworthy indicator of a person's wealth. The peacock's extraordinary tail offers a biological parallel. Producing and sustaining a large flashy tail requires good nutrition and superior survival skills. Dedicating metabolic resources to producing a Las Vegas–style tail might seem illogical. However, the quality of the tail bears honest witness both to the health of the peacock and to its ability to escape predators. A female peahen would therefore benefit by preferentially mating with the peacock with the most "costly" tail.[9]

The idea that signals may incur costs is also evident in the submission or capitulation display in dogs. Although rolling onto one's back may terminate the aggression, this outcome is purchased at the cost of a loss of social status for the signaling animal. The little toy poodle benefited when the Great Dane stopped its attack, but the signal simultaneously established the poodle's subordinate social standing relative to the Great Dane.

These examples notwithstanding, it's easy to overemphasize the concept of honest signaling. Ethologists have noted that many signals appear to involve no apparent cost to the signaler.[10] In order for a signal to evolve, it is only necessary that the transaction benefit both animals.

By way of illustration, consider the case of *stotting* (see figure 2.1). Stotting is observed in ungulates, including several species of deer, gazelles, springbok, and impalas. It can also be observed in young sheep and goats. Stotting features repeated leaping as high as possible into the air with the legs held stiffly. The behavior typically lasts for less than a minute, and several animals within a herd may engage in stotting around the same time.

Stotting is performed only when the animal detects the presence of a predator. When selecting possible prey, predators typically focus on weak, injured, or ill animals. Stotting effectively demonstrates the health and vigor of the animal. Cheetahs and African wild dogs have an instinctive disposition to ignore stotting animals as potential target prey.[11]

Figure 2.1
Springbok stotting in the Etosha National Park, Namibia. This signal is performed in the presence of an observed predator. Stotting demonstrates the health and vigor of the animal and consequently discourages the predator from stalking or chasing the stotting animal. Photo by S. K. Yathin.[12]

Stotting profits both predator and prey. The predator profits by recognizing that pursuing the stotting animal is likely to be futile. At the same time, the stotting animal benefits by avoiding a prolonged chase and allows the animal to return to (vigilant) grazing despite the presence of the predator.[13] It is the mutual benefit for both signaler and observer that provides the selection pressure allowing the stotting signal to evolve.

One might suppose that the jumping behavior represents a significant cost to the stotting animal. However, the energy expended by a brief display of stotting is much less than the energy that would be expended were a chase to ensue, so it is not at all clear that stotting can be considered a costly signal.

As an evolved behavior, ethological signals are, by definition, innate. Both the signal and the response tend to arise spontaneously and automatically in the appropriate circumstance. Moreover, the behaviors themselves tend to be formulaic or stereotypical. In short, signals tend to exhibit two further characteristics: *automaticity* and *stereotypy*. (Lest readers get the wrong impression, I should at this point hasten to add that the concept of automaticity, like honest signaling, can be overexaggerated. Even among animals with tiny brains, there are exceptions when a behavior is not automatically induced. Among big-brained humans, there is notable latitude for inhibiting or modifying such behaviors—a topic that will be addressed in detail in a later chapter.)

An innate or automatic behavior would seem to represent an open invitation for deception, which returns us to the question of honest signaling. If I know that my display is very likely to result in a predictable innate response by some observer, surely I can take advantage of this automatic tendency. Deception is common in animal communication.[14] However, this is not the case with ethological signals.[15] In a comprehensive analysis, the American biologists William Searcy and Stephen Nowicki have identified a number of circumstances where animal communications tend to be reliable with little or no recourse to deception. Back to the case of the rattlesnake, it is difficult to imagine a scenario in which the snake might shake its rattle as a form of deception. Under what circumstance would shaking the rattle benefit the snake while being detrimental to an observer? Especially important are situations when there is no conflict between the signaler and the observer.[16]

By way of illustration, suppose in some circumstance, you are fully aware that I am lying to you. However, as long as you are still in a position to benefit from the interaction, it may nevertheless be useful for you to respond as though you believe I am telling the truth. In such an interaction, honesty is not essential, but mutual benefit is. The critical factor allowing a signal to evolve is simply mutual benefit.

Actually, it's not so much the case that mutual benefit is a property of signals; rather, signals can only exist in situations where mutual benefits are commonly assured. Signals arise in those rare cases where the fitness benefits for both signaler and observer are in persistent happy alignment. It bears emphasizing that most animal interactions offer ample opportunities for deception or exploitation. Situations of assured mutual benefit are less common, which partly explains why, when compared with cues, signals are much less common.[17]

The situation might be helpfully described using the language of game theory. Most animal interactions amount to zero-sum games in which one animal benefits at the cost of the other animal. For example, if one animal commandeers most of the available food, then there is less food available for the other animal. Especially among social animals, however, certain interactions amount to positive non-zero-sum games where both animals benefit from the interaction. An example would be cooperative hunting where two animals are more likely to achieve a successful hunt than the two animals acting independently. In such cases, a signal that leads to cooperative joint behavior need incur no cost for either animal. Once again, not all ethological signals incur costs.

Returning to our initial point, the principal difference between signals and cues is how the information is manifested. Signals are acts that deliberately "push" information into the environment; cues are inadvertent phenomena in which an observant individual "pulls" serendipitous information out of the environment. Formally, we might define signals and cues as follows:

Signal: An act or display that typically alters the behavior of an observing animal, whose consequences benefit both animals and for which the signaling and responding behaviors coevolved because of the conferred mutual benefits.[18]

Cue: A feature of the world that can be used by an animal as a guide to future action and that benefits only the observing animal.[19]

Cues are deciphered purely for the benefit of the observer. Signals, by contrast, initiate a transaction that benefits both the signaling and observing animals.[20]

It should be noted that signals do not benefit the signaler in *every* circumstance. (Indeed, we will see examples later where sadness-related signals are not helpful for the signaler.) However, as we have already noted, signals simply cannot evolve unless there is typically some benefit to the signaler. If a signal persistently proved detrimental to the signaler, selection pressures would conspire to eliminate the signal from the animal's behavioral repertoire. Of course, conditions can change over evolutionary history, so a signal might evolve that is highly beneficial, but over time, the signal may lose its utility: signals may come and go. In general, however, a given signal may be retained, despite occasional or rare circumstances where the signal proves harmful to the signaler.

It's tempting to suppose that signals are innate, whereas cues are learned, but that's an oversimplification. Signals are indeed innate: biologists regard signals as traits that have evolved because of positive selection pressures.[21] Within species, signals tend to evoke robust responses. Signals between species are also likely to be robust when both species have shared the same environment over long stretches of evolutionary history. However, between-species signals may entail an innate disposition for rapid learning rather than some prepackaged response. Moreover, the source of an innate response may be only indirectly sensitive to the signal features. For example, in the long history of human evolution, our African ancestors never encountered rattlesnakes since rattlesnakes are indigenous only to the Americas. Nevertheless, humans tend to respond to rattlesnake rattling by being suitably rattled. What modern humans appear to have inherited is a disposition to quickly learn to fear snakes, combined with a disposition to fear animals that deliberately draw attention to themselves in our presence. Human wariness of a rattlesnake's rattle behavior is not a "hardwired" response but instead arises from a more general sensitivity to a class of behaviors. The human response might be regarded as innate but not specifically tuned to a rattling sound or the shaking of a raised tail. The source of an innate response may be only indirectly sensitive to specific signal features.

In the case of cues, most are indeed learned, but cues can also arise as evolved traits. For example, mosquitos have an innate disposition to follow a trail of carbon dioxide (CO_2) gas. Since animals exhale carbon dioxide, following a CO_2 trail is very likely to lead the mosquito to a food source. In short, CO_2 is a cue that benefits the observer (the mosquito). There is no advantage to the animal producing the cue, but it's impossible to prevent the cue display (stopping breathing is not really a viable option). While most cues are learned, many are innate.

In most cases, opportunities to make the cue less conspicuous are limited. Polar bears have white fur that provides effective camouflage in snowy environments. However, their black noses provide cues that seals and other prey apparently recognize. How do we know that seals use the black nose as a warning cue? When hunting in close proximity to their prey, polar bears instinctively cover their noses with a paw. That is, they appear to exhibit an innate behavior that masks what must otherwise be a salient cue for their prey.[22]

It bears emphasizing that the various attributes that help biologists distinguish signals from cues only represent *tendencies*. There exist cues that are conspicuous, multimodal, sustained over time, and lead to automatic responses that are innate rather than learned. The critical difference between signals and cues is that signals evolve because they typically benefit both the signaling and observing animals.

Finally, we might consider how signals arise in the course of animal evolution. As we have noted, cues offer an advantage only to the observing animal (such as the innate disposition of mosquitos to follow a trail of carbon dioxide gas). Since there is no advantage to the displaying animal, there is no selection pressure to make the cue more conspicuous. In fact, since the cue is detrimental to the displaying animal, selection pressures would endeavor to make the cue *less* conspicuous—as in the case of polar bears covering their noses when hunting.

Over the course of evolutionary history, however, the costs and benefits of particular behaviors can change. For example, a former cue may be transformed so that the cost to the displaying animal becomes negligible, or the behavior may even become beneficial. If a cue offers an adaptive advantage for the animal exhibiting the cue, then selection pressures will favor enhancing or amplifying the communicative properties of that cue. That is, selection pressures will tend to lead to adding features in another modality, sustaining or repeating the display, and, in general, making the display more conspicuous. Over time, the displaying behavior will tend to become innate—transforming a former cue into a bona fide signal. Dutch ethologist Niko Tinbergen cogently argued that all signals evolve from earlier cues.[23] In ethology, this process of a cue evolving into a signal is called *ritualization*.[24] As we will see later, this process of ritualization will prove helpful in understanding the origin of weeping.

Reprise

In this chapter, we have introduced the ethological distinction between signals and cues. Signals are evolved display behaviors whose function is to change the behavior of observers so as to produce a net benefit for both the signaling and observing individuals.[25] When the potential for deception exists, signals may incur an overt cost to the signaler in order to assure the observer of the honesty of the signal. By contrast, cues are incidental sources of information (artifacts or by-products) that benefit observers with no necessary advantage to the individual generating the cue. Both signals and cues are informative to observers, but only signals are overtly communicative.

In distinguishing signals from cues, ethologists look for various telltale features. Signals involve unique displays that are not easily confused with other behaviors (*distinctive*). Signals have little value unless they are observed: consequently, signals tend to be *conspicuous* (rather than subtle). Contributing to conspicuousness, signals are more likely to engage multiple senses (*multimodal*) and tend to be repeated or sustained over time (*redundant*). Unlike cues, signals are more likely to involve *purpose-specific anatomy*. Both the signaling and response behaviors tend to be fixed or formulaic (*stereotypy*), and the motivation to generate a signal or respond to a signal suggests innate or instinctive dispositions (*automaticity*). Finally, the interaction typically results in observer behavior that ultimately benefits both the signaler and the observer (*mutually beneficial*).

3 Emotions and Displays

Returning to the subject of sadness, we might now address the question that would lurk in the heart of any ethologist: Are weeping and melancholic displays *signals* or *cues*?

Melancholy as Cue

Let's begin with melancholy. How conspicuous is a melancholic display? How well do the acoustical, visual, and other behaviors characteristic of melancholy distinguish it from other possible states? Recall the six acoustic features associated with melancholic speech: What—we might ask—do quieter and slower speech, lower pitch, smaller pitch movement, more mumbled articulation, and darker timbre all share in common? The answer is that all of these features are associated with low physiological arousal. In the peripheral nervous system, low arousal is linked to reduced acetylcholine, which in turn reduces both muscle tone and reactivity. Muscles become more flaccid and sluggish.[1]

Reduced acetylcholine will affect all of the skeletal muscles, including the muscles of the vocal folds, tongue, lips, chin, and pulmonary muscles. Reduced muscle tone means that the vocal folds will be less tense, resulting in a lower overall pitch.[2] A slower cricothyroid muscle will produce more sluggish pitch changes—and therefore generate a more monotone prosody.[3] Relaxed pulmonary muscles result in lower subglottal air pressure, causing a quieter voice. Slower reactivity of tongue, lips, and chin will result in a slower rate of speech and more slurred or mumbled articulation. When the zygomatic muscles of the face are relaxed, the lips tend to fall away from the teeth (in contrast to smiling); this results in a longer effective vocal tract length with a concomitant lower resonance—producing a darker timbre.[4] In short, *all* of the acoustical features of melancholic speech can be traced to the effects of low physiological arousal—they are all artifacts of reduced acetylcholine.

Melancholy is not the only state linked to low physiological arousal. Low arousal is most commonly experienced when people are relaxed, tired, or sleepy. This raises the question of whether a melancholic voice is truly distinctive. Does a melancholic voice differ recognizably from the more commonly occurring relaxed, tired, or sleepy voice?

In preparing for a possible experiment, I recruited an actor to read neutral sentences distinguishing a melancholic voice from a sleepy voice. Excluding the telltale sound of yawning, the actor struggled to create distinctive renditions. To my ears, the feigned melancholy and feigned sleepiness sounded only marginally different, and there was nothing to suggest which version was which. I was skeptical that listeners would be able to decipher which was ostensibly the melancholic voice.

Acted or feigned utterances are inevitably suspect. Ideally, we would use recordings of genuine spontaneous melancholic or sleepy utterances. However, assembling such recordings is a formidable practical challenge. Moreover, with spontaneous live recordings, we can't have speakers speak the same (neutral) sentences for comparison purposes. Finally, assembling pertinent recordings is further complicated because, as noted earlier, one of the characteristics of melancholy is the tendency to remain mute: melancholic people simply tend to avoid speaking altogether, so it's difficult to record spontaneous utterances. In light of all the complications, I abandoned my proposed experiment. Nevertheless, informally one could observe that both melancholic speech and tired/sleepy speech exhibit the same quieter sound intensity, slower speaking rate, lower pitch, more monotone prosody, mumbled articulation, and darker timbre.

So what about the visual aspects associated with melancholy? Recall that melancholy is associated with relaxed musculature. This may lead to slumped posture, lowered head, and drooping eyelids. Relaxation of the zygomatic muscles causes the cheeks to flatten, and the relaxation of the mouth causes the chin to descend. The result is the classic "long face" appearance. Examining the morphed photograph shown in figure 1.1 (chapter 1), it's not clear that the "lengthened" face should be judged melancholic rather than sleepy, relaxed, or tired.

So how well do observers correctly infer that a person is feeling melancholy? Apart from laboratory experiments, we might simply ask people about their personal experiences. Since the features of melancholy appear to be artifacts of low arousal, how often are relaxed, tired, or sleepy states mistaken for melancholy? To address this question, I conducted a brief survey in which I merely asked forty-seven people whether they had ever had one or both of the following experiences:

Have you ever had the following experience? Someone you know (or even a complete stranger) approaches you and asks, "Why the sad face?"—despite the fact that you don't feel sad at all.

Have you ever made this mistake yourself—where you thought someone was sad, but they were just tired or relaxed?

Roughly half of the respondents claimed to have had either or both of these experiences. Traveling on public transport in particular, it's common to conclude that many people appear to be sad (see figure 3.1). More plausibly, when people are removed from a social context (such as when traveling alone without a companion), faces are apt to relax. Evidently, we appear to be quick to assume that a relaxed or tired face may be symptomatic of melancholy. At least one female respondent to my survey mentioned the unsettling tendency for male strangers to remark, "Smile, it's not that bad."

By way of summary, we have seen that the features associated with both melancholic speech and melancholic facial expressions appear to be indistinguishable from other low-arousal states—notably sleepy, tired, or relaxed states. Apparently, there are good reasons why some languages (like Tahitian) might use the same word to indicate both sadness and fatigue.[5] It may be that there exist reliable observable features that successfully distinguish melancholic behaviors from sleepy, tired, or relaxed behaviors. However, the apparent ease of confusion is not consistent with the properties of distinctiveness and conspicuousness that ethologists consider important for signaling.[6] Moreover, recall that the most characteristic acoustic feature of melancholic voice is the tendency for melancholic individuals to remain silent. In the absence of any clear visual feature, failing to produce a sound is not what we would expect if the goal is to communicate.

All of the foregoing observations favor the idea that melancholy is an ethological cue rather than a signal. Nevertheless, we'll postpone rendering any final decision until we consider the complete range of evidence. In particular,

Figure 3.1
Buses, trains, and subways appear to be full of seemingly sad people. It is not clear
whether a given facial display is symptomatic of melancholic sadness or simply a
relaxed or tired state. Photo: "Time Passages" by John G. Hoey, with permission.

we need to consider any potential adaptive function of melancholy (chapter 5) that might swing our assessment one way or another. At a minimum, the above observations should cast some doubt about whether a supposedly melancholic display might qualify as a bona fide ethological signal.

Weeping as Signal

Consider now the parallel ethological question related to grief. As we have noted, the behavioral manifestations of grief can range from inconspicuous *choking up* to overt *weeping*. By itself, the globus sensation (feeling choked up) is impossible for observers to recognize.[7] Similarly, the feeling that tears are imminent is not something easily deciphered by some observer. Since the symptoms of choking up are difficult or impossible for observers to recognize, being choked up cannot be a signal.[8] Weeping, however, is a different story. Might weeping be an ethological signal? How conspicuous is weeping? How well do the acoustical, visual, and other behaviors characteristic of weeping distinguish it from other possible states?

Recall that the visual features of weeping include furrowed brow, squinting eyes, downturned corners of the mouth, flushed red face, and—of course—tears.[9] Acoustical features of weeping include vocalized staccato exhaling, long sustained tones (wails), audible gasping, use of falsetto phonation, breaking voice, trembling voice, creaky voice, pharyngealization, and sniffling. When crying, sounds may range from quiet whimpering to loud sobbing or wailing. Tears can be shed without producing a sound, but prototypical crying involves notable spontaneous sound production.

An especially pertinent characteristic of crying is the compulsion to vocalize. As we've noted, people tend not to vocalize when they feel melancholic. In the case of grief, however, the situation is reversed. Whether whimpering or wailing, the grief-stricken weeper experiences a spontaneous inclination to produce sounds. When exhaling, staccato puffs of air are forced through the vocal folds. When inhaling, it's common to hear vocalized gasping (ingressive phonation). Whether inhaling or exhaling, the vocal folds are continuously engaged. In the context of signaling theory, this resolute involuntary compulsion to vocalize is consistent with automaticity, while the sustained sound production over time is consistent with redundancy. The contrast with melancholy is striking: people tend to avoid any vocalizing when experiencing melancholic sadness.

Apart from the tendency to make sound, we might also consider the distinctiveness of those sounds. As noted earlier, the constriction of the pharynx changes the acoustic resonance of the vocal tract, leading to a pharyngealized or "tense" sound. In addition, laryngeal constriction leads to phonatory instability where the voice chaotically switches back and forth between normal (modal) phonation and high (falsetto) phonation. This results in the unique cracking or breaking voice, which—even more than whimpering or wailing—is the quintessential sound of weeping. Finally, we might add one further observation regarding grief-related sounds: weeping sounds commonly occur in the absence of speech. Unlike the sounds associated with melancholy, weeping sounds are not merely modifications to willfully produced speech sounds; they are spontaneous distinctive sounds in their own right.

In considering whether a particular behavior represents a bona fide ethological signal, perhaps the most persuasive evidence would be the reliance on some dedicated anatomical feature or structure. In the case of the rattlesnake, the rattle represents such a purpose-specific anatomical organ. This raises the question of whether weeping involves any purpose-specific anatomy.

As noted earlier, ophthalmologists distinguish lubricant, irritant, and emotional tears. Both irritant and emotional tears originate in the same lacrimal glands. However, irritant and emotional tears are triggered by different neural pathways. Irritant tears arise from the trigeminal nerve (cranial nerve V), whereas emotional tears arise from a branch of the facial nerve (cranial nerve VII). That is, the lacrimal glands are innervated by two distinct efferent nerves. When performing eye surgery, an ophthalmologist can anesthetize the eye so that lubricant and irritant tears are entirely inhibited. However, given a suitably emotional situation, an anesthetized patient can still cry psychic tears because of the separate limbic pathway.[10] When operating on an anesthetized patient, an ophthalmologist must take care not to tell a story that might cause the patient to be emotionally moved. Critically, no other primate shares this separate neural pathway: the anatomy is unique to us humans. The existence of a specialized limbic path for producing emotional tears implies that weeping is an evolved behavior in its own right, rather than being an artifact of some other process.

Recall that the most important feature of signals is that they influence the behavior of others: signals are foremost intended to change the actions of the observer. Of course weeping does have a dramatic impact on observers. Even toddlers will approach a crying individual in order to console or offer help.[11]

Figure 3.2
Weeping facial display: watery eyes, tear traces along the cheeks, furrowed brow, flexed
facial muscles. A weeping facial display is notably conspicuous, even apart from the
distinctive sounds of weeping. Photo by Taremu.[12]

Finally, we need to consider the conspicuousness of any purported signal. Is weeping subtle or obvious? As we have noted, weeping can range in intensity from the inconspicuous globus response (choking up), through subtle tearing, to full-fledged waterworks, sobbing, and ultimately loud wailing. Apart from the globus response, weeping displays are unmistakable—in contrast to a melancholic face. Consider Nigerian celebrity Hilda Dokubo, shown in figure 3.2. The furrowed brow, squinting eyes, downturned mouth, and tears make her expression seem patently clear. Moreover, what the photograph doesn't convey are any of the characteristic sounds of crying. Notably absent are the punctuated vocalized "ah-ah-ah-ah," the sounds of whining or wailing, the inhaled gasping sound, and the telltale "cracking" voice.

Interestingly, the intensity of one's weeping is correlated with the conspicuousness of the display. When simply choked up, it's not possible for anyone to observe unless the displayer chooses to speak. With tears in one's eyes, these can be observed only by a person who has direct sight of the displayer's face. So depending on the direction the weeping person faces, they may be able to intentionally make their face visible to no one or only to a select person or group of people. If the emotional intensity is even greater, then sobbing or wailing may occur. Quiet sobbing can be observed by anyone in close proximity. Wailing, by contrast, can be observed by the widest possible range of observers. In other words, the various display elements associated with increased weeping intensity also exhibit differences in conspicuousness that appear to broadcast the display to an increasingly larger audience.

Apart from the conspicuousness of weeping, there is also evidence that observers are especially sensitive to *recognize* weeping. For example, studies with digitally altered photographs show that even when the addition of a single tear fails to be consciously perceived in a brief visual presentation, we nevertheless unconsciously judge the photographed person to be in a heightened state of emotion.[13] Although a person may endeavor to hide their face or otherwise mask their weeping, we seem to have a "sixth sense" for detecting the slightest hint that someone is crying. It's not simply the case that weeping is usually conspicuous; it's also the case that, as observers, we are especially vigilant and attuned for evidence that a person may be crying.

By way of summary, human weeping exhibits the principal hallmarks of an ethological signal. The features of weeping are manifestly conspicuous, including distinctive visual and auditory elements. Together, the combination of elements makes weeping unmistakable. The existence of a separate neural

pathway for generating psychic tears indicates a purpose-specific evolved anatomical feature. In addition, the involuntary compulsion to vocalize, especially while inhaling, provides further evidence consistent with the notion
that weeping is an ethological signal.

Emotions versus Displays

As we have already noted, apart from grief, weeping has long been observed
in widely varied circumstances. People frequently cry at weddings. Tears can
arise in situations of extreme humor where a person "laughs 'til they cry."
Weeping is regularly observed in the joyful winners of sports competitions.
Crying is also a common response to injury; for example, a child falling off
a bicycle might cry in response to the pain of a skinned knee. A toddler's
temper tantrum is an expression of anger or frustration but also commonly
involves tears. Famously, female fans of The Beatles wept in the presence
of their idolized and adored stars. Tears may well up in the eyes of a soldier
experiencing feelings of patriotism when saluting a flag. And the grandeur
of natural scenery or touching music can lead to nominally "aesthetic" tears.
In short, many emotions might bring a person to tears. Minimally, weeping
might arise due to grief, joy, humor, pain, anger, frustration, patriotism, adoration, or aesthetic beauty. This surely means that weeping itself indicates no
fixed emotion. When we observe someone with tears in their eyes, without
additional information, we cannot assume that the weeping is an indication
of sadness or grief.[14]

The loose relationship between behavioral display and motivating affect
is not limited to weeping. As we also noted earlier, people smile when happy
but also when feeling stressed.[15] Smiling can arise when a person feels uncomfortable, embarrassed, shy or polite, or socially apprehensive.[16] Smiling is a
common form of greeting, even when meeting complete strangers or a person of whom we are suspicious.

These observations suggest that it is essential to distinguish *display* behaviors from the presumed motivating emotion. In order to remind ourselves of
this distinction, it helps to be circumspect about terminology. In English, the
words "grief," "melancholy," and "weeping" are helpful. "Weeping" draws
attention to the display behavior without necessarily implying grief alone.
The words "grief" and "melancholy" more clearly refer to distinctive feeling
states. However, there is no English word for the facial display associated

with melancholy. If necessary, we will use the term "long face" to denote the relaxed facial display that is commonly observed for melancholy—as well as sleepiness, fatigue, and relaxation.[17]

What is important to recognize is that while either choking up or weeping is a necessary feature of grief, neither choking up nor weeping are limited to grief alone. As understood in this book, grief is an emotion, whereas choking up and weeping are behaviors whose motivating emotions may vary. In short, it is not grief, but weeping, that is a candidate signal.

Reprise

In reviewing the physiological and behavioral characteristics of melancholy and grief, we have seen that melancholic displays seem most consistent with an ethological cue, whereas weeping appears to exhibit the hallmarks of an ethological signal. With this background, we might now return to the question posed at the end of chapter 1: Why do we have two main sadness-related emotions rather than one? Do melancholy and grief serve different functions?

If weeping is a bona fide ethological signal, then the induced behavior of observers should benefit both the weeping individual as well as the observer-spectator. This suggests that the question "What is the function of weeping?" might be profitably rephrased as, "What are the benefits of weeping—for both the weeper and the observer?"

Despite the diversity of weeping-related circumstances, for the purposes of this chapter, we'll address this question by simply focusing on *grief*-induced weeping. Although this narrows the discussion, our narrative will nevertheless lay important groundwork that will prove useful later when we broaden the discussion to include the complete range of weeping-related situations—tears of joy, humor, pain, loyalty, adoration, and aesthetic beauty.

So what are the benefits of grief-induced weeping? It bears stating the obvious: weeping commonly leads observers to engage in prosocial helping behaviors. The tears of a person surveying their broken-down automobile will likely lead to offers of transport. The tears shed by an elderly person who is utterly lost in an unfamiliar building will likely result in observers helping orient the weeper. The tears evoked by a child's ice cream cone dropped on the ground may well induce an observer to arrange a replacement.

Across cultures and over the expanse of human history, ancient texts, myths, stories, and modern anthropological reports all suggest that weeping commonly has the effect of encouraging observers to help the weeping individual: grief-induced crying seems to signal an appeal for aid or assistance.[1] Modern experimental research offers plenty of evidence affirming this idea while adding some telling refinements.[2] For example, "helping" may take the form of stopping *un*helpful behavior. Consider the following firsthand account relayed by Jeffrey Kottler:

> A male physician had been verbally abusing a female hospital administrator. The more she apologized, the more he berated her. It was clear he was not accepting her [apology].
>
> All of a sudden, a tear welled up in her eye, just a single tear, and ran down her cheek. He stopped cold. This guy, big time surgeon and all, used to having his way and blustering onward, just stopped dead. This tiny spot of wetness communicated to him very clearly what he otherwise had not seen. He started backpedalling so fast, apologizing like crazy. That single tear had meaning for him a way that nothing else did.[3]

Notice the transformation in the affective state (and consequent behavior) of the surgeon. One minute, he is in "attack" mode; the next minute he is apologetic and consoling. We saw a similar behavior earlier in chapter 2—although in a different species—namely, the canine submission display. In the case of the toy poodle, the act of rolling over utterly transformed the behavior of the Great Dane. The "attack" mode was abruptly switched off. In the case of the hospital administrator, the appearance of a single psychic tear resulted in a similar transformation. In both cases, the display had a dramatic impact on the behavior of the observer.

This single anecdote is not atypical; it is consistent with more representative empirical research. In two experimental studies carried out by Carrie Lane at the University of Texas at Arlington, she found that, in situations of conflict, weeping tends to bring the conflict to a quick resolution; by contrast, if no weeping occurs, there is a tendency for conflicts to escalate.[4]

The benefits of weeping are related to need. Those who have the most to gain from crying are those who experience significant ongoing stress. This includes individuals who are frequently attacked or bullied, people who have little access to resources, and those who lack the ability to acquire resources. In early work at the University of California's Institute for Child Welfare, Catherine Landreth found a significant negative correlation between crying frequency and IQ: individuals with lower IQs cry more than those with high IQs.[5] To the extent that intelligence correlates with a capacity for acquiring resources, this relationship is consistent with an increased value of crying for those individuals less able to provide for themselves.

At this point, we might suppose that weeping appears to serve perhaps three related functions: as an appeal for aid or assistance, as a specific plea to terminate aggression, and possibly as an act of surrender, capitulation, or submission.[6]

The Cost of Weeping

Apart from the benefits for a weeping individual, there are also disadvantages. In the case of the Great Dane and the toy poodle, we noted that the submission display resulted in a loss of social status for the poodle. In human weeping, there is a similar decline in social status for the weeper relative to the observer. These social costs are not the same for all individuals. Age and sex are important factors. In general, high social position is biologically important insofar as it impacts reproductive success.[7] In most species, nearly all mature females successfully reproduce. By comparison, males are much more variable in their reproductive success, with many low-status males failing to reproduce at all. Among mammals, high social rank disproportionately benefits males. A high-ranked male can produce many more offspring than a high-ranked female. Genetic studies, for example, estimate that 1 of every 200 individuals in the world is a direct descendent of Genghis Khan.[8] Genghis is reputed to have said that

> the greatest joy for a man is to defeat his enemies, to drive them before him, to take from them all they possess, . . . and to hold their wives and daughters in his arms.[9]

It's unlikely that most men would regard such a scenario as key to their "greatest joy." But Genghis's remark (and evident behavior) highlights the reproductive chasm separating high-ranked males and females. Apart from one's sex, the negative effect of reduced social status is greatest for those individuals with the capacity to reproduce: infants and children have little to lose when they weep.

In summary, those who pay the greatest costs for weeping are reproductive individuals of highest social rank, with males incurring a greater relative cost than equivalently ranked females. Conversely, those individuals who have the least to lose by weeping are nonreproductive individuals and those at the bottom of the social hierarchy.

These theory-based notions accord strikingly well with empirical observations concerning the frequency of crying. The greatest amount of crying is evident in infancy. Crying decreases around two to three years of age, when toddlers become more socially engaged and so are more likely to suffer from the social penalty incurred by crying. Sex-linked differences in crying frequency are not evident in early life.[10] At puberty, a dramatic reduction in the frequency of crying is evident for both males and females. At the same time, marked sex-related differences appear at puberty, with males much less likely to cry than females.[11] The difference in crying frequency between males and females is evident in virtually every culture observed. To be sure, the disposition for males to cry less is reinforced in many cultures where men are explicitly encouraged not to cry. In the *Republic*, for example, Plato characterized weeping as "unmanly."[12] In addition, in many (though not all) cultures, people of high social station (both male and female) exhibit significantly less crying.[13]

With the onset of old age, sex-related differences are attenuated. Elderly men are slightly more likely to weep than younger men,[14] and this difference is consistent with some loss of reproductive fitness.

Incentives for Helping

In a grief context, the benefits for the weeping individual seem obvious. So what's the benefit for the observer? In attempting to understand behavioral motivation, it is essential to distinguish *proximal* from *ultimate* causes.[15] The biological purpose of eating food is to supply the metabolic resources that sustain life. But the immediate motivation for eating arises from feelings of hunger and the rewarding pleasure of food consumption. Without the proximal allure of appetite, the motivation to eat has little urgency. Similarly, in order to understand why people are motivated to help others, we need to distinguish between proximal and ultimate incentives.

With regard to ultimate causes, the origin of cooperative and altruistic behaviors is well-trod territory in evolutionary biology. Evolutionary theory posits at least six ways by which altruistic acts can be biologically adaptive by enhancing an individual's fitness. First, selfless acts that assist closely related individuals can enhance fitness through *kin selection*.[16] Especially when the cost of assistance is low, assisting one's closest relatives can have a marked facilitating effect in propagating shared genes. A second mechanism was proposed by Robert Trivers, who noted that helping people who are not kin can also benefit an individual through *reciprocal altruism*: the person you help today is more likely to help you in some future hour of need.[17] A related though distinct third benefit is the building of a social alliance, where both individuals are more likely to feel a sense of partnership that may prove useful in dealing with future external threats. Apart from expectations of such quid pro quos, a fourth benefit is an enhanced social status for the observer with respect to the stressed individual. We tend to hold in higher esteem those who come to our assistance.

When helping behaviors are conducted in the presence of an audience, or when gossip is able to communicate stories of altruistic acts to a wider group beyond immediate spectators, a fifth mechanism comes to the fore: creating a positive reputation. Moreover, when we interact with the same people repeatedly over years or decades, it becomes adaptive for individuals to attend to *reputation maintenance*.[18] A reputation for altruism and cooperation has notable long-term social benefits. People who are regarded as reliable team players are more likely to benefit from an elevated social reputation and are more likely to attract assistance from others when in need themselves.

Finally, a sixth mechanism adds *social ostracism* to the equation. Even a single defection or social violation can lead to long-term mistrust and suspicion. Failing to come to a person's assistance can lead to social chastising or punishment.[19] People who could have helped but didn't are often reviled.

Of these six ultimate benefits for encouraging altruistic behaviors, *kin selection* is the most important, followed by *reciprocal altruism*. The preeminence of these two factors is evident in the effect of observer group size on helping behaviors: our propensity to help someone is influenced by the availability of other potential helpers. Tragically, we are less likely to help when we are part of a large crowd of potential helpers. Given the size of the audience, this reluctance to help would seem to contradict the idea that we are motivated by the opportunity to enhance our social reputation. However, the reluctance to help is much less when the audience members are familiar, and the effect disappears entirely when the person needing assistance is a kin relation, friend, or colleague with whom we interact regularly.[20] Moreover, the long-term effects of social ostracism are much greater when the audience members are familiar and the person in need of help is a kin relation or friend.

Apart from the advantages arising from acts of altruism, there are also disadvantages to helping others. Offering assistance depletes one's own resources, either in terms of material assets or time or energy. A hungry individual might not be motivated to share their food. In many situations, however, there are unequal costs involved. The marginal cost to the observer is often much less than for the person seeking assistance. For example, if food and water are readily accessible to the observer, then sharing these resources with a hungry or thirsty individual incurs only a small cost relative to the benefit for the stressed person. In general, one might anticipate many circumstances where the benefits of engaging in acts of altruism would outweigh the individual costs.

Proximal Motivations for Helping

Although ultimate motivations provide the underlying foundation for behaviors, the immediate cause of specific behaviors is to be found in *proximal* motivations. Proximal motivations typically take the form of feeling states or emotions. As already noted, it is the feeling of hunger that motivates eating, not some abstract goal of sustaining one's metabolic energy; it is the feeling of anger that causes us to defend our threatened children, not an abstract assessment that our genetic legacy is in jeopardy. Once again, it helps to consider feelings or emotions as action tendencies or motivational amplifiers.[21]

If the purpose of a signal is to elicit a particular behavior in an observer, then the most effective way to achieve that is by inducing a pertinent feeling state in the observer—as when the rattling of a rattlesnake evokes fear, causing an observer to back away or withdraw.

In the case of altruism, philosophers and commentators have long drawn attention to the pleasurable feelings that can arise when we engage in acts of charity. Saint Augustine and Immanuel Kant, for example, argued that acts of charity are not truly altruistic precisely because they make us feel good. Kant regarded charity as a form of selfish pleasure. So what are the proximal feelings that encourage us to engage in charitable or altruistic acts? We can identify at least six proximal feelings that motivate people to be helpful. We might begin by considering the following story:

Seventeenth-century London was a remarkably grim place. There was no organized sewage system, so the streets were filthy. Disease was rampant, with widespread poverty. In the 1660s, one of the city's residents was the famous philosopher, Thomas Hobbes. Hobbes is generally known for laying the foundations for *ethical egoism*—the conviction that individuals ought to do what is in their own self-interest. One might think that this philosophy is a recipe for insensitivity to the plight of others. Not necessarily. Hobbes himself was not unmoved by the sight of human misery. When asked why he gave sixpence to a beggar, Hobbes explained, "I was in painn to consider the miserable condition of the old man; and now my alms, giving some relief, doth also ease me."[22]

There is something about human misery that touches even the most hardened of souls. Indeed, the inability to feel pity for others is a defining trait of the psychopath. In responding to the beggar, notice how Hobbes drew attention to how it pained *him* to see the old man in misery. Hobbes experienced a sort of contagious emotional connection that allowed him to imagine—and vicariously feel—some of the old man's suffering.

The phenomenon of "feeling another person's pain" is referred to as *commiseration*.[23] A compelling demonstration of commiseration can be found in a classic experiment conducted by German social neuroscientist Tania Singer and her colleagues.[24] Singer first carried out brain scans while she inflicted pain on volunteers. Two brain networks were observed to be active: one associated with physical sensation and one associated with the experience of pain. She then repeated the experiment: this time she inflicted obvious pain on a romantic partner in the presence of the person being scanned. As might be expected, brain areas associated with physical sensation were not active. However, even though no pain was directed at the scanned volunteer, the scan results showed similar patterns of activation in brain areas associated with pain: observing the painful experience of a loved one was echoed in the brains of the empathetic observers. When we see someone accidentally cut themselves or get their fingers pinched in a door, it's hard for us as observers not to cringe. Under many circumstances, it is indeed possible to experience a (muted) version of another person's pain.

If we are able to partake of another person's distress, it follows that any action that reduces that person's distress will consequently reduce our own feeling of distress.[25] As Thomas Hobbes recognized, an important motivation for helping someone is to dampen our own (negative) feeling of commiseration.[26]

If reducing the feeling of commiseration were the only incentive for helping, the best we could hope for would be to simply eliminate our own negative feeling in the presence of a distressed individual. However, there exist at least five other proximal motivations.

An additional motivation for helping is the positive feeling of *social approval*. In discussing his taxonomy of human pleasures, the English philosopher and moralist Jeremy Bentham included "the pleasures of a good name" as one of fifteen general motivating pleasures.[27] As one might expect, research shows that people are more likely to help others when their actions are observed than unobserved.[28] Moreover, an experiment conducted by Japanese researchers Keise Izuma, Daisuke Saito, and Norihiro Sadato has linked the behavioral research with neurological activity. While scanning participants' brains, Izuma and his colleagues gave controlled feedback from (fictitious) judges, who ostensibly assessed the participant's trustworthiness. Their research showed that social approval results in activation of brain regions that are anatomically and functionally related to pleasure.[29]

Social reputation entails both push and pull elements. There is the positive allure of enhancing one's social reputation, but there is also the negative fear of losing one's social reputation. In Shakespeare's *Othello*, for example, Cassio gets into a drunken brawl and behaves in a way that he later recognizes tarnishes his good name. He draws attention to the importance of social reputation when he laments, "Reputation, reputation, reputation! O, I have lost my reputation! I have lost the immortal part of myself."[30] Indeed, empirical research shows that, apart from the positive allure, a parallel motivation is *fear of social disapproval*. When engaged in illicit behavior, for example, research indicates that fear of social disapproval often affects behavior more than fear of incarceration or fear of monetary punishment.[31] Accordingly, we can distinguish two socially related motivating feelings that contribute to altruistic acts: the allure of social approval or positive reputation and the fear of disapproval, embarrassment, or social ostracism.[32]

There are other incentives for engaging in altruistic acts apart from concerns about social reputation or wanting to reduce the pain of commiseration. Foremost is the feeling of *compassion*. Compassion is a vivid feeling of sympathy, pity, care, concern, tenderheartedness, kindness, or benevolence for a person experiencing the distress.[33] As noted in the previous chapter, even toddlers will approach and endeavor to comfort a complete stranger who is crying.[34] We feel compelled to console and assist those who weep.

In a study of charitable acts, William Harbaugh and his colleagues at the University of Oregon conducted brain scans while participants made various choices for the disbursal of a $100 account. They found that deciding to donate money to a local food bank was accompanied by activity in the medial forebrain pleasure circuit. For half of their participants, activation of this pleasure circuit was stronger when donating money than when receiving a monetary gift.[35] Acts of charity can be remarkably satisfying. Importantly, Harbaugh's experiment was cleverly designed so that no one, apart from the participant, was aware of their decision whether or not to donate money. Consequently, the experiment controlled for the possibility that the motivation for charitable acts could be attributed to possible positive feelings arising from social approval.

The pleasure of altruistic acts is recognized in the Bible, where Matthew offers advice regarding charitable giving:

> So when you give to the needy, do not sound a trumpet before you, as the hypocrites do in the synagogues and on the streets, to be praised by men. . . . But when you give to the needy, do not let your left hand know what your right hand is doing, so that your giving may be in secret.[36]

Here Matthew recognizes that charitable acts can lead to public praise or social approval, and he explicitly recommends humility, by giving anonymously so that one's reputation is *not* enhanced. But there is more in this biblical passage than an acknowledgment of the pleasures of social approval. With his "left hand . . . right hand" allegory, Matthew goes further: suggesting that we engage in charitable acts with the minimum conscious awareness of our own good deeds. In effect, Matthew points to yet another (fifth) source of pleasure arising from altruistic acts—namely, the private feeling of *virtue*. Virtue here is a feeling of positive self-esteem that arises when we recognize that our actions conform to some internalized norm of good moral conduct—whether or not our conduct is witnessed by others. In doing good, our sense of self-worth increases.

The feeling of virtue is quite different from the feeling of compassion. The object of compassion is external to oneself, whereas the feeling of virtue is internal. Virtue is all about enhanced positive self-assessment.

For Matthew, acts of charity should happen with a minimum of conscious awareness (where your left hand is unaware of the actions of your right hand). This unconscious strategy is preferred, lest we relish too much the positive feelings. In effect, Matthew recognizes the same two sources of pleasure distinguished in modern brain-imaging experiments: pleasures related to social approval and the feeling of virtue that ensues after performing altruistic acts. According to Christian ethics, we should eschew these pleasures because it is immoral to treat the poor or distressed as merely offering charitable opportunities through which we can feel good. While charity is to be encouraged, it can become morally questionable when others' misfortunes become the means by which we experience personal pleasure.

There remains one further proximal feeling that encourages us to engage in altruistic acts. Of the six *ultimate* motivations for helping described earlier, *alliance formation* plays a major role. Among social animals, building an alliance of mutually supportive friends or comrades is one of the very best ways of enhancing one's survival. Altruistic acts solidify existing bonds and help to establish new bonds. Responding to a weeping individual is one way to expand or bolster one's social network—an ultimate motivation. But what is the proximal motivation that encourages such bonding efforts? That is, what is the proximal *feeling* that encourages us to form bonds of friendship that altruistic acts facilitate?

In English, there is no word for this feeling. It is the warm fuzzy feeling of pleasure one gets from someone else's affection or fondness for you. In pair-bonding, it's the positive feeling, not of love, but of being loved. But this positive feeling state is present even in the much more muted appreciation of colleagues or casual friends. It's the anticipation of a future feeling of being the recipient of someone else's fondness or appreciation that constitutes a sixth proximal incentive for helping others. We'll refer to this pleasurable feeling as *bonding warmth*.

Of course one can help someone without necessarily feeling close to them. One can feel motivated to reduce one's own commiserative distress, experience compassion, feel virtuous, or enhance one's social reputation—all while helping someone you don't particularly like. However, when the aim is to build an alliance, our mutual loyalty is better assured if we experience feelings of warmth, connection, or affection for each other: the feeling of connection, affinity, or warmth should be reciprocated. And indeed, it is rare that we don't appreciate those who appreciate us.

So it's not simply that we anticipate the fondness a weeping person is likely to feel toward us when we come to their assistance but that we also feel some fondness toward the weeper at the outset. Here, pertinent research has been conducted by Janis Zickfeld of the University of Aarhus (Denmark) and his remarkable team of ninety-three worldwide collaborators. In a major cross-cultural study involving more than 7,000 participants from forty-one countries, they probed the feelings evoked in observers when seeing an apparently weeping individual.[37] In all of the cultures examined, they documented increased feelings of "connection" and "warmth" toward the weeping individual. This sense of closeness was observed for complete strangers and included people from divergent ethnic or racial backgrounds. The research also showed that these positive feelings of warmth and connection help dispose observers to initiate benevolent behaviors. Zickfeld and colleagues characterize tears as a "social glue."

Feelings of warmth or closeness are themselves positively valenced. The conjecture here is that these feelings of affinity, camaraderie, intimacy, warmth, social connectedness, or affection can motivate, or at least facilitate, acts of altruism.

Notice incidentally that even in situations where we are incapable of providing help, we are likely to feel positively disposed toward a weeping person. We may offer words of comfort or solace or, depending on our social relationship, engage in physical acts of consolation, such as hugging. If we can't help, we may still console.

Feelings Anticipated

Notice that feelings of virtue and social approval are states that can arise only *after* the completion of some charitable act. This raises the question of how such nominally proximal feelings might cause a person to initiate the behavior. Some readers might be skeptical of the claim that future feelings (such as the anticipation of social approval or of feeling virtuous) might help motivate immediate altruistic behaviors.

Pertinent research can be found in studies related to what might be called the "pleasures of opportunity." An animal that is not hungry can still appreciate the discovery of a stash of food. A person may welcome a job offer, even if they have no intention of switching employers. Someone may appreciate finding a lottery ticket, even though the ticket has no meaning apart from the (highly unlikely) possibility of future gain. Many examples of pleasures are linked to *opportunity* rather than actual rewards.

Ample evidence in support of this idea is to be found in classic research on addiction. Addiction research has long established how, over time, dopamine rewards shift from *consummatory* behaviors to *anticipatory* behaviors.[38] That is, dopamine release is more closely linked to "wanting" rather than "liking." Initially, dopamine is released in response to the act of consumption: we enjoy the act of eating a particular food, for example. With experience, however, the release of dopamine peaks in anticipation of the reward. We experience pleasurable feelings that precede (and so motivate) the consummatory phase.

In light of the study by Harbaugh and colleagues showing that altruistic acts activate the medial forebrain pleasure circuit, we would expect that the mere prospect or opportunity to engage in altruistic acts would itself be positively valenced. In fact, given the extensive existing research related to dopamine, it would be unusual to discover that the mere opportunity to behave altruistically is not experienced as pleasurable. In short, there are good reasons to suppose that the future prospect of a positive feeling (such as anticipation of feeling virtuous) would, by itself, be a positively valenced affective state, whether or not the initial feeling (such as the feeling of compassion) ultimately leads to actual helping behaviors.

In effect, a weeping individual offers observers an *opportunity* to engage in altruistic or charitable acts that we know to result in positive feelings. Although the situation likely entails some negative feelings of commiseration, the situation also amounts to a favorable moment to build social bonds, receive social approval, and experience positive feelings of compassion and virtue. Like the discovery of a lottery ticket, the compassion evoked by encountering a weeping individual offers a positively valenced feeling, even if that feeling doesn't ultimately lead to any altruistic action.

Denial of Good Feelings

As we have seen, the neuroimaging research suggests that altruistic acts and even altruistic thoughts activate regions of the brain consistent with the experience of pleasure. The negative feeling of commiseration notwithstanding, the main incentives motivating prosocial or charitable behaviors are the positive experience of compassion and the allure of future feelings of virtue, social approval, and bonding warmth. Although we may be cognitively aware of the stress and misery experienced by someone who is in the throes of grief, as observers, our prosocial feelings are largely positively, not negatively, valenced.

At first, this claim that observer emotional responses to weeping are largely positive seems questionable. When we witness someone crying, our experience is not necessarily one of delight or elation at the prospect of being able to help. Suppose your beloved pet had died and you are crying in the presence of a visiting friend. Suppose further that your friend relayed the following sentiment to you: "Thanks so much for giving me the pleasurable opportunity to console you in your moment of need." Such an expression would seem nothing short of callous. Nor would anyone do this. The experience for you is clearly negative. Moreover, the event or circumstance leading to this state is highly salient: the specific loss, bereavement, misfortune, or catastrophe will be foremost in everyone's mind—for both the observer as well as the grieving individual. When interacting with someone experiencing stress, admitting to the stressed individual (or to ourselves) that helping is pleasurable would be deemed cold-hearted, if not immoral. Nevertheless, the positive feelings (such as the feeling of compassion and the anticipation of feeling virtuous) are crucial since they are the principal feelings that motivate the observer to offer assistance. If the feelings were wholly negative, there would surely be less incentive to help: helping is an approach behavior.

Our cognitive awareness of the stressful context has the effect of pushing the positive feelings into the mental background, rendering the positive affect less apparent to us. In light of the contagious feeling of commiseration, we mistakenly confuse the stress of the weeper as the overarching emotion felt by everyone. The confusion, I propose, is understandable but not an accurate account of the divergent feeling states between the observer and weeper.

Machiavellian Misgivings

Of course, positive proximal feelings such as virtue and compassion have evolved precisely because altruistic acts commonly improve the fitness of the person who engages in such prosocial behaviors. Regrettably, such thoughts can be deeply uncomfortable. Machiavellian motives to maximize biological fitness are simply offensive when viewed from common moral standards. It's disturbing to think that selfless acts have selfish origins. It helps, however, to recognize that signal behaviors evolve, not because they benefit one individual over another, but because they confer benefits to both.

There is a second source of discomfort that commonly leads people to doubt such biological accounts: the narratives fail to align with our subjective experiences. When we see someone in need, our motivation to help does not arise from feelings of delight—anticipating that our social standing will be enhanced or that the needy person will rack up a future obligation toward us. Rather, our motivation to be helpful is to be found in our sensitivity to the plight of others—as manifested especially in the feelings of compassion and social connection. Conversely, when we ourselves experience grief, our weeping does not feel like some clever way of manipulating others in order to commandeer resources for ourselves. Rather, our subjective experience simply reflects an overpowering sense of loss or distress.

The key is recognizing that, phenomenologically, we live entirely in the world of proximal motivations, not in the world of ultimate causes. Nor is the disconnect between ultimate and proximal motivations unique to weeping or grief. For example, while love may simply be Nature's way of encouraging pair-bonding and procreation, the feeling of love is no less profound an experience despite its prosaic biological origin and function. We must simply make our peace with this moral paradox. We are destined to live our lives in a proximal world of grief, compassion, love, and myriad other emotions—without regard for the ultimate world of biological fitness that enabled the emergence of these feelings.

If, as I now claim, weeping is an evolved ethological signal, then the response of observers to weeping must necessarily be innate. Although humans have a notable capacity for cognitive control that can sometimes inhibit or dampen these responses (a topic we will discuss at length in chapter 6), and although humans differ in their responsiveness (a topic featured in chapter 19), when we encounter someone weeping, the pertinent evoked feelings (such as compassion) are rarely absent. Since the observer responses are innate, the observer's subjective experience will tend to be dominated by the proximal feelings described in this chapter. The very automaticity and vividness of these proximal responses will consequently tend to mask or obscure the underlying ultimate biological motivations.

Catharsis

Before bringing our chapter to a close, it's appropriate to revisit the question of how weeping benefits the weeping individual. We have supposed that the principal benefit is the assistance from observers that weeping commonly induces. However, we might briefly consider a venerable alternative theory of the value of weeping—namely, the theory of catharsis. The theory was famously proposed by Aristotle more than two millennia ago. Aristotle argued that negative feelings such as grief can be beneficial when expressed in a controlled environment.

The word "catharsis" has rather coarse origins. In ancient Greek, a "kathartic" (noun) is a type of laxative that accelerates defecation. In using this word, Aristotle recruited a colorful metaphor to refer to a type of beneficial emotional purging. By allowing people to express strong feelings, a person could, in effect, "blow off steam."[39] For Aristotle, the experience of catharsis "cleanses the mind."

The notion of catharsis found a modern champion in the psychotherapy of Sigmund Freud.[40] Freud suggested that the effectiveness of psychotherapy might be attributed to a cathartic effect. By reliving negative experiences through the controlled environment of the psychiatrist's couch, the symptoms of various psychoses might be alleviated. This notion has spread widely—and not just among psychiatrists, clinical psychologists, and health professionals. In the case of grief, the idea of "working through" an emotion has entered into popular consciousness. The popularity of this idea has been documented by Randolph Cornelius, a professor of psychology at Vassar College. Cornelius carried out a comprehensive survey of English-language media reports spanning 140 years. In popular articles that discuss crying, Cornelius found that 94 percent assume that overt crying will reduce psychological tension and produce a cathartic effect.[41]

Given the popularity of this concept in both the arts and mental health, one might expect that considerable research has been carried out testing the presumed cathartic effects of expressed emotions. Indeed, a large body of research exists, although much has focused on the specific case of the presumed benefits of crying.[42] The extant research includes retrospective studies about past crying episodes, laboratory experiments in which participants have been induced to cry, studies of the effect of crying frequency on short-term and long-term health, and large-scale cross-cultural studies of crying behaviors. The results are surprising.

The received wisdom might lead one to expect a positive relationship between crying and health status or well-being. Instead, study after study points to the reverse relationship. People who cry the least are healthier.[43] In laboratory studies, the research consistently fails to show any cathartic effect from crying.[44] Inducing people to weep in nonmourning contexts—such as watching a sad movie scene—has largely failed to induce physiological changes associated with improved health or well-being.[45] In their

comprehensive review article, "Crying, Catharsis, and Health," Suzanne Stougie and her colleagues summarized the pertinent research literature as follows:

> Crying . . . does not appear to lead to any appreciable decrease in emotional ten-
> sion or distress, or to an improved psychobiological functioning. Indeed, results
> of various studies seem to point in the opposite direction, relating crying to
> increases in arousal, tension, and negative affect.[46]

Curiously, the formal research is directly at odds with survey studies in which
a substantial proportion of respondents report that their crying episode had
beneficial effects.[47] For example, the International Study on Adult Crying
surveyed some five thousand participants from thirty-five countries. About
50 percent of participants reported "feeling better" after crying.[48] Tellingly,
the positive effect of crying turns out to be strongly correlated with the presence of another person. Individuals who cry alone report little benefit and
commonly feel worse. The greatest improvements in reported well-being are
found when crying occurs in the presence of someone else. Specifically, crying is most beneficial when an observer reacts in a consoling or comforting
manner. It's the caring, sympathetic, and altruistic acts of others that appear
to account for the mental and physical benefits of crying.

What the research *does* show is that crying has a beneficial impact on social
relationships. Among women, crying helps to cement close friendships. In
the presence of men, crying transforms male behavior so that indifferent or
aggressive behaviors are replaced by sympathetic and altruistic behaviors. In
short, studies have documented that the principal benefits from crying are
those that relate to the social support brought by others—consistent with
weeping as an ethological signal.[49] Despite broad popularity of the concept of
catharsis, there is little evidence that the act of weeping in itself is beneficial
to the person crying. As Randolph Cornelius has noted, if crying might be
regarded as having any "cathartic" effect, it is principally the result of positive social support and the capacity of crying to suppress the hostile behavior
of others.[50]

All of this suggests that there is little empirical support for Aristotle's venerable theory of catharsis. Specifically, there is little evidence that grief-related
weeping, by itself, has any beneficial effect. However, this doesn't necessarily
mean that all forms of private weeping are antithetical to one's well-being or
enjoyment. As we will see later, apart from tears of grief, there are many other
tear-inducing circumstances that may be experienced positively.

Reprise

In the previous chapter, we reviewed the physiological and behavioral characteristics of grief and noted that weeping appears to exhibit the hallmarks of an ethological signal. In that chapter, we noted that signals evolve only in non-zero-sum conditions in which the signal induces an observer to behave in ways that benefit both the signaler and the observer. If weeping is truly a bona fide signal, then weeping episodes must typically benefit both the weeper and the observer.

In the case of grief-induced weeping, observers commonly respond by behaving in a prosocial (altruistic or charitable) manner toward the weeping individual. The benefits to the weeping individual can include receiving assistance, suspending aggression, or simply inducing a favorable (warm) attitude from observers.

The prosocial behaviors induced in an observer can ultimately benefit the observer in a number of ways, including possible reproductive gains through kin support, the banking of future return favors (reciprocal altruism), increased social status with respect to the person helped, enhanced prosocial reputation among spectators, and the avoidance of social ostracism for failing to assist. In order for a signal to evolve, the behaviors induced by a signal must commonly enhance fitness. However, the impetus to produce the appropriate behaviors is in the form of *proximal* motivations—as manifested in the form of feeling states or emotions.[51] The ultimate or biological benefits are typically opaque to us.

The prosocial behaviors induced in an observer of weeping appear to hinge on six proximal feelings. First is the contagious feeling of personal distress or vicarious *commiseration* that motivates efforts to alleviate the weeper's suffering as a way of reducing the observer's own feeling of distress. Second is the positive feeling of sympathy or *compassion*—a vivid feeling of pity, care, concern, tenderheartedness, kindness, or benevolence for a person experiencing distress. A third motivator is anticipation of the positive feelings arising from an enhanced social *reputation* for engaging in altruistic acts. Conversely, a fourth motivator is the *fear of social ostracism* for failing to act in a helpful manner. A fifth motivator is anticipation of the private feeling of *virtue* or enhanced self-worth that arises following acts of charity. And finally, a sixth motivation is the positive feeling of *bonding warmth* arising from the prospect of expanding or bolstering one's network of friends or companions to include the weeping individual.

In a later chapter (chapter 6), we will consider in more detail the balance of benefits and costs incurred for both the signaler and the observer. In that chapter, we will also discuss the notable human cognitive capacities that can amplify, mask, or inhibit the innate dispositions induced by a weeping signal. For example, we will see situations where an observer can resist and override the feeling of compassion when observing someone crying.

As we have repeatedly emphasized, weeping can be evoked by many emotions apart from grief. In this chapter, we have considered only the grief-related context. We will address other weeping contexts in chapter 7 ("A World of Tears"). Although not evident at this point, we will see that these other weeping situations produce the same prosocial effects in observers and typically lead to behaviors that benefit both individuals. The fact that weeping can arise from a number of different emotions reminds us that weeping itself is not an emotion. Instead, weeping is solely an ethological signal that can be triggered by many different emotions. At the same time, *grief* should not be viewed as a signal: it is a form of sadness emotion that routinely recruits the weeping signal. In some cases, grief results only in a *choked-up* response that is typically invisible to observers and so does not qualify as a signal—although the choked-up symptoms imply the early stages of a possible explicit signal.

In contrast to grief, we saw in the previous chapter that melancholy looks like an ethological cue. If melancholy is indeed a cue, then it follows that melancholy, unlike most grief, isn't designed to be communicated to observers and so isn't designed to change an observer's behavior. Attentive observers might infer that someone is experiencing melancholy, but the "long face" or low-energy voice isn't produced as ways to broadcast, convey, or transmit that state by the melancholic person. This raises an intriguing question: What, then, is the biological purpose of melancholy? In the next chapter, we turn to consider this question.

5 Melancholic Realism

One of the characteristic features of melancholy is the phenomenological feeling of malaise—a glum, forlorn, cheerless, or bummed-out state. Melancholy represents a classic unpleasant emotion.

The idea that positive and negative emotions provide the carrots and sticks of behavior seems obvious. However, this seemingly straightforward story seems questionable once we attend to actual behavior. Many people enjoy riding fear-inducing rollercoasters, watching horror films, eating spicy food, and enduring cold and fatigue to climb tall mountains. So what exactly do we mean by a "positive" emotion? On what basis can we distinguish positive from negative feelings?

People are commonly able to introspect and report whether they find some experience pleasurable or not. We are also adept at distinguishing those aspects of some experience we find pleasant from those aspects we find unpleasant. But apart from introspective reports, a common way to recognize the distinction is that positive emotions are states that a person wants to repeat or continue, whereas negative emotions are states that a person wants to avoid or end. If a music lover listens to music that brings tears to their eyes, then repeated listening to the tear-inducing music suggests that the overall experience must be positive in some sense.

Research (reviewed later) suggests that it's common to experience "mixed" (simultaneous positive and negative) emotions. For example, a mountain climber may experience acute negative feelings such as suffering from extreme cold or muscle pain. Given the opportunity, a climber will take steps to reduce these aspects of the experience, suggesting that these experiences are negatively valenced. At the same time, a sense of impending achievement and the prospect of social accolade is often strongly positively valenced and so encourages the climber to persevere and reach the summit. Indeed, the ability to endure painful challenges typically enhances both individual feelings of pride as well as social expressions of admiration. The same mixture of positive and negative emotions can be observed, for example, in something as simple as eating a calorie-rich dessert. On the one hand, there is the positive gustatory temptation; on the other hand, one might experience feelings of nutritional guilt. The negative feeling of guilt may or may not be sufficient to overcome the positive allure. Whether one engages in a given activity is dependent on the relative balance of positive and negative feelings, a balance that can differ from person to person.

As in the case of pleasurable and painful sensations, happy and sad moods are known to shape behavior in useful ways. Unlike the positive and negative sensations of eating delicious food or stubbing your toe, the benefits of nonsensory states such as happiness and sadness are often less immediate. Instead of directly shaping physical behaviors, these mood states tend to influence how we think. Positive and negative moods shape cognition, which, in turn, shapes future behaviors.

Past emotional experiences also offer teachable moments. Our past experiences help us anticipate positive and negative feeling states and so motivate us to formulate effective strategies for the future.[1] Of course, these proximal feelings arise from ultimate motivations intended to increase biological fitness. So anticipating our future emotions ultimately helps us to behave in ways that are commonly adaptive.

Positive moods are known to confer a number of cognitive benefits. For example, feelings of joy encourage play activities that contribute to long-term physical and social development.[2] The positive feeling of interest or curiosity encourages exploration, which offers new experiences and valuable learning opportunities.[3] Positive mood has been shown to contribute to patterns of thought that are more flexible, integrative, and creative and draw on a wider field of attention.[4] Happiness encourages more lenient judgments.[5] When in a pleasant mood, we tend to simplify complex sources of information, and these simplifications often improve the efficiency with which a task is completed.

American psychologist Barbara Fredrickson has drawn attention to the many cognitive benefits of positive emotions in her "broaden and build" theory.[6] In general, when engaged in problem-solving activities, positive emotions or moods encourage people to take a broad "big picture" approach;

moreover, repeated problem-solving experiences under conditions of positive mood ultimately build long-lasting intellectual coping skills.[7]

At the same time, positive mood is also known to foster *optimism bias* where people tend to hold unrealistic expectations regarding the likelihood of positive future outcomes.[8] Positive mood leads to greater risk-taking and a willingness to gamble even when the likelihood of success is low.[9] For example, Kevin Au and his colleagues at the Chinese University of Hong Kong conducted a series of experiments to determine the effect of mood on traders in financial markets. They employed an internet platform that simulated foreign exchange trading based on historical market data. In one experiment, they used happy and sad music to induce different moods in the traders. They found that traders exposed to the happy music made inferior trades and lost money, whereas those traders exposed to the sad music made better decisions resulting in greater profits. The sad-mood traders behaved in a more prudent manner, whereas the happy-mood traders were overconfident and took greater risks.[10]

Happiness-induced risk-taking is not limited to money market traders. Ross Otto and his colleagues at New York University showed that people are more likely to purchase lottery tickets on sunny pleasant days, especially when preceded by several cloudy days.[11] Similarly, lottery ticket sales tend to spike following major sporting events in which a local team has won.

Apart from overtly positive mood, even normal or typical moods can encourage overly optimistic assessments of the likelihood of achieving certain goals.[12] Randy Nesse at Arizona State University has suggested that the optimism that characterizes normal mental life encourages individuals to strive to achieve goals that might be attainable with effort: nothing ventured, nothing gained.[13]

One might suppose that people tend to become pessimistic when experiencing melancholy. However, the research roundly contradicts this intuition. In fact, the research suggests that we are at our most realistic when sad—a phenomenon called *depressive realism*.[14] Compared with a happy or neutral mood, melancholy promotes more detail-oriented thinking, reduced stereotyping, less judgment bias, greater memory accuracy, reduced gullibility, more patience, greater task perseverance, more social attentiveness and politeness, more accurate assessments of the emotional states of others, a heightened sense of fairness, improved reasoning related to social risks, and a disposition to favor delayed larger rewards over immediate smaller rewards.[15]

When approaching some problem, a person might employ a variety of cognitive strategies. Happiness encourages people to maintain whatever cognitive strategy they have been using, whereas melancholy encourages people to change their approach to a problem.[16] In experimental studies, for example, inducing melancholy in a person tends to cause them to shift between different ways of thinking.[17] Although melancholy tends to encourage a more conservative and less risky approach, it also motivates us to abandon failing strategies and try something new. Sad feelings appear to be especially adaptive for analyzing complex problems.[18]

Randy Nesse has suggested that depressive realism provides a mental "grounding" or "reality check" when our goals prove elusive.[19] That is, low mood is likely to discourage futile efforts that may squander resources. A number of researchers have noted that when normal melancholic sadness is misdiagnosed as depression, the use of antidepressive drugs is actually detrimental to individual well-being.[20]

Klaus Fiedler at the University of Heidelberg has observed how the influence of positive and negative mood on cognition bears a striking resemblance to Jean Piaget's distinction between two contrasting approaches to learning—*assimilation* and *accommodation*.[21] When we encounter new information, that information may either confirm or challenge our existing presumptions or expectations. If the new information is consistent with our existing mental model or schema, then that schema helps us to make sense of the new information. The new information is then simply absorbed or *assimilated* into the current schema. However, if the new information conflicts with what we expect, then the new knowledge can be *accommodated* only by either altering our existing schema or creating a new schema altogether. Both assimilation and accommodation are important forms of learning. Fiedler has noted that positive moods or emotion states tend to be associated with the process of assimilation, whereas negative moods or emotion states tend to be associated with the process of accommodation.[22]

Depressive realism is not limited to human beings. Consider, for example, a clever experiment carried out by Stephanie Matheson, Lucy Asher, and Melissa Bateson at Newcastle University in northern England.[23] The researchers trained birds (starlings) to peck at one of two levers depending on the sound they heard. When the sound was short (two seconds), pecking at the blue lever would result in a pellet of food. When the sound was long (ten seconds), pecking at the red lever would produce a food reward. However, with the long tone, the food reward was delayed for some time instead of being produced immediately. Naturally, the birds preferred instant gratification: they much preferred the short-tone/immediate-food trials over the long-tone/delayed-food trials. Of course, if a bird pecked at the wrong lever, then no food reward was given.

The interesting part of the experiment occurred when Matheson and her colleagues played tones of intermediate duration. How would the birds respond to, say, a six-second tone? Would they peck at the blue lever (interpreting the tone as "short") or at the red lever (interpreting the tone as "long")? And what about a four-second tone or an eight-second tone? An "optimistic" bird might be biased to interpret intermediate-length tones as "short" (with the hopeful consequence of immediate food). A "pessimistic" bird might be expected to interpret intermediate-length tones as "longer" (with the less desirable consequences of having to wait for the food).

There was one other component to their experiment. Matheson and her colleagues divided the birds into two groups. One group (the "privileged" birds) lived in premium large cages, with plentiful water, a birdbath in which to wash, branches and other playthings, and frequent cleaning. The "deprived" birds were housed in small cages, with unpredictable water and bath access and no playthings. The birds who lived in the more privileged circumstances exhibited much greater optimism when interpreting the tones of intermediate duration. Even for an eight-second tone, the privileged birds were significantly more likely to interpret the sound as "short" compared with the deprived birds. Moreover, the deprived birds were more accurate, interpreting the intermediate tones according to their objective nearness to the original two-second and ten-second tones. In other words, the deprived birds weren't more *pessimistic* than the privileged birds; they were more *realistic*. Like people in the throes of melancholy, those birds who lived less happy lives were more accurate in their appraisals.[24]

Interestingly, philosophers have long drawn attention to the value of sadness when engaged in deep thought. For example, philosophers as different as John Stuart Mill, Søren Kierkegaard, and Friedrich Nietzsche have suggested that melancholy is beneficial to the philosophical enterprise. Lord Byron characterized melancholy as "the telescope of truth."[25] A propensity for more detail-oriented thinking, less judgment bias, improved memory accuracy, and a greater willingness to abandon unproductive strategies would appear to be well suited to deep philosophical analysis. Indeed, experimental studies by the Hungarian-Australian psychologist Joseph Forgas have shown that people produce more effective and persuasive arguments when in a sad mood than when in a happy mood.[26] Many of the cognitive characteristics of depressive realism are recognizable components of *skepticism*, perhaps the foundation of all philosophizing. The cognitive benefits associated with positive emotions tend to be rather amorphous compared with the more situationally focused cognitive benefits associated with negative moods.[27]

Reflection, Rumination, and Depression

The apparent cognitive benefits associated with melancholy appear to contradict a classic adverse symptom associated with major depressive disorders—namely, *rumination*. Rumination is a cognitive state in which an individual repeatedly recalls past situations or failures, dwelling on negative thoughts and self-assessments.[28] When depressed, we tend to view ourselves as worthless, doomed, blameworthy, deficient, or unlovable. Clinical research recognizes rumination as broadly destructive and unhelpful. So how do we reconcile the benefits of depressive realism with the harmful effects of rumination?

In a seminal study, Paul Trapnell and Jennifer Campbell of the University of British Columbia recruited a large sample of people experiencing negative (melancholic and depressive) symptoms. In their analysis, they found clear evidence for two different patterns of thought. For one group of participants, their thoughts were dominated by brooding negative self-assessments, a pattern dubbed *rumination*. For the second group, their thoughts emphasized self-awareness and self-knowledge, a pattern dubbed *reflection*.[29] In short, Trapnell and Campbell found clear evidence of a cognitive distinction between (pathological) depression and (normal) melancholic sadness.

The beneficial value of melancholy can be illustrated through a story—what might be called the *Parable of Pat*. In applying for college admission, Pat applied exclusively to the most prestigious universities in the country. After the passage of some months, Pat received a bundle of rejections that caused her to experience an episode of melancholy. Pat withdrew to her bedroom, didn't respond to messages from her friends, and skipped her usual exercise regime. Instead, her principal activity was *reflection*. In this melancholic state, Pat began to question her strategy. She entertained thoughts that she is not as accomplished as she might have supposed. Later in the year, she applied to a number of less prestigious institutions and was pleased to receive several acceptances. As we have seen, research shows that melancholy produces a number of cognitive improvements, including more accurate self-perception. In this case, an episode of melancholy led to a reassessment of her goals, caused the abandonment of a failing strategy, and resulted in a plan for a course of action that had a better chance of success. In the words of Paul Gilbert, melancholy stops us from chasing rainbows.[30]

The beneficial value of reflective thought is evident in several studies related to depression.[31] One might suppose that encouraging depressed patients to think reflectively (rather than ruminatively) might lead to improvement. And sure enough, that's the case. For example, having depressed patients reflect on their condition through expressive writing has been shown to alleviate (rather than amplify) their depressive symptoms.[32] It appears that reflective thought is simply another name for depressive realism.

Overall, the research implies a clear distinction between melancholic sadness and depression. Normal melancholy is an adaptive behavior likely to enhance fitness.[33] Depression, by contrast, is a pathology; it arises when the normal sadness system is broken. Peter Freed has aptly characterized depression as "a sadness disorder."[34] In light of the distinction between depression and melancholy, a better name for *depressive realism* would be *melancholic realism*. A change of terminology is especially warranted in order to prevent depressed patients from regarding "depressive realism" as a justification for their state. Consequently, for the remainder of this book, we will use the term *melancholic realism*.

As in the case of physical pain, the value of psychic pain is most evident when it is absent. I once had a colleague (let's refer to him as Adrian) who experienced several months of general malaise that led his physician to prescribe Prozac. Prozac increases the concentration of serotonin by interfering with the chemistry through which serotonin is broken down. After a couple of weeks, the effect of the Prozac was noticeable. Adrian was his old self, smiling and joking, confident. In private, Adrian was frequently critical of the dean and would also voice complaints about the college president. A month after starting Prozac, we had a faculty meeting with the department heads, the dean, and the president. Adrian took this opportunity to publicly declare precisely what he thought of both the dean and the president. He left the meeting feeling satisfied that after all these years, he was finally able to say what he really thought. Months later, having weaned himself off Prozac, Adrian sat in my office with his head in his hands. "Did I really do that?" he asked soberly.

Of course it was good that Adrian had come out of his period of general unhappiness. The Prozac seemed to have helped him through a rather gloomy time. But psychiatrists are well aware of the downside to increasing serotonin. If you artificially raise serotonin levels in a low-ranked male baboon, he is likely to initiate a fight with the reigning alpha male in the troupe. The low-ranked male is unlikely to win this fight and is in danger of grave physical injury. Moreover, the failed challenge is likely to have long-term negative social effects for the defeated male. Occupying a low position in the social register may not make you feel good, but it is probably better than suffering bouts of retaliation for ineffectively challenging the entrenched leadership.

As Randy Nesse has pointed out, sadness plays an essential role in life.[35] When resources are scarce, melancholy leads to conservation of energy. When a life strategy leads to repeated failure, reflective thoughts typically lead us to entertain alternative approaches and to set more realistic goals. Finally, it isn't just people with low social status whose decision-making benefits from periods of melancholy. People at the highest echelons of society make better decisions when they experience bouts of appropriate sadness. In the same way that physical pain is necessarily for good health, the ability to feel psychic pain is also essential. Feeling melancholy in the right circumstances is part of good mental health.

Purpose of Melancholy and Grief

In chapter 3, we saw that melancholy resembles an ethological cue, whereas grief-induced weeping resembles an ethological signal. Recall that cues are not intended to be communicative. Signals, by contrast, are overtly communicative and commonly change the behavior of observers to the benefit of the signaler. Melancholy is a private, reflective, *self*-directed state, whereas grief is a public *other*-directed state.[36]

When a person faces difficulties in life, several possible resources can be recruited to help deal with the situation. One resource is our own mental capacities. We have seen that normal melancholy is associated with reflection (as opposed to rumination). In addition, we've seen that negative mood favors focused, detail-oriented, skeptical contemplation over the creative, optimistic, "big picture" thinking associated with positive mood. In stressful conditions, there are good reasons to suppose that the cognitive dispositions characteristic of melancholic realism are better suited to addressing the source of the stress. Through focused analysis, less biased memory and perception, more realistic appraisal, consideration of options, and a greater willingness to change directions, an individual can strategize—forming a plan of action that may help resolve, overcome, or ameliorate the stressful situation.

Apart from our own cognitive capacities, a second resource for handling stress is our social network. Here weeping provides an essential tool—an ethological signal—through which the assistance of others can be solicited. Companions, partners, acquaintances, and even total strangers may be induced to intervene and provide crucial support in dealing with a distressing circumstance. As we saw in the previous chapter, significant benefits can motivate an observer to be helpful, so weeping often initiates a win–win interaction.

The theory proposed here is that melancholy and grief are distinct yet complementary states that arise in response to difficult or stressful conditions: melancholy rallies *cognitive* resources to address a challenging situation, whereas grief recruits *social* resources. Both melancholy and grief can contribute to biological fitness.

This theory has repercussions for interpreting the commonly observed mourning cycle. According to the theory proposed here, the alternation between periods of quiescent melancholy and active grieving represents phases of inward-directed and socially directed behaviors. The proportion of time allotted to each behavior is likely to be shaped by the severity of the situation, the ability of the individual to cope with the situation alone, the presence of sympathetic observers, the capacity of others to be able to offer genuinely useful assistance, and the willingness of the stressed individual to incur the social cost associated with appeals for help. Minor stresses are apt to lead to melancholy without grief. Major stresses are more likely to benefit from the help of others and so result in grief. We will consider the relative costs and benefits of various behaviors in more detail in a later chapter.

The suggestion that melancholy is a self-oriented cue whereas grief is an other-oriented signal raises a number of questions. If weeping is intended to change the behavior of observers, why would anyone cry alone? Similarly, if weeping is intended to be communicative, why would anyone attempt to suppress or hide their weeping? Although melancholy is difficult to recognize, why do observers nevertheless feel eager to offer to help a melancholic individual, not just a grief-stricken individual? These and other questions will be addressed in due course.

Seeking Altruistic Opportunities

As we saw in chapter 4, acts of altruism typically result in long-term benefits for the person offering assistance. As noted earlier, from the perspective of the observer, a weeping person amounts to an *opportunity*. If altruistic acts are so valuable, why wait for someone to signal that they are experiencing stress? There are circumstances where a person can benefit from the help of others, even when that person doesn't weep. Accordingly, the benefits of engaging in acts of altruism are not limited to situations where we encounter someone crying. In fact, most helping behaviors arise without recourse to tears.

Of course, offering to help someone has somewhat less value to that person if no help is needed. Offers to help are most valuable when the person is stressed. Weeping aside, we might then expect observers to be vigilant for signs of stress that represent an opportunity to engage in altruistic acts. One such stressed state is melancholy.

Although melancholy bears the hallmarks of an ethological cue rather than a signal, cues nevertheless remain informative for observers. Since any stressful state represents a potential opportunity for an observer, we would expect people to be vigilant to recognize melancholy in others. Knowing that someone is feeling sad (whether melancholy or grief) is generally valuable.

Depending on the circumstance, an observer may be motivated to offer assistance even when a melancholic person isn't aware of having inadvertently communicated their sadness. Moreover, even if a melancholic person doesn't want others to know they are feeling sad, vigilant observers are very likely to benefit from deciphering the saddened state. For example, a boss might delay conveying a criticism or decide not to assign additional work for a melancholic employee. Conversely, if the sad person and the observer are competitors, an observer might take advantage of the situation by compounding the stresses experienced by the melancholic person. Alternatively, an observer may conclude that it's best to leave a melancholic person alone.

In chapter 3, I described a simple survey chronicling the common phenomenon of mistaking a relaxed face (such as can be readily seen on a bus or train) for a melancholic face. If (as we noted in chapter 3) the features of melancholy are simply artifacts of low physiological arousal, it makes sense that tired, relaxed, or sleepy individuals might commonly be misperceived as sad. However, the reverse scenario seems less common: people do not often mistake a melancholic person for a relaxed person. This asymmetry can be plausibly attributed to contrasting social costs associated with different errors. When inferring whether a person is sad, two types of errors are possible: one might falsely attribute sadness to someone who is not sad (what methodologists call a type I error) or fail to recognize sadness in someone who is truly sad (a type II error). There are good reasons to suppose that the type II error would be the more serious mistake. Failing to recognize that someone is sad (especially a loved one) may have onerous repercussions. Accordingly, it is reasonable to conjecture that observers should be biased toward interpreting low-arousal states as suggestive of melancholy rather than relaxation or fatigue. This bias would explain why we tend to interpret the relaxed faces of people riding public transit as symptomatic of a sea of sadness. The passengers with the animated faces are the ones interacting with friends or colleagues.

Apart from the difficulty of recognizing melancholy compared with grief, there are at least three other differences between grief-induced responses and melancholy-induced responses. First, as a cue rather than a signal, the proximal motivations to help are less compelling in the case of melancholy. Our responses to melancholy are more likely to be dominated by cognitive rather than affective origins. So feelings such as commiseration or compassion are typically less strong, less reliable, and less automatic compared with responses to observed weeping.

Second, a stressed person may experience melancholy rather than grief for the very good reason that, given the nature of the stress, other people may be unable to offer genuinely useful assistance. (Very few people would be able to help a melancholic mathematician whose apparent proof of the Collatz conjecture turns out to be flawed.) Accordingly, acts of altruism are less likely to be effective in the case of melancholy compared with grief.

Finally, we need to mention situations where the individual being helped doesn't want any help. Unwanted assistance may nevertheless incur an implied cost of being expected to return some possible favor in the future and a relative loss of social status with respect to the nominally helpful person. The Japanese have a term for this—*arigata-meiwaku*—which is the irritation one feels when someone has done you a favor you didn't want, but you're expected to be grateful anyway.[37]

The problem of unwanted help aside, the important point here is that there are times when offering assistance can benefit an observer, apart from responding to explicit appeals for help in the form of grief-induced weeping. Accordingly, we are sensitive to signs of stress in others, including subtle signs of possible melancholy. Indeed, we may be hypervigilant for such signs—seeing melancholy when only relaxation or fatigue is present.

Reprise

In this chapter, we have considered some of the research chronicling how our moods affect the way we think. The research suggests that positive and negative mood states offer complementary cognitive benefits. Positive moods entail feelings of satisfaction, contentment, and safety. Safe and secure circumstances give us the freedom to engage in more wide-ranging or adventurous thoughts. We are more likely to entertain creative or speculative ideas and to consider plans that may involve notable risk. Negative moods are associated with times when we are stressed—circumstances where a more vigilant, careful, skeptical thought process is warranted.[38]

In the past, psychological research has tended to highlight the value of positive mood on productive thinking. However, ongoing research has established that negative mood offers complementary cognitive benefits that help us to think through stressful situations in constructive ways. In particular, a pattern of thought—referred to as *depressive realism* (preferred: *melancholic realism*)—appears to be admirably tailored to deal with the sorts of stressful circumstances that would lead a person to experience melancholic sadness.

At the same time, we've learned that melancholy is not the same as depression. As we've seen, major depressive disorders are associated with cognitive symptoms of *rumination* (brooding negative self-assessments), whereas melancholy is associated with *reflection* (self-aware and situationally aware contemplation). Depression offers none of the cognitive benefits of ordinary melancholic sadness. In general, happy people are overly optimistic, depressed people are overly pessimistic, and melancholic people are the most realistic.[39]

In recent decades, psychiatrists, psychologists, and other clinical practitioners have reevaluated the experience of melancholic sadness. When melancholy is properly distinguished from depression, it becomes clear that, like the absence of pain, the absence of melancholic feelings may be detrimental to good health. Pharmaceutical interventions that treat normal melancholy with drugs designed for depression are apt to cause more harm than good.

In conjunction with the previous chapter, it should now be clear why there exist two sadness states (melancholy and grief) rather than just one.

When addressing a stressful situation, melancholy and grief represent two different coping strategies. Grief helps us deal with the stressor by enlisting the assistance of others via the ethological signal of weeping. Melancholy helps us deal with the stressor by fine-tuning our cognitive style so as to favor more focused, flexible, and realistic analysis and planning.

In this chapter, we have also noted that helping behaviors can benefit the altruistic person, even if a stressed recipient is experiencing only melancholy rather than making an overt appeal for assistance by weeping. At the same time, we have highlighted the difficulties in recognizing melancholy in others. It's not easy to distinguish melancholy from other low- arousal states. Nevertheless, as observers, we seem to be extraordinarily vigilant (if frequently mistaken) in trying to detect melancholy in others.

6 Automaticity and Control

In the ancient classic *Naturalis Historiae*, the first-century Roman philosopher Pliny the Elder argued that virtually all human behaviors are learned. Pliny argued that not only are language and customs learned, but the very act of speaking—and even walking and eating—are all learned. There was, according to Pliny, only one behavior that is innate to human nature: crying.[1]

Instinctive behaviors are commonly defined as complex behaviors that are not learned and that arise spontaneously in predictable circumstances. When applied to human behavior, the idea of "instinctive" or "innate" behaviors has fallen out of favor.[2] For example, a survey of the dozen most popular English-language psychology textbooks from 2000 found that none of the books refer to human instinctive behaviors except when discussing antiquated obsolete theories.[3]

In studies of nonhuman animals, research suggests a range of behaviors from the highly constrained to the highly flexible. Early research emphasized the predictable and invariant aspects of many animal behaviors. For example, a wet dog will shed excess water by instinctively rotating its torso back and forth in a formulaic and automatic fashion. Ethologists have chronicled innumerable such stereotypic animal behaviors—behaviors variously dubbed "fixed action patterns" or "innate releasing mechanisms."[4] However, subsequent research has played down the notion that many or most animal behaviors are fixed or inevitable.[5] For mammals in particular, some degree of executive control may override or modify behaviors that otherwise appear to be automatic.

Compared with other animals, humans have a much greater capacity for self-control. Large regions of the frontal cortex are known to serve inhibitory functions—suppressing, masking, or modifying otherwise compelling behaviors. Weak neural connections in the frontal lobes of the brain are implicated in impulsive behavior.[6] Despite the reluctance to characterize any human behavior as innate, weeping appears to be a widely acknowledged exception.[7] Our capacity for executive control notwithstanding, even when we don't want to cry, weeping may be inescapable.

Cross-cultural studies of crying have established that both infant and adult forms are highly similar around the world.[8] Infants with hydranencephaly (born without cerebral hemispheres) still cry when experiencing discomfort or stress. Weeping by congenitally deaf individuals includes typical weeping vocalizations, and weeping by congenitally blind individuals includes the usual shedding of tears. You don't ever have to see or hear weeping in order to weep in a stereotypic manner. Moreover, the globus response (feeling choked up) is experienced by all those who grieve, even though there is nothing to see and constricting the pharynx and larynx is generally not accessible to voluntary control. So the experience of being choked up cannot be attributed to learning through exposure or observation. The similarities we see in weeping behaviors across individuals and cultures conform to the property of *stereotypy* that characterizes all ethological signals. The display behaviors are conventional and formulaic.

Weeping can incur costs, such as a reduction in social status, so one might suppose that people endeavor to consciously assess their situation and then choose whether and when to weep. However, this is certainly not a common experience. Weeping is spontaneous and impulsive rather than willful or voluntary.[9] This spontaneity is notably evident in weeping vocalizations. We choose when to speak and when not to speak, but weeping is an exception: when we weep, sounds often just burst forth, seeming to bypass any willful control. The compulsion to vocalize is immediate and unbidden.

The contrast between weeping and melancholy is instructive. There are no uniquely melancholic sounds. Instead, what we regard as "the sound of melancholy" is simply speech whose acoustics are altered as a consequence of low physiological arousal. We still must choose to speak in order for the sound of melancholy to be apparent. In the case of weeping, the characteristic sounds are independent of speech. If we do choose to speak, the effects of weeping are readily apparent. But like laughter, the distinctive sounds of weeping can arise whether we speak or not. We vocalize without choosing to do so.

This does not mean that we are mere slaves to the passions of weeping. If the individual assesses the social cost of weeping as too burdensome, then they may indeed attempt to suppress, modify, or mask their weeping.[10] Such executive control of weeping behavior is evident in several ways. In the first instance, the weeping signal can be physically masked by turning away, hiding one's face, or seeking privacy.

I once had an experience where just before starting a class lecture, a student privately told me something that I found deeply touching. My throat immediately tightened, and I knew that if I were to begin lecturing, my voice would surely crack. I dealt with the situation by burying my face inside my knapsack and rummaged around as though I was looking for something. Only after I had regained my composure did I pull my head out of the bag and begin the class.

A survey carried out by a commercial British airline found that movies shown on long-haul flights frequently cause discomfort for passengers who are moved to tears, especially when they are seated among strangers. In such circumstances, the survey found that women are likely to pretend to have something in their eyes. The same survey found that 41 percent of male respondents buried their heads under a blanket in order to hide their tears.[11]

If physical masking is impossible, various strategies exist for psychological masking. For example, a person moved to the edge of tears may consciously think of something else—typically something mundane—such as the need to wash the car or what to prepare for dinner. Here, prefrontal "executive" control is used to inhibit the propensity to weep. Conversely, an individual might conclude that, under the circumstances, weeping would be advantageous. In this case, a reverse psychological strategy might be employed, such as thinking sad or tragic thoughts—as in recalling the death of a loved one. The intention is to induce, reenforce, or *amplify* the impetus to weep.

Notice, however, that weeping is difficult to generate willfully, de novo. While several aspects of weeping can be voluntarily emulated, that's not the case with tears: we can't willfully spontaneously generate tears. For professional actors, for example, producing tears is one of the most challenging of thespian tasks. In so-called method acting, actors make an effort to identify fully with a role and aim to generate tears through vividly imagined experience. However, all too often, professional actors must resort to various technical aids, such as the Kryolan Tear Stick, which relies on irritant chemicals smeared below the eyes in order to induce tears. In general, it's easier to amplify or facilitate an already latent disposition to cry.

The difficulty of willfully generating tears means that tears, in particular, represent a notably authentic indicator of a highly emotional state. This honesty is recognized by observers. In experiments conducted by Asmir Gračanin and his colleagues, they found that facial displays involving tears are perceived by observers as more sincere. Interestingly, this is the case for both tears of anger as well as tears of grief.[12]

The very fact that people hide their faces, pretend to have something in their eye, or think thoughts that either inhibit or amplify weeping attests to the involuntary origin of the weeping behavior—despite its malleability through executive control. That is, the modifications afforded by executive control notwithstanding, weeping exhibits the stereotypic spontaneous tendencies seen in signaling among nonhuman animals.

Crying Alone

Notice that the foregoing discussion provides answers to two questions posed in the previous chapter: if weeping is an ethological signal, and if signals are intended to influence the behavior of observers, why would anyone weep alone? Moreover, if signals are intended to be communicative, why do people often attempt to mask their weeping by seeking privacy, hiding their face, or surreptitiously wiping away tears? As we have just noted, the very fact that people mask or hide their weeping suggests that the act of weeping is largely involuntary. Circumstances can dispose us to weep, whether we want to or not. It is the involuntary character of weeping that accounts for the phenomenon of private weeping: we can weep, even when there is no audience to witness the signal.[13]

Similarly, the second question has already been answered. Since weeping commonly incurs a social cost for the weeper, cognitive appraisals (including appraisals shaped by cultural norms) might be expected to contribute to a complex calculation as to whether the cost is prohibitively high. If the cost is assessed as excessive, then voluntary efforts to mask or disguise weeping may be expected. Standing in front of my class, I'd rather bury my head in my knapsack than risk my voice breaking in front of my students.

Response Automaticity

Apart from the question of automaticity in generating weeping, we can also consider the automaticity of responding to weeping by observers. Working at the University of Tilburg in the Netherlands, Martijn Balsters and his colleagues conducted a revealing experiment. They exposed participants to photographs of neutral faces, some of which had been modified by the addition of tears. The images were presented to viewers for just fifty milliseconds—which is barely enough time to register that an image represents a face and is not sufficient for observers to be able to accurately report whether tears are present or not. Nevertheless, when asked which images suggest a person may be "in need of social support," participants selected the tear-doctored images. The extremely short presentation time indicates that the effect of tears is evident at a preconscious level. Tears influence our responses, even when we are not consciously aware of their presence.[14]

When we witness someone weeping, cross-cultural studies of crying show a consistent pattern around the world: there is a strong tendency to want to help, or at least suspend any aggressive action toward the weeping individual. Adult responsiveness to crying (especially maternal responses to infant crying) is highly stereotypic.[15] The emergence of compassion in child development also shows a conventional pattern. As noted earlier, at roughly two years of age, toddlers commonly exhibit spontaneous helping behaviors toward someone crying. With the exception of psychopaths, most of us find it nearly impossible not to be affected by the sight or sound of someone weeping.

Nevertheless, our responses to signals are not merely involuntary reflexes. The role of assessment is particularly evident in situations such as responses to conspecific aggression or in mate choice. Whether or not a threat display succeeds in inducing fear in an observer depends on the observer's assessment of pertinent qualities of the displayer, such as size, musculature, or social status. (One shouldn't be afraid of someone who appears to be much less powerful.) Similarly, the importance of assessment is evident in female responses to male courtship signals. Much mating behavior is dominated by female choice. While male courtship signals may attract the attention of females, in most species, whether females elect to mate with a particular male is dominated by an assessment of features linked to reproductive success, such as animal size, vigor, and quality of feathers or fur.

In the same way that cognitive processes can mask, inhibit, or amplify the production of a signal, cognitive processes can also shape our spontaneous *responses* to signals. Before we consider some of the ways human cognition shapes how we respond to weeping, it's appropriate first to see an example of how cognition can mold the responses of nonhuman animals in response to signals. An accessible example of such assessment is evident in the peculiar phenomenon of "loud scratching" in our great ape relatives.

Loud Scratching in Primates

Scratching is a common spontaneous behavior in response to various skin irritants. However, conspicuous scratching has been observed in specific social contexts in several ape species, including chimpanzees, bonobos, gorillas, and orangutans. In particular, great ape mothers have been observed engaging in what primatologists call "loud scratching."[16]

As the name implies, loud scratching differs from ordinary scratching insofar as it is, well, conspicuously loud. Human observers report hearing loud scratching in noisy rainforest environments at distances greater than fifteen meters.[17] Tellingly, the scratching tends to occur just prior to the mother (or troupe) departing from their current location. The behavior is notable in that it attracts the attention of the mother's infant.

Primate mothers commonly carry their infants. Young infants tend to stay close to their mothers. Older offspring may stray farther afield, but these more mature offspring are likely to move with the group without being carried by their mothers. There is an intermediate age, however, when offspring tend to wander away from their mothers but are still carried by their mothers when moving from location to location. It is these intermediate-aged infants who are most attentive to maternal loud scratching; they respond by approaching and joining their mothers. In short, loud scratching tends to cause infants to reunite with their mothers in order to coordinate joint travel.[18] Loud scratching is a maternal message that beckons infants to "come hither and climb aboard—it's time to move on." The behavior is, in fact, a classic ethological signal that benefits both mother and offspring.

The fact that loud scratching has been observed in orangutans is especially noteworthy since orangutans split from other hominids roughly 10 million years ago[19] That a species as phylogenetically distant as orangutans exhibits the same social loud scratching observed in chimpanzees, bonobos, and gorillas suggests that the behavior is very old and so surely innate.

What's important in our discussion here is that the loud scratching signal of apes is intended for such a small audience (a single infant offspring). Other members of the social group, including older offspring, are not expected to respond to the signal, and if they did, the mother would reject their efforts to hitch a free ride. Individual infants must recognize that the signal is intended for them only when issued by their own mother. Adolescent and adult troupe members must recognize that the signal has no value for them. This suggests that, although the compulsion for mothers to make a loud scratching display may be innate, it's quite likely that the attentive response of infants is learned—as is the *un*responsiveness of more mature offspring, unrelated infants, and adults. What is instructive about loud scratching is that the response of observers to an innate signal may be strongly shaped by an assessment of the observer's relationship to the signaling.

Human responses to weeping exhibit a similar flexible pattern. Although weeping evokes feelings of prosocial compassion in human observers, in certain circumstances, we can impede or override such feelings via cognitive assessment. Of all the cognitive factors influencing responsiveness, none is more important than the assessment of one's social relationship to the displayer. Foremost is whether the displayer and observer are affiliative or antagonistic—friend or foe. Seeing your worst enemy weep from grief is more apt to evoke feelings of glee or gloating than feelings of compassion.

History is replete with sobering examples where the weeping of victims did little to evoke feelings of pity or mercy. Horrific examples include the behaviors of Nazi concentration camp guards and perpetrators of the Rwandan genocide. Indeed, in the presence of a hostile adversary, weeping can actually lead to *increased* aggression. A failure to evoke compassion is especially likely if the observer regards the weeping as a form of psychological manipulation. Of course, these are extreme examples. You don't have to be a psychopath in order to resist responding to some signal.

Strangers represent an interesting third category—distinct from friends and foes. Strangers might not be soulmates, but neither are they necessarily opponents. In contrast to most enemies, strangers offer potential opportunities for useful cooperation. Consequently, we are much more likely to respond compassionately to the weeping of strangers.

In general, whether an observer of weeping endeavors to inhibit a prosocial response depends on a number of factors. These include the perceived honesty of the signal, whether the individual is assessed as friend or foe, one's kin relationship to the weeping individual, the probability of repeated future interactions,[20] the capacity of the individual to offer useful reciprocal help in the future, the presence of other observers who may be able to help (or provide more effective help), the presence of an audience to observe either reputation-enhancing actions or reputation-destroying inaction, the seriousness of the stressful situation, our capacity to offer effective aid, and the marginal cost of offering assistance, including time, resources, and effort. It bears reminding that these executive control factors simply mediate an innate prosocial disposition toward the weeping individual. When we encounter someone crying, our initial spontaneous experience is dominated by feelings of commiseration and compassion.

Individual Differences

Apart from the various contextual factors that shape cognitive assessment, what we feel is also shaped by individual differences. As one might expect, people differ in both their responsiveness to the sadness of others as well as their own personal susceptibility to sadness.

Notably, people differ with regard to their capacities to experience compassion and commiseration. At one extreme, there are some psychopaths for whom observing the plight of others has little psychological or emotional impact—which is the reason why they are so potentially dangerous. At the other extreme are people who are deeply moved by the slightest suggestion of distress in others. We might suppose such sensitivity to be a wonderful blessing—and, indeed, who wouldn't want to live among people who are sensitive to the plights of those around them? However, Yale University psychologist Paul Bloom has offered compelling descriptions of how such unbridled sensitivity can sometimes subvert moral behavior.[21] Later (in chapter 19), we will see how such personality traits help to explain why people differ in their enjoyment of tragic arts and entertainments, such as the enjoyment of sad music.

Apart from how people respond to grief-stricken or melancholic individuals, some also differ in their own propensity for sadness. In the extreme case of clinical depression, it has long been known that some people (and families) are more susceptible to bouts of depression than others.[22] Even for nonpathological melancholy, it's also the case that some individuals are more susceptible to sad feelings than others. In the case of grief, modern research on bereavement has shown a wide range in how well people adapt to such stresses. For example, some bereaved people are overwhelmed at their loss and experience chronic grief that may burden them for years. Other bereaved individuals are able to recover from the death of a loved one and return to normal life in a matter of days or weeks.[23]

Columbia University psychologist George Bonanno has pioneered the study of how and why people differ in their resilience when experiencing various stresses. There are a variety of behavioral and dispositional factors, including both cultural and genetic aspects. For example, certain gene variants are implicated in how well a person fares when exposed to serious stress.[24] Cultural factors also can shape individual resilience in the face of stress. For example, Bonanno has described at length traditional Chinese mourning rituals and how certain elements of these rituals better mitigate long-term suffering for those experiencing loss compared with Western bereavement practices.[25]

Criticisms

The concept of automaticity or prepotency in emotion-related displays has not been without its critics. Alan Fridlund has argued that deception and the need to mask one's motivations makes automaticity impossible. That is, the temptation to deceive an observer and efforts to avoid conveying one's motives would cause any automatic disposition in production or reception to be selected against. Automaticity would necessarily disappear.[26]

Fridlund is right that communicating one's emotional state (and hence likely motivations) to an observer is commonly detrimental. And indeed, most human emotional states are covert or private and not communicated. Of all the many emotions that people experience, only a handful are manifested in overt displays like facial expressions. What Fridlund ignores is that ethological signals evolve only in non-zero-sum conditions where the interaction typically benefits both displayer and observer.[27] If deception were commonplace in signaling behavior, then indeed, the display behavior would be selected against and disappear from the behavioral repertoire. Similarly, if conveying one's emotional state were *mostly* detrimental to the signaler, then no display would arise. The fact that a signal is retained testifies to its overwhelming value to both individuals. In this regard, selection pressures would promote (rather than impede) automaticity in both the displaying behavior and the evolved observer response.[28]

This is not to say that responses are not shaped by cognitive assessment. As we've seen, automaticity should not be regarded as involuntary "reflexes" that are devoid of cognitive involvement. We'll have more to say about the role of cognition and executive control in display behavior in chapter 15, "Cultural Cues and Signals."

Reprise

In this chapter, we have addressed three key concepts related to the innateness of ethological signals: *stereotypy*, *automaticity*, and *executive control*.

Stereotypy is the tendency for both display behaviors and observer responses to be standardized or formulaic. Many of the archetypal patterns evident in weeping can be traced to anatomy and physiology, such as the unique efferent nerve innervating the lacrimal glands responsible for producing psychic tears. At least in the case of grief, weeping behaviors appear to be highly stereotypic and universally present across the panoply of known human cultures.

Automaticity is the tendency for signal displays and signal-induced observer behaviors to arise from innate or instinctive dispositions. Evidence of automaticity includes the fact that weeping is easily observed in individuals whose opportunities for learning are limited, including infants with hydranencephaly and congenitally deaf and congenitally blind individuals. Pliny the Elder's claim that virtually all human behaviors are learned may or may not be correct. But his claim that crying is not learned appears to be warranted. Whether or not one is willing to use words like "innate" or "instinctive," weeping is clearly a biologically prepared behavior.[29]

Executive control refers to the cognitive ability to temper or circumvent these involuntary tendencies. When generating signals, executive control is evident in *masking, inhibition,* and *amplification.* Masking occurs when a displayer endeavors to hide the display by turning away, obscuring the display by placing a hand in front of their face, or seeking privacy. Inhibition occurs when a displayer endeavors to avoid producing the spontaneous display by thinking thoughts that confound or suppress the feeling state—such as thinking happy thoughts to avoid weeping or sober thoughts to avoid laughing. Amplification is the reverse process by which a displayer endeavors to facilitate or encourage the display, such as thinking unhappy thoughts that might foster or fuel weeping.

Executive control can also occur when observing and reacting to a signal. Indeed, executive control appears to be somewhat more flexible when responding to weeping. For example, if a weeping person is assessed as an enemy or foe, it's much easier to inhibit or suppress any induced feelings such as commiserative distress or compassion. Nevertheless, weeping has a tendency to induce prosocial feelings in observers, whether or not those feelings are acted on.

Compared with other animals, humans are endowed with a much greater capacity for executive control that can mask, inhibit, or amplify many otherwise compelling behaviors. But this capacity is largely limited to efforts aimed at modifying involuntary spontaneous tendencies that have already been unconsciously initiated. Our ability to turn on or off weeping behaviors or responses is quite limited.

So far, our discussion has been intentionally restricted to *grief*-induced weeping. In the next chapter, we broaden our discussion beyond grief and address the many other situations in which tears can arise.

7 A World of Tears

Apart from grief, crying is commonly observed in many other circumstances. Weeping can arise from physical pain or mental frustration. People commonly become teary-eyed at wedding ceremonies or when receiving an award or praise. We may weep with pride or satisfaction at the success or accomplishment of someone close to us. A father may cry when holding his newborn child for the first time. Tears may well up in the eyes of a pious believer in the presence of a religious relic, or in the eyes of a patriot when singing a national anthem. Tears commonly arise in situations of extreme amusement, and fans may weep in the presence of a deeply admired celebrity. Tears accompany innumerable experiences of being moved or touched, including while viewing tragic dramas or when listening to music.

If, as claimed here, weeping is an ethological signal, then we need to show how all of these various weeping scenarios are consistent with an ethological interpretation. Why, precisely, is it that experiences as diverse as grief, pain, joy, pride, piety, patriotism, humor, adoration, being moved, or aesthetic appreciation might bring a person to tears?

Recall that the purpose of a signal is to change the behavior of the observer so as to benefit both the signaler and the observer. Also recall that emotions act as motivational amplifiers. If the purpose of a signal is indeed to change the behavior of an observer, and if emotions act as motivational amplifiers, then we would expect signals to routinely induce specific emotions in observers that act as the proximal motivation that ultimately results in the desired behavior. In this chapter, my claim is that, whatever other emotions might be evoked, weeping in a wide variety of circumstances does indeed tend to induce a single common emotional response in observers and that this response typically benefits both individuals.

To anticipate our story, I propose that independent of whatever causes someone to weep, the result is the same: *weeping is a signal whose purpose is to induce a broadly prosocial disposition in observers—feelings that commonly lead to behaviors such as reduced aggression, the offering of altruistic assistance, a feeling of connection or bonding, or simply a favorable attitude toward the weeping individual.* Notice that this ethological account will necessarily focus on the behavior of observers.

Of course, the feelings that lead a person to weep also require explanation. Apart from a few passing remarks, we will postpone discussing the emotions experienced by the weeper until the final section in this chapter ("Phenomenology"). The following accounts begin the conversation by focusing on functional explanations.

Grief and Pain

It should come as no surprise that weeping often arises in situations where the weeper may benefit from the help of onlookers. Grief and pain represent precisely such situations. Grief can arise in many different circumstances of high stress. Most obviously this occurs with the departure or death of a loved one. But grief can also arise when resources are lost (house burning down, crop failure, food spoilage, theft), threats to personal well-being (serious illness, bullying, personal attack), injury (intense pain), threats to livelihood (unemployment, bankruptcy), social threats (loneliness, social exclusion, shame, loss of reputation), reduced behavioral effectiveness (handicap, physical restraint, inability to help friends or loved ones), or reproductive challenges (infertility, parenting difficulties). Provided that the observer does not assess the weeper as a competitor or foe, in all of these circumstances, weeping tends to induce a prosocial disposition in observers, notably by evoking the proximal feelings described in chapter 4: compassion, commiseration, the anticipation of virtue, the appeal of enhanced social reputation, and the warmth of social bonding. These feelings commonly lead to altruistic interventions that often help the stressed weeper.

Tears of Joy

Weeping is a common response to moments of joy. The cliché example of weeping for joy is that of the weeping beauty pageant winner. Why would someone who is likely experiencing "joy" weep?

The effect of such weeping is best understood if we contrast weeping with gloating celebration. Suppose the response of a beauty pageant winner was akin to the triumphant boasting of a boxing champion. Turning to the other contestants, she thrusts her arms into the air and declares "I am the greatest" or "Tough luck suckers!"

In a world where gender-related expectations exhibit a double standard, personality counts for something in women's beauty pageants. In contrast to the boxing champion, we tend to prefer pageant winners to be magnanimous, generous, self-effacing—not self-centered or egotistical. As the winner of the competition, our victor could adopt an imperious, aloof, or self-satisfied demeanor. On the contrary, instead of gloating, her weeping amounts to a powerful declaration of humility and the absence of vanity.

In our ethological account, the effect of weeping is to evoke prosocial feelings in observers. As observers, we experience feelings of compassion, of wanting to be helpful and supportive. We feel more connected to the weeping pageant winner (bonding warmth) and minimally simply hold a favorable attitude toward her. Indeed, when interpreted as an act of modesty, her weeping is likely to enhance our admiration for her.

When a culture expects a winner to be magnanimous, gracious, and grateful, there is arguably no more powerful expression of these traits than to effectively downgrade oneself in the social hierarchy through the act of weeping. The words of self-effacing thanks she utters are heartfelt: "I feel so honored; I've grown so close to the other participants, etc."

The weeping of the pageant winner is matched by equally remarkable behavior on the part of the pageant losers. If anyone has cause to weep, one might suppose that it's the losers who share the same stage with the winner. So why don't we witness a background chorus of women bawling their eyes out? As a signal, weeping is explicitly intended to draw the attention of observers and evoke prosocial feelings. However, in the context of the beauty pageant, the declaration of a winner is intended to be a celebratory occasion. The cultural script calls for us to attend to the joy of the winner rather than drawing attention to the misery of the many losers. Accordingly, the losing contestants will feel obligated to mask or inhibit any compulsion to cry since weeping on stage would be viewed as a socially inappropriate commandeering of attention away from the winner. Tears of grief will surely be shed once the losing contestants depart the stage. However, on stage, the cultural script calls for the losing contestants to exercise executive control, doing everything in their power to suppress or stifle the compulsion to cry.

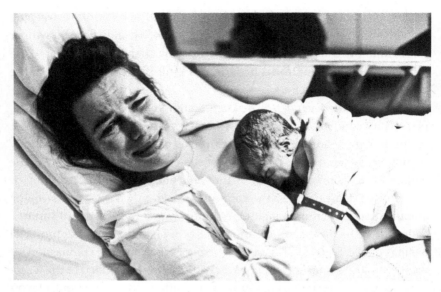

Figure 7.1
Without contextual information, it is commonly difficult to distinguish joyous weeping from grief. The purpose of a signal is not to convey the emotional state of the signaler. Instead, the most reliable emotion associated with weeping is the prosocial emotion induced in observers.[1]

A striking example of joyful weeping is shown in figure 7.1. If the photograph were cropped so that only the left half of the image were visible, one might conclude that the woman in a hospital bed is in a state of grief. (Perhaps her oncology surgeon has just informed her that her cancer has metastasized and nothing further can be done.) However, the right half of the image provides a crucial context that suggests the weeping expression is anything but grief. Here again, her facial expression may be interpreted in different ways. One might conjecture that her infant has been still born (not the case); perhaps she has learned that her infant has a serious heart defect (again, not the case). Instead, she is overwhelmed with happiness—a moment captured by her husband's camera.

Perhaps her weeping arises from a combination of happiness mixed with the acute stress of childbirth. The combination of stress and happiness seems especially plausible when interpreting the celebratory tears of victorious athletes—as commonly seen in Olympic contests. In Japan, the most famous example of a crying athlete is the public weeping of sumo champion

Tokushōryū Makoto after beating Takakeishō Mitsunobu at the 2020 New Year Grand Sumo Tournament in Tokyo. Photographs of Makoto's weeping are renowned in Japan (and alas, copyright permissions for such famous images proved too expensive to allow inclusion in this book). Like Western wrestlers and boxers, public weeping is extremely rare in sumo. It's not the behavior one expects of a physical champion, especially among males. Entering the competition, Tokushōryū was the lowest ranked of the eighteen competitors. He certainly did not expect to prevail over all of the others. Despite his voluminous build (and near naked attire), his weeping suggests that he is not arrogant, self-obsessed, boastful, or pompous.

Curious about people's reactions, I conducted a simple survey where I polled students and colleagues to respond to a celebrated photograph of Tokushōryū weeping. When asked to speculate about his personality, my Western survey respondents described him as hardworking, conscientious, sensitive, empathetic, kind, and agreeable. They also described him as vulnerable and lacking in self-confidence. When asked whether they responded positively or negatively to Tokushōryū, with only a couple of exceptions, my respondents all declared a positive or very positive response.

As with the beauty pageant winner, public weeping can amount to an act of modesty. In witnessing the tears, observers are commonly captivated by the apparent vulnerability and humility of someone who has just been promoted to a higher social position. It is the appeal of someone who is accomplished but is nevertheless deferential, eager to please, and (apparently) does not take their superior social standing for granted.

It's certainly possible that tears of jubilation might arise from a mix of joy and stress. However, many joyous tears appear to be shed in situations with little or no stress. In my collection of crying photographs, I have a photo of a mother wiping away tears when posing for photographs at her daughter's college graduation. Many people weep at weddings, even as mere spectators with only a moderate acquaintance with the bride or groom. These are hardly stressful circumstances for the weeping individual.

In any event, signaling theory tells us that the emotional state of the signaler is not the principal concern. Instead, our interest is the effect of the signal on behavior of the observer. Our point here is that, as in the case of grief, the effect of the tears is to induce a prosocial attitude toward the jubilant weeper, an attitude that would be less positive in the absence of the tears. As observers, we might cognitively interpret the tears as communicating a feeling of humility or gratitude (as opposed to entitlement), which, for most observers, makes the joyful weeper more appealing rather than less appealing.[2]

Tears of Laughter

So what about tears of laughter? We mostly associate laughter with humor, but the extant research has shown that most laughter occurrences arise in contexts without any obvious humor. University of Maryland laughter researcher Robert Provine has noted that only about 10 to 20 percent of events or utterances that precede laughter are even remotely humorous.[3]

Laughter is strongly associated with social interaction. Although people laugh when alone, we are thirty times more likely to laugh in the company of others.[4] Indeed, laughter itself exhibits the hallmarks of an ethological signal.

Although typically associated with social play, as with weeping and smiling, laughter can arise in a variety of circumstances. At least three types of laughter can be distinguished: playful laughter, nervous laughter, and mocking laughter. Playful laughter is associated with situations of social entertainment, joke-telling, and physical engagement such as tickling and rough-and-tumble play. Nervous laughter is common when in the presence of someone of significantly higher social station, such as meeting a famous personality or head of state. Nervous laughter is typically subdued—in the form of quiet giggling, rather than boisterous guffaws or bellowing laughter. Mocking laughter is a form of aggression where laughter is directed *at* someone with the goal of humiliation or ostracism. The most common form is playful laughter—laughter strongly associated with positive social interaction—with mocking laughter being the rarest form.

Of the three laughter contexts, tears are most likely to appear in the case of playful laughter. One does not typically see tears accompanying nervous laughter, and mocking laughter appears to never include tears.

Playful laughter offers an especially powerful bonding experience. Friendships are cemented through joint laughter. Since the principal effect of weeping is to induce prosocial dispositions in observers (notably feelings of bonding warmth), the appearance of tears in playful laughter is certainly consistent with an association function—a function that supplements the social bonding arising from laughter alone. It would appear that tears serve to augment or amplify the positive prosocial consequences of playful laughter.

As noted, tears sometimes appear with nervous laughter but rarely or never in the case of mocking laughter. If tears tend to induce prosocial disposition in observers, one can see why tears would not accompany mocking laughter. It would make little sense to attempt to induce feelings of bonding warmth in a person one is scorning.

Incidentally, recall from chapter 1 that laughter and weeping are physiologically nearly identical. Both involve staccato vocalizations ("ha ha ha" or "ah ah ah"). The principal difference between laughing and crying is, in fact, the presence or absence of tears. A number of actors have commented on the difficulty of distinguishing between laughing and crying. If crying occurs without tears, it is often impossible to differentiate crying from laughter. It's very likely that weeping and laughter share a common ancestor signal.

The Promise of Affiliation

Among the possible benefits for observers of weeping, one is reciprocal altruism where the assistance offered by the observer may lead to a returned future favor by the weeper. In order for reciprocal altruism to arise, weepers must generally be willing to incur the implied debt. To weep in someone's presence, then, is to effectively propose a form of friendship or bond. Weeping carries an implicit appeal for the observer to trust the weeper.

In the following scenarios, we will see examples of weeping behaviors whose function hinges on the weeper's implicit promise or declaration of alliance or coalition. These scenarios typically relate to situations of personal loyalty or group bonding.

Tears of Loyalty

In my collection of weeping photographs, there is a picture of a soldier, standing at attention, his right hand raised in salute, his erect posture matched by a stoic facial expression with tears clearly visible on his face. It is certainly possible that he is experiencing grief—perhaps reminiscing about the death of a fellow soldier. Whatever the motivating emotion in this particular case, the photograph reminds us that it's not uncommon for people to weep merely from feelings of patriotism.

Patriotism is commonly defined as a feeling of commitment or devotion to one's homeland or nation. It's associated with feelings of loyalty and dedication. Who would you trust more to defend your nation—the soldier who salutes the flag with dry eyes or the soldier who salutes the flag with tears in their eyes? It may be that the more sober soldier proves to be a more effective fighter. But the teary-eyed soldier certainly suggests someone with a stronger sense of group identity and commitment and therefore a greater willingness to engage in costly behaviors on behalf of that group.

The scenario is consistent with weeping as an implicit promise or declaration of affiliation. The main effect on observers is the evoking of a more positive (prosocial) disposition toward the weeper, including regarding the weeper as loyal or trustworthy.

Similarly, consider the case of a father's tears when holding his newborn child for the first time. It makes no sense to suppose that the tears constitute an appeal to the baby to come to the assistance of the father. For any new father (or mother), the foremost feelings are those of devotion and commitment to the new child. The preeminent thought for any new parent is that they will do anything in their power to help their child thrive in the world.

Now suppose that the weeping amounts to a promissory note: the weeping represents a pledge or covenant, an expression of loyalty and deep affiliation. But, we might ask, who is the target observer whose behavior might be influenced by this display of tears? As a new mother assessing the parenting dedication of your husband or partner, who would you trust more to help with future caregiving: the spouse who holds his newborn child with dry eyes or the spouse who holds the baby with tears in his eyes?

The most important target observers here include the mother as well as other adult members of the mother's family. The effect of the tears is to induce a positive disposition toward the teary-eyed father. A weeping father suggests he is less likely to be haunted by doubts as to whether the child is truly his offspring and so less likely to harm the child. Like the weeping patriot, a weeping father warrants greater trust.

Mutual Weeping

As a final example of weeping behaviors, we might consider the case of contagious weeping. It's not uncommon for the weeping of another person to bring tears to our own eyes. This is most commonly observed in grief situations, but mutual or tandem weeping can arise in other circumstances, such as joint tears of loyalty or joint tears of laughter.

A useful place to begin is by noting the remarkable consequence of mutual weeping. Mutual weeping is known to lead to especially strong social bonding. Especially among women, joint crying has been shown to deepen friendships.[5]

Suppose you are in a state of grief, and my response is to join your weeping. If, as we've argued, weeping is an ethological signal, then we might focus on the pertinent behavioral consequences: What is the effect of my weeping behavior on you, and what benefit accrues to both of us?

In chapter 2, we illustrated an ethological signal by describing the capitulation display in dogs, where a submissive dog rolls over on its back. Suppose that a dog's response to another dog's capitulation display was similarly to roll over and also display capitulation. In effect, the dog is communicating "no, no, I don't accept your declaration of submission toward me; we're equals here."

Although weeping tends to induce a prosocial disposition toward the person weeping, it nevertheless incurs the cost of reduced social status. By engaging in mutual weeping, the observer effectively declares that the situation is not grounds for the observer's social advancement at the expense of the weeper. The observer might still be motivated to engage in helpful acts of altruism, but the observer's tears effectively say, "There's no cost to you here in terms of our social relationship; you are not incurring any debt."

Taken in this spirit, mutual weeping might be interpreted as the ultimate compassionate response since it declares that the observer not only holds prosocial feelings toward the weeper but also does not accept the weeping person's implicit reduced social status.

At the same time, the observer's weeping is likely to evoke prosocial feelings in the weeper toward the observer. Both individuals are likely to be consumed by prosocial feelings toward the other. Moreover, knowing that one is receiving help while incurring little debt is likely to be experienced as an especial positive state.

This interpretation is bolstered when we consider situations where the individuals differ significantly in social status. Recall that the cost of weeping is greatest for those of high social station. Suppose you were in the private presence of a high-status person (perhaps your boss) who was weeping. As an observer, you might feel considerable discomfort—certainly compared to the weeping of someone of lesser social standing. Engaging in tandem weeping would likely lessen the joint discomfort. The suggestion here is that the reason why the discomfort would be lessened is because your own weeping weakens the loss of social status for the weeping person of high social station. Your boss is apt to feel far more embarrassed if she or he weeps alone in your presence than if you join in.

By way of summary, observer weeping communicates to the weeper that the weeping is incurring little or no cost. At the same time, mutual weeping amounts to a declaration of reciprocal prosocial dispositions. Accordingly, mutual weeping effectively cements friendships as a bond between two equals. There may be no more effective path to becoming soulmates than crying together.

Social Status

Earlier, we noted that weeping typically incurs a cost—a relative loss of social status. This might make sense in the context of grief-induced tears, but what about those whose tears arise from laughter, joy, or loyalty? Do these tearful situations also incur the cost of a decline in status relative to the observer? Alas, there currently exists no pertinent empirical studies. Instead, consider the following hypothetical scenarios.

Suppose I am introduced to a high-status individual—say Japanese Emperor Naruhito. Suppose further that I tell him a joke. Not only does he laugh, but it brings him to tears of laughter. I propose that, as a consequence, the difference in social status between us is reduced. In this regard, it's not so much that my success in inducing tears of laughter has raised my status but that his tearful response has lowered his status relative to me.

In the case of the joyful tears of the sumo wrestler, we would say his weeping "makes him more human," by which we mean that we regard him as more approachable and less of an aloof celebrity compared with the sumo champion who does not weep when he wins. Similarly with the weeping beauty pageant winner. I propose that the winner who doesn't weep is perceived as less approachable—that is, someone who is asserting their superior social status.

It's important to understand that a loss of social status may sometimes be a *desirable* result for the tearful person. It's difficult to build true friendship between individuals with contrasting social status. If a friendship is to be established between people of highly contrasting status, then it is common for the higher-status individual to take steps that intentionally mark themselves down. Consider two further scenarios.

Suppose I am an alpha male chimpanzee. Despite my alpha status, I might still worry that my circle of supporters is too small or weak and that I need to expand my social alliance. In soliciting possible recruits, acts of intimidation ("join my group or else") are unlikely to produce a true feeling of alliance or comradeship. I need to make an effort to lower my superior social status—but only with respect to the individual I want to recruit. I can do that, for example, by grooming the lower-status individual.

Suppose I am King Henry VII, eager to raise an army to defeat Richard III, but I have insufficient resources. I must court and recruit several allies to my cause. Rather than simply command the Earl of Oxford to deliver his support, I will endeavor to establish our friendship. I invite him to join my court and otherwise take steps to seem less aloof. I need to show vulnerability as part of lowering my status so the Earl feels a genuine connection. I laugh at his jokes (even the lame ones) perhaps even to the point of tears.

By way of summary, I contend that, whatever the motivating emotion, weeping as an ethological signal is indeed associated with a decline in social status relative to the observer, but while in most circumstances, this represents a "cost" to the weeper, there are also situations in which lowering one's social status may be beneficial.

Other Tears

There remain other common contexts in which tears are routinely observed. Of particular interest are those tears associated with the arts and entertainment—notably through poetry, literature, drama, film, and music. Considerable empirical research has been conducted, especially in the case of music. Indeed, the theory of sadness presented in this book arose from efforts in my lab to understand the seemingly paradoxical enjoyment of nominally sad music. We will address what might be called "the aesthetics of sadness" in two later chapters (chapters 18 and 19) where we consider how sad arts might be enjoyable.

Polygenic Proclivity of Signals

Figure 7.2 offers a simple illustration summarizing our discussion of weeping. Although weeping can be induced by multiple motivating emotions, weeping tends to evoke a much narrower range of responses in observers. The induced observer emotions act as proximal motivations that typically

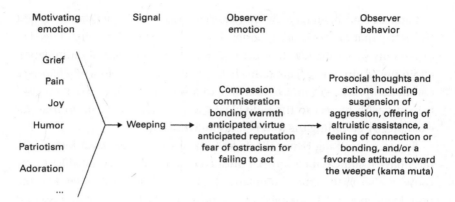

Figure 7.2
Weeping can be induced by many emotions, including grief, joy, pain, humor, patriotism, and so on. Observers of weeping tend to experience feelings such as compassion, which encourage prosocial responses toward the weeper.

result in prosocial behaviors, although these behaviors can be masked,
inhibited, or amplified through executive (cognitive) control.

Although figure 7.2 pertains only to weeping, it turns out that this pattern
is characteristic of signals in general (including smiling, threat displays, etc.).
The range of emotional responses induced in observers of a signal tends to be
much more constrained than the many emotions that might lead someone
to generate the signal.

We tend to think of displays such as weeping or smiling as reliable indicators of a person's emotional state. However, our examination of weeping
raises serious doubts about this common presumption. When we encounter
someone with tears in their eyes, in the absence of other contextual information, the tears tell us little about what the person is feeling. A reasonable
initial guess might be that the person is experiencing grief but that conclu-
sion might be entirely wrong.

By contrast, we have a better chance of correctly inferring that an observer
of weeping is likely to experience a prosocial feeling such as compassion. As
we will see in the next chapter, the emotional state that is most reliably
predicted by a signal is that of the observer, not that of the signaler. Here we
can simply note that signals like weeping can arise from multiple causes—an
observation we might refer to as the *polygenic proclivity of signals.*

Phenomenology

So what indeed does the *weeping* individual feel? Sigmund Freud was a prolific author whose collective writings span some twenty-four volumes. Freud produced a wealth of interesting and compelling ideas, many of which are utter nonsense. But Freud was notably insightful in his observations concerning the unconscious mind.[6] For Freud, everyday thoughts, memories, aspirations, and feelings constitute the conscious mind. These experiences are accessible to our awareness and subject to rational contemplation. The unconscious mind, by contrast, consists of underlying thoughts, feelings, and urges that lie outside of our awareness. They often include desires, impulses, or cravings that, to our conscious mind, are unacceptable or even immoral. As many of our behaviors originate in the unconscious, we are often strangers to ourselves, trying in retrospect to comprehend why we acted in a particular way.

Psychologists have shown that people regularly make up stories about their behaviors that are demonstrably false. For example, clever experiments can force someone to "choose" a particular option purely by chance. Yet when asked why they had chosen that option, people will nearly always fabricate a story "explaining" their behavior in what seems like a logical account. University of Michigan psychologists Richard Nisbett and Tim Wilson refer to this as "telling more than we can know."[7] In this regard, modern research supports Freud: we simply don't have the sort of access to our underlying motives that we think we do. Our self-understanding is remarkably shallow, and we often deceive ourselves when attempting to account for our own actions. When we introspect about our own behaviors, many of our self-reports are consummate fabrications.

The distinction between conscious and unconscious motivations is the psychological parallel of the proximal/ultimate distinction in biology. As noted earlier, we live our subjective lives entirely in the domain of proximal feelings, not in the domain of ultimate causes. We consume food, not because it's essential to sustain metabolic energy (the real purpose), but because eating is pleasurable and hunger is unpleasant.

In the above accounts of various forms of weeping, our attention was focused exclusively on ultimate rather than proximal accounts. My aim was to explain how various forms of weeping might serve biologically useful functions. For most readers, however, such functional accounts will ring hollow because they don't match our subjective phenomenal experiences.

As noted in chapter 4, when we experience grief, our subjective experience is not one of endeavoring to hoodwink observers into providing us with assistance. Instead, our experience is one of profound capitulation to an unhappy condition. When we cry from grief, it's because of a tragic sense of loss. Similarly, when we cry in response to some injury, our subjective experience is one of pain, not one of soliciting assistance. We cry because "it hurts."

When we shed tears of laughter, our experience is not one of wanting to establish a close bonding relationship. Instead, our proximal experience is that the tears have appeared simply because the situation is hysterically funny. In the case of the weeping patriot or the weeping father holding his newborn infant, the subjective experience is not one of establishing one's bona fides as a trustworthy patriot or parent. Instead, the subjective experience is a sense of profound connection and a sense of heartfelt dedication, responsibility, or duty.

In many other crying situations, our proximal feelings seem utterly inscrutable. We find it difficult or impossible to put into words what has brought us to tears. We resort to such nebulous phrases as "being touched" or "being moved." Indeed, these seemingly cryptic descriptions may represent the most accurate or veracious introspective reports.

Anthropologist Alan Fiske at the University of California, Los Angeles and psychologists Beate Seibt and Thomas Schubert at the University of Oslo have adopted the Sanskrit term *kama muta* ("moved by love") to describe the general amorphous positive feelings of being touched or moved, stirred, smitten, infatuated, entranced, or transported.[8] They identify many situations in which *kama muta* might be evoked, including situations of patriotism, loyalty, religious rapture, nostalgia, and when observing heartwarming or poignant acts, such as acts of self-sacrifice, forgiveness, generosity, heroism, feelings of collective pride, or the tender feelings induced when encountering an adorably cute infant or animal.[9]

Fiske, Seibt, and Schubert's conception of *kama muta* includes many situations beyond our more restricted focus on weeping or feeling choked up. Nevertheless, what all of these experiences share in common is the feeling of a broadly prosocial disposition.

Whether we call this feeling *kama muta* or something else, Fiske, Seibt, Schubert, and colleagues are right, I propose, to point to the wide range of situations in which positive prosocial feelings characterize our subjective experience. Whether weeping is overtly present or not, the induced feeling is a generally benevolent disposition in observers—feelings that lead to reduced aggression, compliance to the explicit requests or desires of the weeper, the offering of altruistic assistance even in the absence of overt requests, a feeling of connection or bonding, or simply a favorable attitude toward others.

I would propose that the feeling of *kama muta* amounts to a default all-purpose proximal feeling that shields us from the more cognitively uncomfortable awareness of the often Machiavellian goals that provide the ultimate motivations for many of our nominally altruistic acts. That is, *kama muta* (being touched or moved) is the feeling accessible to consciousness that masks the underlying unconscious self-benefiting motives that have evolved over the course of human existence. Like feelings of romantic love, our biology harbors ulterior motives. Like romantic love, our proximal feelings allow us to experience prosocial behaviors as overwhelmingly positive and noble.

Yet again, it bears reminding that we live our lives exclusively in the world of proximal motivations, even when these motivations owe their existence to underlying ultimate goals. In our day-to-day living, proximal emotions occupy the driver's seat and so it is possible for proximal emotions to override and even subvert the ultimate motivations. People do take vows of chastity and poverty, and no greater love is evident than when self-sacrifice entails giving one's life for another. People do often act in truly selfless ways. Indeed, it's possible that some truly selfless behaviors arise because of a person's cognitive awareness and moral opposition to the Machiavellian foundations for common altruistic acts.

Reprise

In this chapter, we have considered a range of familiar weeping scenarios apart from grief—including tears of pain, joy, laughter, and loyalty.

The functional analyses presented in this chapter highlight the independence of displays and motivating affects. Weeping by itself is a poor indicator of the weeper's emotional state. Signaling theory reminds us that the purpose of a signal is to induce particular behaviors in observers—behaviors that are motivated by inducing a predictable proximal emotion. The suggestion here is that the induced emotion in observers is a broadly prosocial disposition that commonly leads to reduced aggression, the offering of altruistic assistance, a feeling of connection or bonding, or simply a favorable attitude toward the weeper. What the various weeping scenarios share in common is not the emotion of the displayer but the emotions of the observers.

In discussions of emotion-associated displays such as facial expressions, emotion researchers have generally been looking at the wrong person. As we've seen in this chapter, it is the behavior of observers that represents the commonality across these otherwise contrasting tearful situations.

Inspired by the field of ethology, this book is premised on a rather different understanding of emotion-associated displays, one that contrasts strikingly with long-standing emotion theories. At this point, we might pause and consider the problem of how observers decipher the emotional states of other people. How do we make sense of emotion-related displays?

Covert Emotions—Emotions Concealed

The foremost function of emotions is to act as motivational amplifiers. We are more likely to behave in a particular way in the presence of some emotion than in its absence. For observers, there is obvious value in trying to infer the emotional states of others. Since emotions motivate behaviors, deciphering someone's emotional state helps us anticipate their likely actions and so sidestep potential dangers and better prepare to take advantage of possible opportunities. Consequently, we should be perpetually vigilant observers—trying to discern as best we can the emotional states of those around us.

Historically, much emotion-related research has focused on examining facial displays. Facial displays appear to be associated with only a handful of emotions such as fear, disgust, grief, anger, and joy. Most human emotions have no associated distinguishing expression. There are no characteristic displays for emotions such as regret, pride, loneliness, nostalgia, indignation, relief, compassion, boredom, devotion, or dozens of other affects.

Even vital feeling states like hunger and thirst are opaque. When my friend comes to visit, he is much too shy to mention that he's hungry. And if I neglect to ask him whether he's eaten, there is no nonverbal sign testifying to his hunger—much to his regret (and to my embarrassment as host).

Despite the critical importance of feelings like hunger and thirst, they are not evident from looking at someone's face or listening to their voice. For many of our most basic feeling states, there exist no recognizable displays.

Biologists note that there can be significant costs incurred when communicating one's feelings. In many situations, an individual benefits most by masking or hiding their emotional state. If a monkey discovers a cache of food, there may be considerable value in not advertising one's excitement to others. Similarly, if I am playing poker and have been dealt an especially good set of cards, it would be foolhardy for me to communicate to observers my obvious glee. Now it may be difficult or impossible for me (or the monkey) to mask our delight from others. Perhaps there is a glint in my eyes or a lilt in my voice that a sharp-eyed poker player will recognize. But such cues benefit only the sensitive observer—to my detriment.

In the business world, it's understood that one should not give away information unless it's to your advantage. Although the information may seem innocent, you never know how your competitors might be able to make effective use of it against you. Businesses will sometimes announce their intentions in advance, but only if they conclude that making the announcement offers more benefits than keeping the information secret.

In summary, communicating one's emotional state should occur only when there are good reasons to suppose that the result will be to one's advantage. There is no obvious value in indiscriminately broadcasting our emotional states to others. In the same way that expert poker players learn to sustain a blank "poker face," selection pressures always aim to nix any displaying behavior that typically fails to benefit the signaler.

The different benefits of concealed and displayed emotions can be usefully illustrated by considering the phenomenon of anger. It's appropriate to briefly digress from our topic of sadness in order to better understand the underlying ethological principles.

Anger—Hot and Cold

It has long been observed that feelings of anger can motivate an aggressive facial display ("hot anger"). But it is also often the case that people experience feelings of intense anger without exhibiting any outward signs of that anger ("cold anger").[1] For the purposes of the following discussion, let's use the word "anger" to refer to the emotion or feeling, "threat" to refer to the commonly associated facial display, and "aggression" to refer to any potentially harm-inducing action or behavior such as striking at someone. Threat displays can be spontaneous and involuntary or willful and posed. In chapter 15, we'll discuss at length voluntary, feigned, or acted emotion-related displays. In the case of hot anger, both the feigned and spontaneous threat displays have similar effects (although that's not the case for all displays).

If a threat display is a signal, then it must be beneficial both to the person making the display as well as the person observing the display. Consider the case of two children in a school playground—Allan and Bert. Allan is angry because he just discovered that Bert has stolen the cookies from his lunch. Allan's threat display might presage the possibility of an attack. The purpose of an attack, were it to occur, would be to retrieve the stolen cookies, punish Bert for the theft, or both. Now any altercation or fight has the potential to cause either one or both individuals to become injured. So even if Allan is stronger than Bert, it would be better if he could avoid having to fight.

If Bert assesses the situation as one where Allan's ready to fight, and if Bert thinks Allan is stronger, then Bert is likely to return Allan's cookies and express regret for his transgression. The benefit for Allan in making a threat display is the return of his food (and coincidentally affirming that he is socially dominant over Bert). The benefit for Bert is that Allan's threat display forewarned him of the possibility of a fight and allowed him to avoid potential injury—which would have been especially likely if his assessment is correct that Allan is stronger. In short, Allan's threat display benefits both him and Bert. If Allan didn't make the display and simply launched into attacking Bert, both of them would be taking the risk of injury.

Now consider the contrasting scenario where Allan assessed Bert as being the stronger individual. In this situation, it would be foolish for Allan to make a threat display since that might invite Bert to hurt him. For example, Bert might consider Allan's threat display as a challenge to his dominant status, so it might be beneficial to attack Allan simply to avoid future challenges that would erode Bert's relatively higher status.

Also notice that failing to make a threat display wouldn't necessarily mean Allan doesn't *feel* anger. Having assessed that Bert is stronger than he is, Allan could very well feel angry, without showing any overt sign associated with that feeling.

A widely popular idea is that cold anger contributes to poor health.[2] The idea is that it is detrimental to keep one's feelings "boxed in." However, the pertinent clinical research is equivocal. Despite decades of study, there is little direct evidence that cold anger has negative health consequences due to increased stress.[3] The research suggests that we might consider the possibility that the presumed cause and effect have been mixed up. The reason why we don't generate a threat display when angry is precisely because we conclude that we are unlikely to prevail should we have to fight the person who has perpetrated or incited our anger. We typically make a threat display only in those circumstances where (1) we are convinced that we can prevail should a fight actually ensues or (2) where we are convinced that the observer is likely to cave in to our aggressive overture—whether or not we are the stronger individual. In other words, we are more likely to make a threat display under conditions of lower stress and more likely to avoid making a threat display when we experience higher stress. This suggests that it is the level of stress that determines whether or not we make a threat display, not the other way around. That is, it's the stress we feel—rather than failing to display our anger—that might be confounding research related to possible negative health consequences.

Now it would be wrong to suppose that cold anger is useless. Emotions are important insofar as they tend to induce adaptive behaviors, and even cold anger has behavioral consequences. Although we might feel cold anger toward someone who is more powerful than we are, the anger itself might motivate us to seek out others who have similarly been abused by that person and build a collaborative alliance. Among social animals (like humans), even the most powerful individuals can be overthrown by a suitably large coalition. So although cold anger may not be displayed, a covert feeling of anger can still motivate future behaviors that ultimately serve an adaptive purpose for the person who experiences the anger.

The distinction between hot and cold anger reinforces the distinction between affect and display: an affect can be experienced without necessarily generating any observable display. Moreover, our discussion of threat displays reinforces the idea that signals are intended to change the behavior of the observer to the mutual benefit of both the signaler and observer. Once again, the main purpose of a signal is not to communicate one's emotional state.

The Emotion Expression Model

The idea that our emotional states may be better kept covert contradicts the most widely assumed model of emotion-related displays—what might be called the *emotion expression model*.[4] The idea is that there exist "emotional expressions" such as smiling, frowning, weeping, and laughing whose evolved purpose is to communicate our emotions to others. The emotion expression model might be illustrated by the following logic:

I feel happy.

→ This causes me to smile.

→ You observe my smile.

→ And infer that I am happy.

(And perhaps this also causes you to feel happy as well.)

Although the emotion expression scenario is widely presumed in most theories of emotion, it's problematic for a number of reasons. The key question to bear in mind is, What is to be gained by revealing one's emotions?

Researchers have long supposed that there exist a handful of emotions that are commonly communicated. Paul Ekman, for example, famously proposed six "basic" emotions—anger, disgust, fear, happiness, sadness, and surprise— on the basis of what appear to be unique facial displays. Ekman conducted cross-cultural tests that showed above-chance recognition of these various facial displays, leading him to propose that these basic emotions and their displays are universal.

On the one hand, people in different cultures do seem to agree on some displays—such as associating weeping with feelings of grief or sadness. However, the cross-cultural studies carried out by Ekman and subsequent researchers did not produce the robust results that are commonly recounted in the secondary literature.[5] For example, high rates of agreement depend on the use of tasks where participants must select their answers from a limited list of possible emotions. If we are asked to choose whether a photograph of a weeping face represents *anger, fear, disgust,* or *sadness,* almost everyone will choose sadness. When participants are given an open-ended task where they simply describe the presumed associated emotion, "recognition" of the associated facial expressions is barely above chance levels.[6]

The equivocal results from the cross-cultural studies seem troublesome. Surely if these facial expressions are universal, we would expect nearly perfect recognition scores. Moreover, how do we explain the absence of distinctive displays for the many other emotions people experience apart from the so-called basic emotions? As already noted, most emotions we experience are entirely covert. We do not wear our emotions on our sleeves.

Over the decades, a considerable volume of research has been conducted examining the relationship between behavioral displays (such as facial expressions) and their purported motivating emotions. Juan Durán and José-Miguel Fernández-Dols of the Autonomous University of Madrid and their colleague Rainer Reisenzein at the University of Greifswald conducted a major meta-analysis examining the relationship between emotions and facial displays. They reviewed a large number of individual studies in order to measure the degree of agreement or co-occurrence between a given emotion and its purported facial expression. How often is the feeling of surprise associated with the prototypical surprise facial display (or vice versa)? How often is the feeling of anger associated with the prototypical anger facial display (or vice versa)? They examined the level of agreement for 158 individual studies relating to seven emotions: amusement, happiness, surprise, disgust, sadness, anger, and fear. With the exception of amusement, they found surprisingly weak links between the feeling state and the presumed associated facial display.[7] The research suggests, for example, that it is common to experience fear without a fearful facial display. And conversely, it is common for a facial expression (like fear) to arise without the presumed (fearful) emotion. As readers of this book will appreciate, they found that weeping displays were not always linked to sadness.

In light of the poor results, Durán and his colleagues considered possible reasons why there would be such a low correspondence between displays and the presumed motivating emotions. They pointed to several potential confounds. For example, laboratory studies (in particular) may use stimuli that fail to have sufficient intensity to evoke the target emotion in research participants. In the case of field studies, it is possible that people deliberately or involuntarily tend to mask or inhibit the expression of the target emotion. In both laboratory and field studies, the research may employ less than optimal designs, so the effect size is understated or attenuated. Another potential problem is that the methods used to measure the emotion or the facial display are not sufficiently sensitive. Durán and his colleagues went on to suggest that none of these potential confounds is likely to explain the generally low agreement or co-occurrence between a given emotion and its purported facial expression.[8]

When combined with the research showing generally weak associations between emotion and display in cross-cultural studies, the idea that there is a one-to-one correspondence between display behavior and motivating emotion is deeply suspect. With the notable exception of Alan Fridlund's work, emotion theorists continue to wrestle with this apparent anomaly.[9]

An Ethological Approach

Alan Fridlund was the first to suggest that an ethological approach offers a better account of the situation. The key claim is that the notion of "emotional expression" is biologically questionable.

Earlier, we noted the ambiguity evident in various displays. Does weeping indicate grief or joy? Does laughter reveal playfulness or nervousness? Does a smile denote happiness or stress? The problem with these questions is evident in verbs like *indicate*, *reveal*, *denote*, or *express*. The problem arises from assuming that weeping, smiling, disgust, and other overt displays are intended to convey or communicate our emotional states. Ethological signals do not *indicate*, *reveal*, *denote*, or *express* some presumed emotion of the signaler. Once again, the purpose of a signal is to change the behavior of the observer to the mutual benefit of the signaler and the observer. The purpose of a signal is *manipulation*, not *revelation*.[10,11] That's not to say signals aren't informative for observers. Signals do indeed tell us that the displayer is feeling something, and that the feeling is likely drawn from a limited subset of possibilities.[12] That observers are sometimes able to decipher the emotional state of someone by attending to such displays is not what is being challenged here. What ethology tells us is simply that that's not the purpose of the displays.

As we noted in chapter 3, it's important to distinguish display behaviors from motivating affect. The conflation of display and affect is already evident in the names Ekman gave to his inventory of basic emotions. The labels *anger*, *disgust*, *fear*, *happiness*, *sadness*, and *surprise* identify affective states. Yet Ekman defined these states by explicit reference to particular facial displays. The presumption is that weeping denotes sadness, that smiling denotes happiness, that a threat display denotes anger, and so on.

In contrast to the emotion expression model, an ethological approach favors the *polygenic proclivity of signals* I described in the previous chapter. This relationship might be summarized as follows:

Some (but not all) emotions . . .

→ Might evoke an innate signal, such as weeping, laughing, or smiling. The same signal can be evoked by more than one emotion.

→ The signal tends to induce a predictable emotion in observers—an emotion that serves as the proximal motivation encouraging a particular behavior by the observer.

→ The resulting observer behavior commonly benefits both signaler and observer.

Notice that if facial displays such as weeping, smiling, and disgust are signals, then we can explain why studies of emotion recognition tend to produce equivocal results. Researchers are simply asking the wrong question when they ask what emotion is represented or conveyed by a particular facial display.[13] They should be asking what emotion the display tends to induce in observers. As we noted in the previous chapter, the emotion that's most reliably predicted by a signal is the emotion evoked in the observer, not the emotion experienced by the signaler.

Stotting Revisited

The problem of inferring the emotional state of a signaler can be illustrated by reconsidering the phenomenon of stotting described in chapter 2. Suppose you are a philosophically inclined cheetah observing a stotting gazelle. You're interested in what the stotting animal might be feeling.

On the one hand, the stotting behavior arises because of the observed presence of a predator. The presence of a predator might be expected to induce a feeling of fear. So it seems reasonable to suppose that the stotting gazelle is feeling fear.

However, the whole purpose of stotting is to demonstrate the health and fitness of the particular gazelle compared with other animals in the herd. Those animals who are *unable* to stott (the old, lame, or sick) are the animals that should feel truly afraid. The stotting animal is the animal that should feel least afraid. If the animal is able to make a convincing display of its physical vigor, the behavior might well be accompanied by feelings of relief rather than fear. Indeed, the stotting animal might very well be "jumping for joy" (as David Attenborough once speculated).

The point of this thought experiment is to illustrate the difficulty of inferring the underlying emotion that motivates a given signal. In the case of the stotting gazelle, we have no idea whether the associated feeling is fear or joy or something else (including feeling nothing at all). It's not easy to decipher the feeling motivating a signal, even when you know the context.

The main effect of a signal is to influence some behavior in the observer, not to broadcast the displayer's emotional state. The one feeling that can be predicted by stotting is the feeling of indifference induced in the cheetah; for the cheetah, the stotting gazelle will hold little or no allure as a possible dinner item.

Ageing and Emotion Recognition

The problem of inferring someone's emotion is highlighted by research comparing older and younger human observers. A considerable volume of research has documented age-related differences in "recognizing" emotional expressions. When asked to identify various emotions (from photographs or sound recordings), it turns out that younger people are much more "accurate" than older folks.[14] This is the case whether viewing facial expressions, observing body posture, or listening to nonspeech emotional vocalizations. A summary analysis of seventeen studies conducted by Ted Ruffman and his Australian and New Zealand colleagues found that older adults perform consistently worse than younger adults in recognizing expressions assumed to be indicative of sadness, fear, and anger.[15] This apparent deficit is more marked for some expressions than for others. For example, in some studies, older adults appear to be more consistent and accurate in recognizing disgust.

Researchers have suggested a number of possible reasons for the apparent poor performance of older adults, including declining sensory acuity, cognitive decline, positive affect bias, and other possibilities. However, careful subsequent research has rejected these conjectures. For example, César Lima and his University of Porto colleagues conducted a study where Portuguese and English listeners were asked to rate eighty nonspeech vocalizations representing eight feeling states: achievement, amusement, pleasure, relief, anger, disgust, fear, and sadness.[16] They recruited eighty-six younger and older adults who were carefully matched for educational attainment, general health, and hearing acuity. They also measured twenty-two traits via nine cognitive tests. The average age was twenty-two years for the younger group and sixty-one years for the older group.

Participants were asked to rate each sound example according to how well it suggested each of the eight target feeling states. The results are worth examining in detail. Age-related differences are readily apparent in table 8.1, which shows the percentage of rating points assigned to each of the eight target emotions for a given type of vocalization. In every case, the younger listeners assigned a higher proportion of rating points to the assumed target emotion than did the older adults.

So what explains the dramatic difference between younger and older adults? Lima and his colleagues found that both older and younger adults were equally sensitive to the acoustic features of nonspeech emotional sounds and that age differences in emotion recognition could not be attributed to differences in measures of general cognitive abilities, emotion regulation, current emotional state, or personality traits.

Examining table 8.1, it's evident that the older adults spread their ratings across many more emotion categories. For example, when hearing laughter, older adults assigned greater points to fear than younger adults. Similarly, older adults were more likely to characterize "sadness" (i.e., weeping sounds) as possibly indicative of fear, anger, disgust, or achievement.

The ostensibly poor performance of older people contradicts established research showing that recognition of emotion-associated facial displays increases with the amount of a person's social exposure.[17] How can older people with a lifetime of experience encountering a variety of emotional situations do so poorly at "emotion recognition?" In light of our claim that emotion-related displays are actually ethological signals, readers will surely suspect that rather than displaying a deficit in emotion recognition, the older adults are actually displaying their superior skills. With increasing age, we have many more opportunities to observe nervous or mocking laughter (not just amusement-induced laughter), stress-induced smiles (not just happy smiles), and tears of joy and achievement (not just tears of grief).

Emotions are not explicitly communicated by signalers; instead, they are deduced by careful observers. Observers learn to *decipher* emotional states, not to *recognize* them. For bona fide ethological signals, what is "automatic" is the evoked feeling in the observer. What is not automatic is the observer's

Table 8.1

Comparison of emotion perceptions for younger and older adults hearing nonverbal emotional vocalizations, based on Lima et al. (2013, table 3). The table highlights the ostensibly poor emotion recognition of older adults. Rows identify stimulus type; columns identify participant responses. Table values represent the percentage of all rating points assigned to each emotion category for a given stimulus type segregated by participant age. For example, for younger participants, 20.7 percent of the rating points for "achievement" vocalizations were assigned to "amusement." Diagonal values (highlighted bold/asterisk) identify nominally correct identifications. Rows may not sum to 100 percent due to rounding.

		Achievement	Amusement	Pleasure	Relief	Anger	Disgust	Fear	Sadness
Achievement	Young	**50.6***	20.7	7.5	20.7	0.2	0.1	0.0	0.0
	Old	**34.8***	12.9	16.3	24.8	3.6	2.5	3.8	1.2
Amusement	Young	18.2	**55.8***	5.7	10.3	0.2	0.3	1.7	7.7
	Old	21.0	**32.3***	11.1	18.2	3.4	3.2	3.9	6.9
Pleasure	Young	8.5	6.0	**76.7***	6.9	0.7	0.5	0.3	0.3
	Old	10.8	5.2	**42.4***	16.5	5.5	8.9	6.4	4.3
Relief	Young	8.8	1.7	6.2	**73.3***	2.7	0.8	5.4	1.1
	Old	14.0	3.1	10.2	**39.6***	6.0	7.2	13.6	6.3
Anger	Young	5.1	0.3	0.3	0.6	**61.1***	19.0	9.2	4.4
	Old	13.7	2.7	4.0	8.4	**32.6***	21.7	12.5	4.5
Disgust	Young	0.7	0.0	0.5	0.7	10.4	**84.2***	2.5	1.0
	Old	6.8	2.6	3.3	5.8	13.0	**53.1***	11.3	4.1
Fear	Young	4.6	0.7	0.6	6.1	2.0	6.9	**65.0***	13.9
	Old	11.0	3.4	8.1	12.2	7.6	10.1	**32.3***	15.1
Sadness	Young	0.4	0.7	0.3	0.4	1.3	2.1	19.6	**75.1***
	Old	4.6	3.8	3.1	6.0	7.6	7.6	21.7	**45.4***

ability to discern the emotional state of the signaler leading to the signaling display. What the signaler is feeling is hardly transparent for the observer.

Once again, this is not to say that observers don't often correctly infer the emotional state of someone. Much or most weeping does indeed arise from the feeling of grief. Similarly, most smiles are indeed associated with positive feelings such as happiness. So it makes sense that we initially assume that weeping signifies grief and smiling signifies happiness. However, with experience, we learn to pay attention to other clues such as the situation and subtle behavioral signs such as smiling mixed with a furrowed brow or laughter combined with an icy stare.

Of critical importance is the context.[18] A man's wife has left him so we assume that his weeping is due to grief. A woman receives a coveted award so we infer that her weeping is due to joy. A child has fallen and skinned her knee so we conclude that her weeping is due to pain. A friend is reduced to tears following hysterical laughter and we presume that he is feeling extreme amusement. The presence of weeping minimally indicates that the person is feeling some strong emotion. But further work is needed in order to decipher the emotion that motivated the signal.

In resolving the signaler's emotion, we are effectively treating the display as an ethological *cue*. We use whatever evidence is available to us in an attempt to deduce the underlying emotion that motivated the signal. Our success in inferring emotional states is possible because we are attentive to subtle clues— not because the display is intended to reveal the other person's inner feelings or motivations. Once again, we do not wear our emotions on our sleeves; if anything, our emotions are hidden under multiple layers of opaque clothing.

I would propose that what researchers have interpreted as evidence of a deficit in older adults is actually demonstrating a deficit in younger (less experienced) adults. Younger adults (and apparently emotion researchers as well) tend to interpret display behaviors as representing a single stereotypic emotion. They are assuming that weeping indicates grief and grief only, that smiling indicates happiness and happiness only, that laughter only indicates amusement, and so on. That is, they are assuming that the display represents the most common emotion that tends to lead to the display.

In the end, Lima and his colleagues concluded that the apparent weakness of the older adults cannot be attributed to any form of cognitive decline. After discounting the most popular conjectures for this difference, they finally concluded that older adults used different "inference rules" to process the vocalizations. I would suggest that this is indeed correct and that the "rules" for deciphering someone's emotional state from their display behavior are more sensitive and sophisticated for experienced adults. The recognition "rules" used by older adults lead to a broader range of responses precisely because they have encountered these emotion-associated displays in many more circumstances than their younger selves. The purported "deficit" in emotion recognition by older adults is better viewed as evidence contradicting the emotion expression model. Instead, the research is consistent with the polygenic proclivity of signals and an ethological interpretation of emotion-associated displays. The research highlights the role of learning and cognition in deciphering the emotions of others. Emotions are not "expressed"; they are covert and cryptic. Emotions are not recognized; they are decrypted.

Reprise

The foremost purpose of emotions is that they motivate behaviors that are typically adaptive for the situation in which the emotion is evoked.[19] Even if emotions don't result in an immediate change of behavior, they often result in changes of attitude, mood, or disposition that may ultimately influence future behaviors.

When others recognize our emotions, it helps them to anticipate our future actions. This allows them to tailor their own actions so as to better serve their particular goals. How they respond might also benefit or prove costly to us. In cooperative situations, we will benefit if the other person can anticipate our behavior. But in competitive situations, it is best if the other person cannot anticipate our behavior, and so it is best to hide or mask our emotions.

Rather than concerning ourselves with emotional communication, it would be especially useful if we could actually induce an emotion in the other person—an emotion that would dispose them to act in a way that benefits us. However, the other person would accede to such a manipulation only if their induced behavior also benefited them. Although not ubiquitous, such cooperative circumstances do regularly arise. The mutual benefits of such interactions provide positive selection pressures that encourage the emergence of instinctive displays and responses. Such cooperative interactions are initiated by displays that ethologists call a signal.

The purpose of a signal is to encourage the observer to *behave* in a particular way. The aim of a signal is not to communicate one's emotion. There are typically many situations in which the ensuing behavior can be beneficial to both the signaler and the observer—apart from whatever emotion might have motivated the signal. By itself, a given signal doesn't tell observers what the signaler is feeling. Signals such as weeping, smiling, or laughing might be expected to arise from many different feeling states.

At the beginning of this chapter, we posed the question, How do we decipher the emotional states of others? Our answer is "with difficulty." With experience, we learn to attend to the context in which the signal is issued and learn to recognize subtle clues in the signal itself that help us discern the likely motivating emotion. Emotions are not recognized; they are deciphered by experienced and sensitive observers.

It bears noting that our story of emotion-related communication is not yet finished. As we will see in later chapters, signals themselves also provide the raw materials for many culturally defined modifications that can indeed serve overtly communicative functions.

9 The Immune Connection

Melancholy isn't the only state in which we feel glum, cheerless, or blue. Malaise is also characteristic of feeling sick. In the late 1920s, a young Hungarian medical doctor named Hans Selye was conducting rounds in a Prague hospital. Selye was struck by how patients with different illnesses tended to exhibit the same symptoms. It didn't seem to matter whether the illness was caused by a virus (like the flu), a bacterium (like typhoid), or a parasite (like malaria); patients all seemed to look similarly "sick." Sick people commonly exhibit an elevated temperature or fever (*pyrexia*) and frequently suffer from a headache. Sick people often experience muscular aches and exhibit an increased sensitivity to pain (*hyperalgesia*).[1] Sick people are commonly fatigued or lethargic: when we are sick, we mostly just want to curl up in bed and sleep (*anergia* and *somnolence*). Many activities that are normally enjoyable lose their allure, including food, sex, play, and socializing (*anhedonia*). Importantly, Selye also observed that sick people experience low mood or *malaise*. For all intents and purposes, sick people also appear to be "sad." In general, sick people commonly experience a seven-ingredient cocktail of pyrexia, headache, hyperalgesia, anergia, somnolence, anhedonia, and malaise.[2]

Selye's clinical observations inspired his remarkable career in medical research. He spent the next five decades unraveling the complex mechanisms involved in stress. Selye provided the first detailed description of the so-called hypothalamic-pituitary-adrenal (HPA) axis—a major neuroendocrine system that plays a critical role in stress reactions and regulates many body processes, including the immune system as well as emotions and moods. He wrote a lovely (though now dated) book entitled *Stress without Distress*[3]— just one of twenty-two books and an astonishing 1,700 research studies he published in his lifetime. Selye, it turns out, was one of the most prolific scientists of his era.[4]

Sickness

When we are physically injured, infected, or poisoned, the immune system springs into action. A cornucopia of research in recent years suggests that many of the physiological, psychological, and behavioral changes associated with sickness facilitate immune functioning and assist recovery. For example, for some illnesses, people experience reduced appetite (*anorexia*). It turns out that reduced food intake is especially helpful in bacterial infections, since the reduction in food denies nutrients needed in order for the bacteria to reproduce.[5] When sick, many of the drugs we take make us feel better but actually interfere with immune functioning. For example, children with chicken pox who are given acetaminophen (such as Tylenol) feel better than children who are given an inert placebo, but the chicken pox infection lasts on average a day longer for children who receive the acetaminophen.[6] University of Michigan physiologist Matt Kluger has chronicled the benefits of fever in response to infection. Drugs that suppress fever reduce the effectiveness of the immune response to infection and can make people worse off.[7] Feelings of anergia, anhedonia, hyperalgesia, fever, headache, changes of appetite, purging, depressive mood, and sleepiness have all been shown to aid recovery from various illnesses.[8] Feeling sick may take much of the fun out of life, but these feelings play vital roles in giving us more life to live.

When fully engaged, the immune system is an energy hog: it places considerable demands on metabolic resources.[9] In fact, an active immune system is comparable to the brain as a high-demand energy consumer.[10] The effectiveness of the immune response is reduced if the immune system must compete for metabolic assets. Consequently, when the immune system is active, it's important to minimize any unessential energy drains. Of the major energy-consuming systems, voluntary motor movements can be singled out as the most optional. By refraining from movement, more metabolic resources are available for the immune system.[11] In order to achieve this helpful state, the individual must lose their motivation to move. Coincidentally, four feelings contribute to this desirable goal: *hyperalgesia* (in the form of muscle soreness or pain sensitivity), *anergia* (fatigue or lethargy), *malaise* (low mood), and *anhedonia* (the loss or reduction of pleasure). Anergia, hyperalgesia, and malaise discourage movement itself. Anhedonia reduces the incentive or motivation to engage in (what otherwise) might be rewarding behaviors.

The classic symptom of virtually every kind of pathology is inflammation. Inflammation is associated with two important classes of compounds—*prostaglandins* and *proinflammatory cytokines*. Prostaglandins cause blood vessels to expand (vasodilation), increasing the blood flow to regions of injury. Enlarged blood vessels also produce the warm/red/swollen appearance at an injury site. At the same time, prostaglandins[12] cause pain sensors (nociceptors) in the skin and muscles to become hypersensitive—leading to increased pain sensitivity (i.e., hyperalgesia).[13] Prostaglandins also act on the thermoregulatory center of the hypothalamus, causing fever. Aspirin and aspirin-like drugs act to inhibit the production of prostaglandins, thereby reducing both pain sensitivity and fever.

Proinflammatory cytokines, by contrast, are implicated in feelings of anergia and anhedonia—although indirectly. Lucile Capuron and her colleagues at Emory University found that administering proinflammatory cytokines produces a decrease in dopamine.[14] Dopamine is a critical hormone and neurotransmitter involved in both movement and hedonic rewards. Low levels of dopamine interfere with a person's ability to move effectively. It is the paucity of dopamine that characterizes Parkinson's disease with its tragic robbing of both the patient's ability and will to move. At the same time, dopamine plays a preeminent role in reward-motivated behavior. Many (pleasure-inducing) addictive drugs increase dopamine concentrations, either directly or indirectly. Overall, when proinflammatory cytokines suppress dopamine, the result is that we find movement more draining, are less motivated to engage in normally pleasurable activities, and experience less pleasure if we nevertheless pursue those activities.

An example of a proinflammatory cytokine is interferon-α, a powerful antiviral used to treat various viral infections. Injecting interferon-α into a healthy individual results in a series of changes, including a rise in body temperature, feelings of fatigue, muscle pain, and possible headache. In short, an injection of interferon-α will cause the person to feel sick, even when there is no illness to combat. Moreover, the effect of interferon-α doesn't stop there. There is a notable psychological effect: nearly half of the patients who receive interferon-α will also exhibit symptoms of depression.[15]

Remember that prostaglandins and proinflammatory cytokines are the *good* guys—coming to our rescue when we are attacked by viruses. Yet both

classes of immune-related molecules are associated with the classic symptoms of feeling sick: anergia, anhedonia, and hyperalgesia. Moreover, these
same "good guys" commonly lead to sad or depressive feelings.

People usually assume that it is illness that makes us feel sick. In fact, it's
our own immune systems that produce the sick feelings. A virus is most successful when it rapidly disseminates throughout some population. Dissemination is facilitated when infected individuals feel well enough to continue
with their daily activities, interacting with family and friends, shaking hands,
sharing air, and spreading the contagion far and wide. The virus is not served
by us feeling bad and staying at home in bed. If a pathogen itself produced a
headache, fatigue, fever, anhedonia, or other unpleasant symptoms, it would
be doing itself no favor. In fact, these symptoms would tend to be selected
against in the evolution of pathogens. Once again, it's our own immune systems that produce these symptoms—symptoms intended to enhance the
effectiveness of the immune system itself. Said another way, rather than
health-*robbing* feelings, these are health-*giving* feelings.[16]

Melancholy

Apart from feeling sick, notice that some of the same sickness symptoms
are characteristic of melancholy. Notably, melancholy includes feelings of
anergia, anhedonia, and malaise. That is, the main symptoms of melancholy
resemble downstream effects of prostaglandins and proinflammatory cytokines released by the immune system.

Over the past decade, research in the field of depression has witnessed two
related revolutions. The first of these revolutions was described in the chapter 5: researchers are now better able to discriminate (normal) melancholy
from (pathological) depression and have recognized that melancholic sadness
serves to enhance cognitive processing and so represents a positive (and normal) response to stress.[17] Both researchers and clinicians have observed that
the experience of melancholy is commonly beneficial and that it's important
not to misdiagnose normal sadness as depression.[18]

The second revolution is a body of research drawing attention to the
importance of the immune system in melancholy and depression.[19] A telling example of the close relationship between depression and the immune
system is provided by the effects of sleep deprivation. For centuries, sleep
deprivation has been used as a form of punishment or torture. If a person is

deprived of sleep long enough, the result will be death. You might suppose that the cause of death is exhaustion, cardiac arrest, or perhaps neurological dysfunction. Instead, the cause of death is a catastrophic bacterial infection. Bacteria that are normally held in check by the immune system run riot when a person is deprived of sleep for a long period. Ultimately, death is caused because of the collapse of the immune system.[20] Sleep confers many benefits, but one of the most important is ensuring a well-functioning immune system. Especially pertinent to our discussion, however, is that sleep deprivation has been used as an effective therapy for short-term treatment of major depressive disorders.[21] It has long been known that depriving a depressed person of sleep will cause the depression to go away. Enhance the immune system by administering interferon, and depression can be induced; suppress the effectiveness of the immune system through sleep deprivation, and depression is diminished.[22]

It makes sense that the immune system would be active when handling physical stressors such as injury and illness. But much melancholy or depression in humans arises from nonphysical stressors, such as social stress. Why should your immune system get involved if a close friend moves away or your boss makes a nasty comment about your work? Surely an inflammatory response makes no sense when dealing with cognitive forms of stress. However, the research suggests otherwise.[23] It turns out that the physiological mechanisms involved in cognitive or social stress are elaborations of immune responses. For example, proinflammatory cytokines have been shown to target a region of the brain associated with social pain, such as when we experience social rejection.[24] Moreover, the greater the feeling of social rejection, the greater the inflammatory response.[25] In addition, antidepressants that are effective for cognitively induced depression have been shown to inhibit the production and release of proinflammatory cytokines and stimulate the production of anti-inflammatory cytokines.[26] Altogether, the research implies that ancient immune responses provide the foundation for addressing cognitive and social stressors: immune responses are not limited to physical injury or microbial infections.

Although ordinary melancholy may commonly enhance fitness, a portion of the population is prone to suffer from major depressive disorders that are clearly pathological. The origin of depression remains a complex problem addressed by ongoing research. One part of the puzzle appears to involve a genetic component: depression often runs in families. Both depressive disorders and innate immune traits are known to exhibit moderate to high heritability.[27] Interestingly, many of the genes implicated in depression are also known to enhance immune effectiveness.[28] Andrew Miller and his colleagues at Emory University have proposed that depression-related genes have been preserved in the human gene pool precisely because of their adaptive value in protecting against pathogens.[29] People who suffer from major depressive disorders are typically better protected from infection. However, their reduced susceptibility to infection is directly linked to a genetic endowment that coincidentally renders them more likely to suffer from depression. In this regard, some forms of depression appear to constitute a form of autoimmune disorder.

By way of summary, recent research suggests a tight linkage between immune responsiveness and the negative feeling states associated with melancholy and depression. In particular, feelings of melancholy and depression are closely linked to the release of proinflammatory cytokines.

Grief

So what about grief? Is the immune system similarly implicated in grief? Consider again the main symptom of grief—namely, weeping, with its watery eyes, nasal congestion, constriction of the pharynx (choked up feeling), and erratic breathing. When we cry for an extended period of time, we typically also experience red swollen (i.e., inflamed) tissue around our eyes. In addition, some individuals experience hives in response to sustained weeping.[30] As we noted earlier, these are classic symptoms of an allergic response. The response originates in the immune system's efforts to expel an irritant or allergen: watery eyes attempt to flush the allergen from your eyes, nasal congestion endeavors to prevent the allergen from entering through your nose, pharyngeal constriction is intended to discourage swallowing the allergen, and staccato exhaling (coughing) attempts to expel allergen particles from your lungs or windpipe. Altogether, the act of weeping itself resembles yet another immune response.

In fact, the relationship goes deeper. The allergic response itself is known to be caused by the release of *histamine*. These are the same histamines that encourage an allergy sufferer to reach for a bottle of antihistamines. Tellingly, the ability of emotional stress to evoke watery eyes and a stuffy nose can be inhibited or eliminated by administering an antihistamine such as Benadryl.[31] At this point, you might not be surprised to learn that histamine is yet another type of proinflammatory cytokine. If you are planning to attend a tear-jerker movie and want to avoid the embarrassment of weeping in the presence of your friends, taking an over-the-counter antihistamine in advance will help to keep the tears at bay.

So what about the relationship between weeping and feelings of malaise? Known pharmacological effects point to a close relationship between tears and feelings of sadness. Notably, the drug dextromethorphan has proved to be an effective treatment for pseudobulbar affect (pathological weeping and laughing).[32] Yet this same drug is also known to reduce major depressive disorders.[33] So a drug that inhibits weeping can also inhibit sad feelings.

Histamine

As we will see shortly, the relationship between sadness and histamine warrants a somewhat more detailed discussion. Histamine is a neurotransmitter and hormone known to be involved in at least two dozen physiological functions. Its principal role is immunological. However, histamine participates in many other functions, including arousal, circadian rhythms, body temperature, pain sensitivity, appetite, stress, learning, memory, motivation, motor movement, and cognition.[34]

There are at least two sources of histamine within the body. One source is a group of specialized brain cells located in a region known as the tuberomammillary nucleus.[35] In addition, histamine is stored in mast cells that are diffusely distributed throughout the body. Among its many functions, histamine serves several roles related to disease and injury. For example, when we experience tissue damage, mast cells release histamine, which causes dilation of the blood vessels, leading to the distinctive symptoms that characterize inflammation. When we inhale or ingest an allergen, it's the disgorging of histamine from mast cells that is responsible for the classic symptoms of an allergic reaction, including watery eyes, nasal congestion, runny nose, hives, rash, and itchiness.

Four physiological effects are especially pertinent to our discussion of melancholy and grief. First and foremost is the effect of histamine in generating allergic symptoms—the physiological precursor and foundation for weeping.

Second, histamine is known to interfere with the dopamine reward system: histamine reduces reward incentives, so contributing to anhedonia.[36] Third, histamine tends to interfere with movement.[37]

Most suggestive is a fourth physiological effect: histamine has been implicated as playing a major role in cognitive processing.[38] Since the middle of the past century, it has been known that intravenous administration of histamine induces more rational thinking in psychiatric patients.[39] Many antipsychotic medications work by increasing histamine production.[40] Low levels of histamine have been found in individuals suffering from impaired cognitive functioning, such as in Alzheimer's dementia.[41] Moreover, antihistamines, often prescribed for treatment of allergic disorders, are known to sometimes induce cognitive deficits.[42] Numerous studies have shown that histamine is involved in memory and learning.[43] Paradoxically, histamine facilitates both forgetting as well as memory consolidation and retrieval. As any clinical psychologist will tell you, some memories are best forgotten. Forgetting is not always a "failure." When thinking through a problem, some memories should be weighted more heavily than others.[44] That is, cognitive performance can be improved through a judicious combination of memory facilitation and memory suppression. Most people know about histamine because of its role in allergies. However, recent review articles concerning histamine have all highlighted its influence on cognitive processing as a major or even preeminent function.[45]

The above observations notwithstanding, research investigating histamine function remains somewhat rudimentary. The neurophysiology is complicated, and there remains much to learn. Nevertheless, in light of our discussion of melancholy and grief, the existing findings regarding histamine are suggestive. In summary:

1. Histamine is the common cause underlying both the allergic response and the characteristic features of weeping, including watery eyes, choked-up pharynx, nasal congestion, and so on.

2. A well-documented effect of histamine is that it inhibits reward. This suggests that histamine may contribute to feelings of *anhedonia* that are prominent symptoms of both sickness and melancholy.

3. Histamine confers notable cognitive benefits. The release of histamine enhances the sort of rational reflection that is characteristic of *melancholic realism*—an essential function in melancholy.

It bears emphasizing that histamine function is complicated. For example, although histamine release is a fundamental part of the allergic response, bouts of crying can also arise due to chronically *low* histamine levels (a condition known as histapenia). Also, recall that depressive symptoms are linked to other proinflammatory cytokines, not just histamine. In accounting for the various stress-related symptoms like anhedonia, melancholic realism, and watery eyes, histamine alone will surely not provide a sufficient account. Histamine is simply one of the suspects, whose effects may or may not help to explain many of the mechanisms underlying the physiological, cognitive, and affective features of melancholy and grief.

Triadic Theory of Immunological Responses to Stress (TTIRS)

There exist many forms of stress and many responses to those stressors. For a handful of critical stressors, evolution has fashioned stressor-specific responses. For example, the experience of fear can lead to the classic behaviors of fight, flight, or freezing. Touching a hot object will cause a reflexive response so fast that the hand will be withdrawn before the sensation of heat has registered in consciousness. Similarly, bad smells or tastes can lead to a stereotypic disgust response characterized by rapid retreat from an offensive stimulus, profuse salivation, and possible spitting. In chapter 12, we will encounter another stressor-specific response where the separation of an infant from a caregiver results in a separation distress call.

Apart from such stress-specific responses, we can also identify responses that address a wider and more diverse range of stressors such as injury, infection, confusion, embarrassment, suspicion, frustration, and low social rank. In chapter 5, we identified two general resources for dealing with stress— namely, our thoughts (a cognitive response) and other people (a social response). In light of our discussion of the immune system, we can now add a third general resource: our bodies. As proposed here, this triad of resources is linked to a triad of distinct responses: *sickness*, *melancholy*, and *grief*. Each state offers a different strategy for dealing with stress. Each state is associated with a characteristic combination of feelings or affects whose purpose is to motivate us to engage in behaviors that are usually adaptive in dealing with the pertinent stressor. Feeling *sick* (the corporeal response) functions principally as an energy-conserving state whose purpose is to enhance the effectiveness of the immune system by minimizing competition for metabolic energy and impeding the growth of pathogens. Feeling *melancholy* (a cognitive response) functions principally through the phenomenon of melancholic realism where the individual benefits from enhanced reflective thinking. Feeling *grief* (a social response) functions principally by recruiting the assistance of those around us. Together, these three states help us deal with a huge range of stressors—from malaria parasites and broken bones to boredom, loneliness, or the departure of a loved one.

In this chapter, I have suggested that the immune system is implicated in all three responses. In the case of sickness, useful feelings of anergia (lethargy), anhedonia (loss of pleasure), and hyperalgesia (pain sensitivity such as muscle soreness) originate in proinflammatory cytokines and prostaglandins. In the case of melancholy, the feeling of malaise is minimally linked to the release of interferons. Most important, melancholy is characterized by improved cognitive processing (melancholic realism) for which histamine is a prime facilitating candidate. In the case of grief, the allergic response is recruited and embellished to form the distinctive ethological signal we call *weeping*, a signal whose value lies in the evoking of prosocial behaviors from observers by inducing a bouquet of proximal feelings such as compassion and commiseration. All three affective states originate in the immune system, with proinflammatory cytokines (such as interferon-α and histamine) playing prominent roles.

In chapter 5, we observed that the long-face display of melancholy resembles an ethological *cue*, whereas weeping (the main symptom of grief) bears the hallmarks of an ethological *signal*. We can now see that these observations align perfectly with the proposal that melancholy is principally a private cognitive response to stress, whereas grief is principally a public, observer-directed social response.

Reprise

In this chapter, I have introduced what I call the *triadic theory of immunological responses to stress* (TTIRS, pronounced "tears"). The TTIRS theory posits the existence of a general stress response system that complements stressor-specific evolved responses such as fear and disgust. TTIRS entails three affective states—sickness, melancholy, and grief—and proposes that these states exist to rally corporeal, cognitive, and social resources, respectively. The theory further proposes that all three of these states originate in the immune system. Although we call it "the immune system," in light of its wide-ranging responses, it might better be called the stress–response system.

In the next two chapters, we will expand on the TTIRS theory by considering how melancholy and grief might have evolved from sickness behaviors. In particular, we will consider in detail how the common allergic response could have been recruited and transformed into the ethological signal of weeping.

How did feelings of sickness, melancholy, and grief emerge over the course of evolutionary history? In this chapter and the ensuing chapter, I offer a speculative evolutionary story describing how these affects might have emerged over time.

Evolutionary stories rightly invite skepticism. Evolutionary accounts are susceptible to all kinds of empirical, methodological, and ethical misadventures.[1] Not all biological change arises from evolution by natural selection, nor are all phenotypic characteristics adaptive. Most genetic changes are due to genetic drift rather than arising from selection pressures.[2] When applied to human behavior, the concept of "instinct" is thorny and problematic due to the notable human capacity for cognitive reflection, allowing the inhibition or modification of otherwise compelling behaviors.[3] When existing social norms are mistaken for "human nature," purported evolutionary accounts can invite restrictive moralizing that merely reinforce entrenched cultural practices limiting human freedom. When certain behaviors are portrayed as inevitable, efforts to curb or eradicate them may be abandoned or disparaged. Methodologically, evolutionary storytelling is plagued by the fact that most adaptationist conjectures are untestable.

These (and other) dangers notwithstanding, theoretical attempts to reconstruct biological history are not without value. While evolutionary theorizing appears to be an open invitation for unbridled speculation, as the Norwegian political theorist Jon Elster has noted, "The first step toward finding a positive answer is telling a plausible story."[4]

In the following account, I will suggest that each of the three target stress responses—sickness, melancholy, and grief—arose by building on the physiological, cognitive, and behavioral foundations of already existing stress responses. I will suggest that melancholy emerged as a distinct affective state by elaborating and tweaking features of sickness and that grief, in turn, emerged by elaborating and tweaking features of melancholy. In chapter 3, we reviewed evidence suggesting that weeping conforms to an ethological signal. My evolutionary story will include a causal chronology suggesting how a weeping signal emerged from an existing melancholy cue. In order to help in telling our tale, I will propose precursor or intermediate states (now biologically defunct) that formed bridges between sickness, melancholy, and grief. Consequently, my evolutionary scenario will describe six hypothetical states that emerged in the following sequence: *illness, sickness, protomelancholy, melancholy, hypermelancholy*, and finally *grief*. It should be noted that these six posited states are descriptive conveniences. They represent momentary signposts in a continuous process that would have unfolded over a very long period of time. Finally, even if my proposed origin story happens to be broadly correct, the actual evolutionary history will surely have followed a more convoluted path.

Sickness versus Illness

Sickness and melancholy address quite different stressors and recruit different resources in their efforts. Sickness focuses on physical stressors (tissue damage, pathogens, etc.), and it does so by rallying various corporeal resources, notably inflammation. Melancholy grapples with mental stressors (loneliness, anxiety, loss of social status, etc.), and it does so by marshaling cognitive resources, notably melancholic realism.

The corporeal/cognitive dichotomy is echoed in an important distinction physiologists make between *central* (brain) and *peripheral* (rest of the body) phenomena. These two realms are concretely defined by a hermetic border in the form of the *blood–brain barrier*. This is a sophisticated chemical barrier that allows very few compounds to pass from one realm to the other. Even the tiny histamine molecule is unable to make the journey in either direction—either body to brain or brain to body.[5]

An important consequence of this barrier is that the same chemical typically has different effects in the brain and the body. For example, in chapter 1, we learned that serotonin (in the brain) influences mood, with low serotonin associated with social stress, sadness, or depression. Elevated serotonin is associated with self-confidence and assertiveness. In the periphery (outside of the brain), serotonin has dramatically different effects. Among other effects, serotonin influences the speed of movement of food through the intestines. An injection of serotonin is likely to result in a case of diarrhea. Serotonin cannot pass from the brain to the rest of the body or vice versa.

Many states (like sickness) involve a coordination of central and peripheral processes. When we are sick, we experience both distinctive corporeal and mental symptoms. Examples of corporeal (peripheral) phenomena include inflammation and warmth around an injury site, gastrointestinal turmoil associated with purging some ingested toxin, itchiness in response to a topical parasite, and coughing in response to viral- or bacterially induced respiratory congestion. Examples of central phenomena include feelings of lethargy (anergia), sleepiness (somnolence), a loss of appetite (anorexia), fever (pyrexia),[6] and the absence or attenuation of pleasure (anhedonia).[7]

Headache is an interesting case. One might suppose that headaches are central phenomena since we localize the pain in our heads. However, the brain itself has no pain sensors so the pain of a headache is not actually a "brain pain." The pain associated with headaches originates either immediately outside the brain (such as in the scalp, neck, eyes, or sinuses) or in large cranial veins or cranial nerves that traverse the brain but reside outside the blood–brain barrier. For this reason, physiologists regard headache as a peripheral rather than a central phenomenon.

Two central phenomena contribute directly to combating disease. High body temperature (fever) disrupts the functioning of many viruses and bacteria, and loss of appetite (anorexia) can directly hinder bacterial infection by denying nutrients needed in order for bacteria to reproduce. However, the remaining four central responses contribute only indirectly to fighting injury or disease: anergia, somnolence, anhedonia, and malaise are useful only insofar as they discourage voluntary movement and so boost the effectiveness of immune operations by reducing competition for metabolic resources.[8]

Feelings of anergia, somnolence, anhedonia, and malaise are really *enhancements* that amplify or boost the effectiveness of the peripheral responses. This implies that these four central components of sickness behavior were Johnny-come-latelies to the immune response arsenal. That is, central feelings of lethargy, sleepiness, a decline of pleasure, and low mood were likely later additions in the evolutionary history of biological tactics for fighting disease.

By way of analogy, consider Anne, a promising world-class athlete. Providing her with high-quality nutritious food and a pleasant and comfortable place to live might well contribute to her success. Anne's friend, Bonnie, however, is a poor athlete. For her, good food and superior lodging will do little to help her win competitions. Similarly, without the active peripheral immune response, these central responses would have little or no value. By itself, anhedonia won't get rid of a virus or heal a broken bone. The immune system must have already had in place some effective tools for fighting injury or pathogens before reducing metabolic competition from other systems could have any utility.

All of this suggests that there was an earlier version of "sickness" behavior that relied exclusively, or almost exclusively, on peripheral responses to infection, injury, or disease. Let's refer to this hypothetical earlier response as *illness*—as distinct from *sickness*. I propose that this earlier illness state is evolutionarily quite old and was shared by virtually all ancient animal species.

Modern *sickness*, by contrast, is experienced by the majority (although not all) animals.[9] In particular, the combination of peripheral and central phenomena means that sickness requires a somewhat more sophisticated brain. Sickness, as defined here, can be observed in mice. For example, if a mouse sustains tissue damage, one can directly observe peripheral effects such as inflammation and warmth around the injury site. If the mouse is infected, one can also directly observe reduced activity (less movement) and reduced food consumption—suggesting that the animal is experiencing the central effects of lethargy and loss of appetite. The mouse is experiencing *sickness*, not merely *illness*—as defined here.

Although it's common to see both central and peripheral symptoms in diseased animals, zoologists have also observed that some physically stressed animals can exhibit peripheral symptoms without exhibiting any central symptoms. For example, some diseased birds and reptiles fail to show any evidence of reduced activity, increased sleep, or reduced food consumption, suggesting that the animal does not feel lethargic, sleepy, or less hungry. Such differences can even be observed within a single species. For example, in contrast to sick females, male animals in some species will continue to engage in mating displays and behaviors, even when they are near death due to disease or injury.[10]

Biologists have proposed several theories in order to explain why there is no reduced activity in some diseased animals. When seriously ill, it may make sense for males to engage in mating behaviors since mating may represent a reasonable last-ditch effort to enhance one's biological fitness despite the animal's impending demise.[11] Other possibilities relate to predator–prey interactions. In general, predators are attentive to prey that are sick or injured, so diseased or injured animals may benefit by suppressing or hiding readily observable behaviors that are symptomatic of their stressed state.[12] (Said another way, lethargic movement and reduced appetite amount to ethnological *cues* that might benefit an attentive hunter.)

With the exception of fever, notice that the central components of sickness behavior (anhedonia, anergia, somnolence, anorexia, and malaise) are also characteristic symptoms of melancholy. Insofar as the function of melancholy is to address nonphysical stressors, there is no need for melancholy to include peripheral responses such as inflammation. This suggests that melancholy might have arisen from sickness behavior by simply jettisoning the peripheral responses. This might also explain, for example, why headache is not a common feature of melancholy.[13]

At this point, it's helpful to consider some of the neurochemistry. Recall from the previous chapter that proinflammatory cytokines are implicated in both central and peripheral responses to stress. Specifically, proinflammatory cytokines (including histamine) are involved in inflammation as well as the various feelings associated with being sick. If both peripheral and central phenomena are caused by the same chemical compounds, we need to consider how it is possible that central symptoms might appear without the concurrent appearance of peripheral symptoms—and vice versa. For example, how is it that one could experience tissue inflammation without also feeling lethargic? How is it that one could experience anhedonia without also having a headache? In short, how is it possible to be "ill" without being "sick?" And similarly, how would it be possible to feel melancholy without feeling "sick?"

As noted earlier, the blood–brain barrier presents a formidable obstacle for chemicals. Consequently, the blood–brain barrier makes it possible to have independent central/peripheral responses. "Illness" occurs when we only experience peripheral symptoms. When we experience only central symptoms, our state looks very similar to melancholy—including lethargy, sleepiness, a decline of pleasure, and low mood.

Now if the brain and body are two truly independent realms, we also end up with the opposite problem: How do we explain the coordination of brain and body when responding to a diseased state as in sickness? And why would both the peripheral and central symptoms of sickness rely on the same endogenous molecules?

Research has shown that the release of peripheral cytokines can often induce a corresponding synthesis or release of central cytokines.[14] Ongoing research has identified at least three mechanisms by which peripheral cytokines are able to induce parallel chemical production or release in the brain. For example, peripheral cytokines can activate the vagus nerve, which in turn induces cells in the brain to produce cytokines (including histamine).[15] Moreover, a variety of complementary mechanisms allow central histamine releases to induce parallel releases in the periphery. So despite their independence, there also exist mechanisms that enable coordinated action by the brain and the body. However, it's important to recognize that these parallel activation systems are specialized adaptations. Not every molecule can induce its siblings to be mobilized on the other side of the blood–brain barrier, nor do these systems operate for a given molecule in every situation.

From Sickness to Melancholy

Now melancholy is not simply sickness stripped of its peripheral symptoms. Melancholy also includes *melancholic realism*. Indeed, in chapter 5, I suggested that the principal value of melancholy is that it encourages more realistic thinking. That is, melancholy has adaptive value because it ultimately shapes behavior in ways that commonly better address a person's cognitive stressors. However, this explanation relies on melancholic realism—a response that is notably absent from sickness when stripped of its peripheral symptoms. Melancholic realism, I propose, was added later in our evolutionary story. Stripped of the peripheral symptoms of sickness and devoid of melancholic realism, let's refer to this intermediate state as *protomelancholy*. To be clear, what we're calling *protomelancholy* is a hypothetical, biologically defunct state (at least in humans) that was characterized by feelings of lethargy, sleepiness, reduced pleasure, low mood, and possibly reduced appetite, without any of the mechanisms leading to melancholic realism.

Before we consider how melancholic realism was added, we need to consider whether protomelancholy itself offered any utility for addressing cognitive or social stressors. Was protomelancholy itself stable enough to exist for a period of time while the mechanisms needed for melancholic realism got their ducks in a row? Without melancholic realism, just what do feelings like lethargy, low mood, and attenuated pleasure get us when dealing with cognitive or social stressors?

When things aren't going well, one of the first remedies is to stop persisting in futile efforts. As the popular maxim goes: the definition of stupidity is doing the same thing over and over again and expecting different results.[16] Now the reason why we persist in particular behaviors is that there are proximal rewards that continue to attract us. Classic examples include addictions such as gambling and alcoholism. People can persist in behaviors they consciously recognize as futile or destructive simply because the proximal positive rewards are too compelling.

What feelings of anhedonia, malaise, and anergia accomplish in protomelancholy is to reduce the proximal allures that have led to repeated past actions that have a consistent history of failure. Simply encouraging inactivity (through feelings of lethargy, low mood, and reduced pleasure) can prove beneficial in many stressful situations. Of course, not all behaviors warrant being suppressed. However, the onset of acute cognitive or social stress has a high likelihood of being preceded by unsuccessful actions.

My proposal here is that protomelancholy (i.e., melancholy without melancholic realism) was useful in its own right. Consequently, protomelancholy emerged—and was preserved in the behavioral repertoire—because of positive selection pressures. Accordingly, protomelancholy was not simply some happy accident that merely provided an inert steppingstone leading to the subsequent emergence of melancholy. Protomelancholy could have enhanced one's fitness by itself.

In order for protomelancholy to morph into melancholy proper, we need to consider how *melancholic realism* might have emerged. In the previous chapter, we reviewed evidence linking improved cognitive performance to proinflammatory cytokines—with histamine being particularly implicated. We saw how melancholic realism confers a host of cognitive benefits tailored for stress-related strategizing. These include improved attention, memory, and reasoning; reduced stereotyping and judgment bias; and reduced gullibility—among other benefits. Histamine was already being released in the periphery as part of the inflammatory response. But histamine was unable to cross the blood–brain barrier and so powerless to affect the brain. The conjecture here is that at some point in evolution, histamine released in the periphery was echoed by histamine released in the brain and that central histamine took on the cognitive effects we see today. Where protomelancholy relies on a single strategy (the suspension of actions likely to be associated with failure), melancholy avails itself of a much-expanded palette of cognitive tools. Apparently, over time, cognitive functions were optimized so as to better deal with stressful situations. It is the addition of mechanisms promoting melancholic realism that I propose would mark the hypothetical transformation from protomelancholy to melancholy proper.

Reprise

In this chapter, I have outlined a speculative scenario by which melancholy might have emerged in evolutionary history. I have proposed that sickness emerged from illness and that melancholy emerged from sickness through a series of rather straightforward steps. In the next chapter, we continue the story by presenting a theory of how grief and weeping might have emerged.

11 Evolution of Grief

Grief is an acute sadness state that includes weeping; grief is also evident via the globus sensation (feeling choked up), a state that is best regarded as a prelude, harbinger, or arrested precursor of weeping. Although both weeping and the globus sensation can be induced by other emotions, grief can be distinguished from other weeping scenarios because it is always linked to the subjective experience of acute sadness.

In previous chapters, we reviewed evidence suggesting that weeping is a bona fide ethological signal whose purpose is to induce a prosocial response from observers. In this chapter, we address the question of how grief-related weeping emerged in evolutionary history. Recall that ethologists argue that all signals evolve from earlier cues through a process called ritualization.[1] In the following scenario, I suggest how a weeping signal might have evolved by recruiting the common allergic response.

Altruistic Opportunities

As we noted in chapter 5, if altruism is so beneficial to the person engaging in altruistic acts, then we might expect that the motivation to help others would not be limited to explicit requests for help. And indeed, there is ample evidence in human behavior that we are eager to offer help, even in the absence of a weeping signal.

Although our disposition to engage in helping behaviors can be greatly enhanced by cultural norms and expectations, as we saw in chapter 6, humans have an innate disposition to engage in helping behaviors. Helping stressed individuals is commonly observed in our closest ape relatives as well as more distant animal relations.[2] The biological benefits enabling and promoting altruism mean that helping behaviors must already have been present in some quite remote nonhuman ancestor.

At least two problems can be identified regarding help-related behaviors. As mentioned in chapter 5, one problem arises when the individual being helped doesn't want any help. Unwanted assistance may nevertheless incur an implied cost of being expected to return some possible favor in the future and a relative loss of social status with respect to the nominally helpful person. However, the more serious problem arises when a stressed individual fails to attract any help when help is truly needed.[3] From the perspective of a stressed individual, there is obvious value in having an explicit signal that effectively says, "Now I need help."

Notice that the clearest "win–win" situations occur when we help someone who is truly in need of help. The assisted individual is much more likely to feel gratitude and to not begrudge either the implied future reciprocal debt or the reduced social status with respect to the altruistic helper. The benefits to both individuals are maximized when help is truly needed. It is this mutual benefit that provides the selection pressure favoring the emergence of an ethological signal.

Mechanism

As noted in chapter 1, the causes (etiology) of melancholy and grief are quite similar. Both emotional states arise from the same or similar sources of stress. The main difference is that grief is more likely to occur in situations that are especially stressful—and, consequently, situations where the benefits of receiving assistance are more likely to outweigh the costs. In considering how grief arose from melancholy, our first hint is that grief might have originated as an exaggerated form of melancholy. As part of my evolutionary story, I propose a long-defunct state of especially acute or intense melancholy that might be called *hypermelancholy*. Hypermelancholy represents an intermediate state between melancholy and grief.

In chapter 9, we considered at length the possible role of histamine in the phenomenon of melancholy. We reviewed a considerable volume of research establishing the notable effect of histamine in enhancing a wide variety of cognitive processes. The evidence suggests that histamine represents a plausible causal agent for melancholic realism. As proposed here, a characteristic of hypermelancholy was exceptionally high levels of histamine—elevated levels that might be expected in an exaggerated or intense form of melancholy.

Although recent research highlights the role of histamine in cognition, historically, the first discoveries related to histamine pertain to its role in the allergic response. As already noted, histamine released by peripheral mast cells leads to classic allergic symptoms, including watery eyes. Like most compounds, histamine produces very different effects in the brain compared with the periphery. One might suppose that weeping arose because the histamine released by the central nervous system associated with melancholy somehow spilled over into the periphery. However, as we learned in the previous chapter, even a very small molecule like histamine is unable to cross the blood–brain barrier. The central and peripheral realms are independent when it comes to histamine.

Yet also in the previous chapter, we learned that various specialized mechanisms sometimes allow molecules released on one side of the blood–brain barrier to induce a parallel release of the same molecule on the other side of the blood–brain barrier. In this regard, I propose that an important step in the evolution of weeping was the appearance of a new mechanism allowing melancholy-related histamine released in the brain to be echoed by histamine released in the periphery. Such echoing is typically sensitive to concentration level, so the greater the amount of histamine released centrally, the greater the amount of histamine released peripherally.

Now suppose an archaic human ancestor was experiencing an especially stressful situation resulting in a notably high concentration of centrally released histamine. A parallel high-volume release of peripheral histamine would trigger classic allergic symptoms, including watery eyes. However, the watery eyes associated with an allergy are generally less profuse than the waterworks evident in modern crying, so these symptoms would surely have been less conspicuous than weeping. Nevertheless, these allergic symptoms would have been more noticeable to observers than the inconspicuous symptoms of melancholy alone.

As a consequence of the allergic symptoms, observers could now be somewhat more confident in recognizing the stressed state of a teary-eyed person. Said another way, compared with melancholy, hypermelancholy would have been notably more conspicuous. Nevertheless, this hypothetical archaic state would still be a cue (like melancholy) with the allergic symptoms such as watery eyes arising as mere physiological artifacts.

Recall that ethologists regard cues as beneficial only for the observer while offering no benefit to the displaying individual. As a comparatively conspicuous cue, an observer of hypermelancholy might benefit simply by staying away from a stressed individual. Or if the observer regarded the stressed individual as a foe, then an observer might take actions to magnify the stress. In such cases, the observable hypermelancholy cue would indeed serve no useful purpose for the stressed individual.

However, as noted earlier, the value of altruistic acts is maximized for both individuals when the stressed person is especially in need of help. It's this mutual benefit that would have provided the selection pressure needed to create the coevolved display-and-response behaviors characteristic of a bona fide signal. The mutual benefits arising from altruistic interactions would encourage the transformation of an allergic cue into a weeping signal (i.e., ritualization). Specifically, the resulting selection pressure would favor changes that (1) increase the conspicuousness of the display and (2) encourage automaticity in the response of the observer. We might now turn to these two facets of ritualization.

Conspicuousness

In general, conspicuousness is a two-edged sword. If you are sick or injured, it's not always a good idea to draw attention to yourself. As noted earlier, predators (like lions and wolves) are constantly on the lookout for animals that appear sick or lame. When surveying a herd, it is these health-compromised animals that carnivores target as the most promising dinner prospects. For many species, there are excellent reasons not to broadcast that you are stressed. On the contrary, every effort should be made to hide one's stressed condition.

The situation changes somewhat among highly social animals. Among social animals like humans, survival strongly depends on mutual assistance rather than individual encounters with predators. Of course, recruiting assistance is much easier when conspecifics are readily able to recognize that you are in trouble. So communicating one's stressed state may be vital for securing help. At the same time, even among our closest companions, there can be good reasons for observers to be wary of someone who is stressed, especially if they are sick with a communicable disease. If the stressed individual is infectious, there are excellent reasons to keep your distance rather than approach with an offer of assistance. Consequently, it would be important for others to know that the source of the stress you are experiencing is not communicable.

If a social animal wants to advertise that it is stressed (with the hope of attracting altruistic assistance), then one should avoid behaviors that suggest the stressor is contagious. Ostensibly, there are lots of stress-related behaviors whose elaboration or ritualization could have been contenders for a "please help me" signal. For example, ritualized limping might have provided an effective signaling behavior. Many other behaviors, however, have strong associations with stressors that are indeed communicable. These include coughing, retching, vomiting, defecation, shivers, sweating, scratching, sneezing, drooling, and foaming at the mouth. Unlike limping, there are excellent reasons for us to give a wide berth to anyone displaying these symptoms. If we limit our potential ritualized cues to behaviors that clearly represent no threat of contagion, then the possibilities are more limited, including various facial displays, blushing, ritualized movements, or vocal calls. Another is the allergic response: allergies are not infectious.

Even in modern societies, most of us can offer only marginal assistance to someone who is ill with an infection or disease. By contrast, we can often offer considerable assistance in the case of noninfectious stressors, especially mental stressors such as cognitive or social stresses. Most of us are better able to help someone who is lonely, confused, embarrassed, hungry, or fearful, compared with someone suffering from a cold, measles, diarrhea, arthritis, or warts.[4] Moreover, unlike many forms of illness, melancholy is not physically contagious. Coming to the assistance of someone experiencing melancholy involves much less risk for the observer and affords opportunities for more effective altruistic intervention than is the case for someone who is sick. The risk of contagion, as well as the limited ability to offer genuine help, is the reason, I propose, why sickness remains a cue rather than a signal—at least in the case of humans. We may sympathize with someone who is throwing up, but compared with tears, vomit has little power to induce feelings of compassion.

In order to transform hypermelancholy to grief, our first task is to make the display more conspicuous. With weeping, the modest quantity of tears associated with the allergic reaction is transformed into a veritable flood. As noted in chapter 4, the profusion of tears is linked to the appearance of a new branch of cranial nerve VII that innervates the lacrimal glands—a branch not found in other primates.

Most important in the transition from cue to signal was the introduction of an acoustic component that increased conspicuousness by adding another sensory modality. With the exception of occasional sniffling, sneezing, or nose-blowing, the allergic response itself exhibits no unique sonic features. As we saw in chapter 1, however, weeping is associated with a compulsion to vocalize. Weeping vocalizations exhibit several distinctive acoustical characteristics that can all be traced to intense muscular contractions in the throat whose original purpose was to block or expel possible allergens. Specifically, these sonic features include vocalized coughing (i.e., sobbing), the gasping sound caused by inhaling through a constricted windpipe, sustained high-pitched (falsetto) wailing, and the sound of cracking voice—caused by phonatory instability and a pinched or pharyngealized timbre. Altogether, these unique acoustic features amount to an *exaptation* where the physiology of an existing allergic response has been repurposed—in this case, recruited as components of a bona fide weeping signal.[5]

Overall, grief usually exhibits higher arousal than melancholy. The difference is evident in several behaviors. For example, melancholy is usually characterized by a compulsion to remain mute. Even when we speak, melancholic vocalizations are less energetic—quieter and slower. Although grief can also be silent, it's much more likely to involve the production of sounds, sometimes very loud sounds. The difference in arousal levels is also apparent in the flexed versus relaxed facial and neck muscles evident in grief versus melancholy, respectively. In this regard, it may be pertinent to note that, along with histamine, mast cells concurrently release adenosine triphosphate (ATP), a compound that is the universal biological energy source. ATP is critical for such energy-consuming functions as muscle contraction and nerve conduction. It is possible that the coordinated ATP release is responsible for, or contributes to, the generally higher arousal levels often seen in grief.

Automaticity

Apart from conspicuousness, ritualization also entails reshaping both the displaying behavior and the observer behavior so they become stereotypic and automatic. We already reviewed the evidence for the automaticity of weeping behaviors in chapter 6. With regard to *generating* the behavior of weeping, we noted the similarity of weeping behaviors cross-culturally and anatomical changes such as the addition of a separate nerve connected to the lacrimal glands for generating psychic tears—an efferent nerve activated only by intense emotions. With regard to *responding* to weeping, in chapter 7, we noted the prosocial responses evident across multiple weeping contexts, including weeping induced by grief, joy, pain, loyalty, and extreme laughter. In short, the weeping behaviors characteristic of grief are indeed consistent with increased automaticity.

Ritualization

Altogether, the narrative presented above is consistent with a classic process of ritualization by which a cue is transformed into an ethological signal.

In suggesting how grief emerged from melancholy, one can't help but notice the apparent serendipitous role of histamine. In sadness-related behaviors, histamine exhibits three very useful features. First, as reviewed in chapter 10, central histamine is implicated in many cognitive benefits that plausibly form the core of melancholic realism. Second, in the periphery, high levels of histamine induce allergic symptoms that provide the physiological basis for weeping. And third, as a signal inviting prosocial responses from observers, allergy-like symptoms are notably fortuitous since allergies are not contagious.

At this point, astute readers might rightly wonder whether this chapter has described the evolution of *grief* or the evolution of *weeping*. As we've repeatedly emphasized, among modern humans, weeping can arise in many different circumstances apart from grief. The claim here is that grief and weeping were originally inextricably linked and that only later in human history did weeping became a distinct behavior independent of grief and so became available to be deployed in other situations such as in tears of joy. How and why this process occurs will be addressed in chapter 16 (section "Whence Polygenic Signals"). There we will learn that it is common for signals to become generalized by gaining independence from their original evolved function.

Reprise

In this chapter, I have extended the speculative story begun in the previous chapter, tracing the evolution of sickness, melancholy, and grief. In order to help clarify the logic, table 11.1 summarizes the various symptoms associated with each of the six conjectured stress states in our evolutionary chronology.

Our story suggests that the early immune system focused solely on physical and physiological stressors arising from injury or infection. An early

Table 11.1

Summary of symptoms associated with six conjectured stress states tracing a hypothetical evolutionary chronology from illness to grief. States dubbed *illness*, *protomelancholy*, and *hypermelancholy* are considered defunct, with only *sickness*, *melancholy*, and *grief* remaining for modern humans.

	Illness	Sickness	Protomelancholy	Melancholy	Hypermelancholy	Grief
inflammation	✓	✓				
fever	✓	✓				
headache	?	✓				
anorexia	✓	✓	?	?		
hyperalgesia	?	✓				
anergia		✓	✓	✓	?	
somnolence		✓	✓	✓	?	
anhedonia		✓	✓	✓	✓	✓
malaise		✓	✓	✓	✓	✓
melancholic realism				✓	✓	
social withdrawal				✓	?	
weeping/choked up						✓

state, deemed here *illness*, was limited to peripheral responses such as tissue inflammation, headache, coughing, sneezing, and purging. Only two central responses were involved—fever and reduced appetite—responses that contribute to recovery by interfering with the metabolisms of pathogens.

The effectiveness of immune responses could be enhanced by reducing metabolic competition from unnecessary behaviors, notably by curtailing energy-consuming voluntary movement. As suggested here, a new state deemed *sickness* enhanced the *illness* state by adding feelings of hyperalgesia, anergia, somnolence, anhedonia, and malaise. That is, voluntary motor movement could be reduced through feelings of muscle soreness, reduced energy level, sleepiness, diminished motivating rewards, and low mood. Prostaglandins and proinflammatory cytokines (such as interferon and histamine) offer plausible neurochemistry that may account for all of these affect-related responses. These behavioral changes imply that immune compounds were synthesized in the brain itself or able to cross the blood–brain barrier, setting the scene for further future immune-system influences on cognitive behaviors.

Our speculative story continues with the suggestion that inhibiting motor action itself may be useful in dealing with various stresses that don't originate in injury or infection—such as feelings of loneliness, confusion, or hunger. When lost or frustrated, responses such as inflammation, fever, or headache serve no useful purpose. However, feelings of anergia, malaise, and anhedonia can prove useful insofar as they tend to discourage repeating recent behaviors whose ineffectiveness may be the source of the stress. As suggested here, a new state, deemed *protomelancholy*, emerged in which peripheral responses such as inflammation and hyperalgesia were jettisoned, and only the central affective responses (feelings of anergia, somnolence, anhedonia, and malaise) were retained.

Over time, further cognitive refinements were added to this protomelancholic state. Most notably, unhelpful optimistic and risk-tolerant thinking was replaced by more detail-oriented, flexible, and risk-sensitive patterns of thought (i.e., melancholic realism). Recent research suggests that histamine offers a plausible neurochemical contributing to these characteristic cognitive changes. In our evolutionary story, it is the emergence of melancholic realism that distinguishes protomelancholy from (modern) melancholy. Note that the emergence of melancholic realism broadly expanded the scope of cognitive and social problems that a stressed individual could fruitfully address.

In circumstances involving especially intense stress, our evolutionary story suggests that a high concentration of central histamine associated with melancholy was able to induce a parallel peripheral histamine release resulting in classic allergic symptoms. The resulting allergic symptoms constituted an ethological cue that would have been more readily recognized by attentive observers. This *hypermelancholy* state would have offered little or no benefit to the stressed individual beyond the benefits of *melancholy*. However, over time, observers discovered the value of responding to the more noticeable hypermelancholic cues by engaging in altruistic acts—the mutual merits of which were reviewed in chapter 4.

In situations where both a stressed individual and an observer typically benefit from an altruistic interaction, selection pressures would be expected to transform the cue into a signal. According to our scenario, selection pressures would have reshaped an intense-stress *hypermelancholy* cue into the *weeping* signal characteristic of a modern *grief*.

In ethological theory, all signals are thought to arise from an earlier cue through the process of ritualization. Ritualization entails two major changes: increasing the conspicuousness of the display and shaping both the displaying behavior and the observer response so they become stereotypic, automatic, or involuntary. In the case of conspicuousness, allergy-related symptoms were magnified and augmented. The volume of tears was amplified through the appearance of a novel branch of cranial nerve VII innervating the lacrimal glands—an efferent nerve activated only by intense emotions. Conspicuousness was further increased by adding distinctive vocal sounds. Interestingly, the unique sonic features of weeping—such as sobbing, cracking voice, and pharyngealization—are direct consequences of allergy-related physiological changes in the vocal tract.

Regarding automaticity, this is most evident in the grief-stricken individual's involuntary compulsion to vocalize: sounds issue forth without any willful intent. Automaticity is also evident in the development of a characteristic feeling evoked in the observer when witnessing weeping: although the ultimate motivation to offer assistance originates in the benefits of enhanced fitness, it is the proximal psychological feelings that dominate the subjective experience of the observer. These include feelings of compassion, bonding warmth, anticipated virtue, the allure of social approval, the fear of social ostracism, and reduced negative feelings of commiseration. Although humans are able to exercise a degree of executive control to inhibit such feelings, they nevertheless tend to arise spontaneously and involuntarily.

Due to the risk of contagion and the limited capacity for observers to provide effective help, displays of *sickness* have failed to evolve into a signal. Similarly, since private reflection can often successfully address stressful circumstances without the need for assistance (and its attendant social cost), *melancholy* has been retained as a separate affect state (independent of grief) for dealing with cognitive and social stressors. Consequently, modern humans exhibit a triad of basic stress-related affects: *sickness*, *melancholy*, and *grief*—only one of which (grief) is linked to an ethological signal (i.e., *weeping*).

While sickness, melancholy, and grief are typically adaptive responses, it bears reminding that any biological system can assume pathological states. In the case of melancholy, pathologies can lead to major depressive disorders. In the case of grief, conditions like pseudobulbar affect can lead to pathological weeping. In the case of feeling sick, pathologies include a range of autoimmune disorders from type 1 diabetes to rheumatoid arthritis.

At the beginning of chapter 10, I emphasized that hypothetical evolutionary accounts are inevitably suspect for a variety of reasons, not least because they are usually difficult or impossible to test. It is hoped that the narrative presented in this (and the previous) chapter is sufficiently detailed with respect to behaviors, physiological effects, and conjectures regarding the ordering of events, and that some of the main claims might ultimately prove susceptible to future empirical testing.

12 Childhood Crying

No discussion of human crying would be complete without addressing those individuals who cry the most—namely, children, especially infants. There are many valuable lessons to be learned from considering infant bawling, colic crying, and that dubious apex of childhood weeping—the temper tantrum. As we will see, research on the crying of infants and children can be remarkably useful in clarifying the relationship between crying and sadness.

Infants and Caregivers

Few sounds are as distressing for adults as the sound of a crying baby. When faced with a seemingly inconsolable infant, parents are eager—often desperate—to discover the "off" switch. As one might expect, the sonic suffering endured by parents has been a major incentive motivating research related to infant crying.

The proper way to stifle crying is to address its root causes. Is the infant hungry, sleepy, sick, wanting to be held, or in need of venting a good burp? Since simple conversation with preverbal infants is not possible, deciphering the cause (if any) of infant crying can prove challenging. Caregivers and researchers look for telltale clues, both in the acoustic character of the crying itself, as well as in the context in which the crying occurs.

Considerable research has endeavored to distinguish possible "cry types" that might differentiate the source of the apparent distress.[1] Parenting magazines are full of optimistic and reassuring articles; however, research efforts to reliably discriminate different cry types (indicative of different sources of distress) have largely proved disappointing.[2] Attempting to decipher the nature of the presumed precipitating distress is not straightforward. The problem is confounded by the phenomenon of colic, in which infant crying arises for no apparent reason.

In describing infant crying behaviors, a useful place to begin is to consider the simple frequency of crying. In cross-cultural studies of infant crying, Ronald Barr and his colleagues at McGill University have documented an apparently universal trend for crying to increase during the first few months of life.[3] At the same time, the overall amount of crying is known to differ from culture to culture. Drawing on anthropological field reports, Herbert Barry and Leonora Paxson at the University of Pittsburg conducted a large-scale study comparing 183 contrasting cultures. They found that infants cry more in Western cultures. They also found that infant crying occurs less frequently in cultures where caregivers respond quickly and attentively in response to the crying—suggesting that the greater frequency of infant crying in Western cultures is a consequence of less rapid responsiveness by parents.[4]

One factor influencing the amount of crying is infant/caregiver sleeping arrangements. In Western cultures, most infants do not share a bed with a caregiver, and the goal of many parents is to provide a separate room for their baby. In their cross-cultural study, Barry and Paxson found that Western culture is unique in this regard. They found no other culture in the world where babies sleep alone. One might suppose that the reduced responsiveness of Western caregivers to infant crying also contributes to the apparent higher incidence of unconstrained crying (colic) among Western infants. However, evidence suggests that colic also occurs regularly in all cultures studied, no matter how attentive the caregiver.

Beyond the age of two, there may be good reasons why caregivers in Western cultures have been more reluctant to respond to crying. One might suppose that affluent caregivers have more opportunities to give in to the resource demands of toddlers and children and so inadvertently reward and reinforce crying-fueled demands. In less affluent societies, the relative inability of caregivers to submit to offspring resource demands may quickly discourage some crying behaviors.[5]

In their cross-cultural study of crying behaviors, Barry and Paxson explicitly tested this theory by examining the relationship between crying frequency and parent responsiveness in forty-four contrasting traditional cultures for which they had adequate data. In early infancy, they found that the frequency of crying was negatively correlated with the speed of caregiver responding.[6] That is, prompt attention by the caregiver reduces the overall frequency of crying. This result is what would be expected if the crying truly arises from some sort of infant stress or discomfort and that the caregiver's actions are typically effective in dealing with that stressor. They reported that

> an opposite tendency for any type of rewarded behavior to be strengthened and expressed more frequently (Miller, 1969) apparently is not manifested at this early stage, indicating that crying by infants is primarily a reflex response to discomfort rather than a learned, instrumental response.[7]

Barry and Paxon's observations here pertain specifically to *early* (i.e., infant) crying rather than the crying of toddlers and older children. As we will see shortly, age and type of crying play important roles.

Bawling

In chapter 1, we distinguished two different forms of crying—*bawling* and *weeping*. The most noticeable difference between bawling and weeping is evident in the rhythm of the sounds. Bawling is characterized by relatively long vocalizations (*waaaah, waaaah, waaaah . . .*) with individual cry sounds ranging from 0.5 to 1.5 seconds in duration.[8] Each "bawl" coincides with a full exhale, with no vocalization made while inhaling.

By contrast, the vocalization most characteristic of weeping is the sound of *sobbing*—the distinctive cough-like convulsions involving punctuated vocalized exhaling (*ah-ah-ah-ah-ah . . .*). With sobbing, the "puffs" of air occur at a rate of roughly five per second. As noted earlier, these staccato vocalizations are shared with laughter. In fact, the main acoustic difference between laughter and weeping is the slight change in timbre caused by smiling. If you smile while sobbing, the resulting sound will be transformed into the sound of laughter.

Compared with weeping, bawling is generally much louder. Weeping can occur without producing any sound at all. However, bawling is rarely quiet, and the absence of sound usually arises only when the lungs are depleted. Apart from the sound, bawling commonly exhibits a gapping or wide-open mouth, squinting or closed eyes, a furrowed brow, and (often) flushed face. The flushed face is likely a consequence of high blood pressure. Curiously, much bawling occurs without any wetness in the eyes. Although not all bawling is tearless, much infant crying occurs without the shedding of tears—a phenomenon that Charles Darwin chronicled from observations of his own children.[9] When tears are digitally added or removed from photographs of infants, their presence has little capacity for eliciting sympathy compared with similar manipulations of photographs of adults and older children.[10] Another difference between bawling and weeping is that weeping commonly includes nasal congestion and pharyngeal constriction; these symptoms are rarely or never observed in neonate bawling. Finally, bawling is more energetic and so metabolically more costly.[11]

Bawling is the predominant crying behavior throughout the first year of life. By twelve months of age, infant crying gradually shifts from bawling to weeping. By late childhood, bawling is rare, and it is virtually nonexistent among adolescents and adults. As noted earlier, immature pulmonary development might account for the delay in the onset of adult-style weeping. Specifically, the staccato breathing or rapid punctuated exhaling characteristic of weeping depends on sophisticated thoracic muscle coordination that requires many months of postnatal development.[12]

At birth, crying is known to be most frequent in the absence of a caregiver, suggesting that the main purpose of infant crying is to maintain infant–caregiver contact.[13] However, by two years of age, crying is maximum in the *presence* of a caregiver, suggesting that the main purpose of toddler crying is to promote caregiver investment.[14]

Infant crying has notable effects on observers, especially on caregivers. Mothers and fathers are remarkably accurate in recognizing the cries of their own infants and respond physiologically most strongly to those cries.[15] Caregivers are strongly motivated to approach, hold, and comfort the infant. Bawling also has notable effects on the mother's physiology, causing an increase in breast temperature and leading to the milk letdown reflex.[16] It is possible that this is simply a learned response, but it appears to be physiologically prepared, consistent with the stereotypy and automaticity of an ethological signal.

The automaticity of these responses is nicely illustrated in studies of interactions between deaf parents and their babies. In the first instance, deaf babies of deaf parents vocalize in a manner indistinguishable from hearing babies.[17] That is, neonates exhibit automatic and stereotypic bawling behaviors even when they are unable to hear themselves or others. It is common for deaf parents to install sound detection devices near their infants that produce a flashing light when the infant makes a sound exceeding some threshold intensity. However, deaf parent reactions to the flashing light are notably muted compared with hearing parents' reactions to the sounds. Hearing parents typically exhibit considerably more distress and react with greater urgency in response to their crying infant. Eric Lenneberg and his colleagues concluded that the emotions and reactions evident for hearing parents "cannot be called forth through the surrogate of a flashing light."[18] For deaf parents, the visual-only facial cues of distress accompanying their infant's crying is not sufficient to evoke a strong response—which is to suggest that the acoustical component is by far the most important communicative aspect of bawling. Bawling is predominantly an acoustical signal, and in the absence of this signal, we see that the automaticity of the observer response is largely lost.[19]

Distress Calls

Several researchers have suggested that infant and adult crying derive from so-called *separation distress calls*.[20] Separation distress calls are evident in thousands of species. They are issued by abandoned or isolated offspring and serve to help reunite offspring with caregivers. Separation distress calls appear to be classic ethological signals: there is an innate disposition for offspring to produce the calls when separated from a caregiver, and caregivers have an innate disposition to be attentive to and respond earnestly to these calls. The calls typically benefit both offspring and caregiver insofar as the biological fitness is enhanced for both.

For many or most species, the same call is produced in other circumstances, such as where the offspring is hungry, injured, or otherwise stressed. Accordingly, ethologists commonly drop the "separation" adjective and refer to the signal simply as a *distress call*. Nevertheless, the call is very commonly associated with offspring/caregiver separation.

Nearly all distress calls are exclusively acoustical without any unique visual component. The absence of a visual component is understandable since the inability to see a caregiver is the main incentive for issuing the call in the first place. Unfortunately, any call is apt to attract predators, so a distress call is most effective when it is assured of attracting the caregiver's attention. A modest call that fails to reach the caregiver is potentially more dangerous than no call at all. As a result, separation distress calls across species tend to be loud.

Although human weeping is sometimes loud, compared with bawling, human weeping is generally quieter. We can recognize that someone is weeping even when the weeping entails no sound at all. But bawling without sound hardly qualifies as bawling. In short, bawling would seem to offer a better candidate for a separation distress function than weeping.

Cache or Carry

Not all species produce separation distress calls. Ethologists have observed that distress calling is related to the pattern of food foraging.[21] For many species, young offspring are left in a burrow or nest while a caregiver goes away to forage. Examples include groundhogs, wolves, and most birds. In other species, animals will forage accompanied by their offspring, as in the case of ducks, moose, and monkeys. For example, a baboon mother will carry its baby on its back while foraging.

Offspring of species that remain cached in a burrow or nest during caregiver foraging are typically more self-sufficient. For example, they are commonly better able to regulate their body temperature than offspring of species that forage accompanied by their offspring.[22] Also, offspring of species that are abandoned during foraging typically remain silent. Since foraging caregivers are apt to be some distance away from the nest, calling out from the nest or burrow is likely to be entirely ineffective and simply court the attention of possible predators.

By contrast, offspring of species that accompany foraging caregivers typically have a prominent separation distress call as part of their signaling repertoire. The separation distress call is simply intended to attract the attention of caregivers when, for whatever reason, infant and caregiver become separated.

Separation distress calls are evident in all primates. Primate infants are not abandoned in a burrow or lair but accompany the troupe when foraging. In the case of humans, caregivers in traditional hunter-gatherer societies also tend to carry their offspring while foraging for food. In light of the pattern observed across many species, one would then expect that humans also have some form of separation distress call.

Indeed, several researchers have drawn attention to the acoustic similarities between early infant crying and separation distress calls in our closest primate relatives.[23] Primatologists have observed that the separation distress calls of nonhuman primate infants are sometimes indistinguishable from the cries of human infants.[24] Critically, the observed similarities relate to *bawling* rather than *weeping* displays.

In fact, acoustic analyses show clear similarities across a much wider range of species.[25] For example, working at the University of Winnipeg, Susan Lingle and her colleagues have chronicled highly similar infant distress calls for fallow deer, fur seals, bighorn sheep, ground squirrels, silver-haired bats, and human neonates—all from recordings of calls produced less than seven days after birth. In particular, duration, sound envelope, and pitch contour are highly similar. The calls differ mainly with respect to the overall pitch height, which is species specific. As you might expect, bighorn sheep produce calls that are much lower in frequency than those produced by squirrels.[26]

Following up on these similarities, Lingle and her Midwestern University collaborator, Tobias Riede, conducted an ingenious experiment where they transposed the overall frequency of different neonate distress calls to the same region of distress calls for the offspring of white-tailed deer and mule deer. They then played recordings of diverse vertebrate distress calls via a loud-speaker placed in fields where the two deer species visit. They observed deer mothers (but not males or female nonmothers) approach the loudspeaker in response to distress calls from all of the various species tested—including human infant bawling. Deer mothers did not approach the loudspeaker in response to various control sounds having the same frequency range but otherwise involving a different acoustical pattern.[27] Evidently, neonate distress calls are very similar across a wide range of diverse mammals and evoke similar responses in caregivers. At least in the case of deer, transposed human infant bawling has all the earmarks of a distressed Bambi.

The idea that bawling (rather than weeping) is phylogenetically linked to separation distress calls is reinforced by the complementary observation— namely, that the sounds of human sobbing and laughter are much more similar to the so-called *pant-grunt* of modern apes (think of a chimpanzee's *oo-oo-oo-oo* . . . or *ah-ah-ah-ah* . . . call). This suggests that weeping (more specifically, sobbing) has a plausible independent phylogenetic precursor distinct from separation distress calls.

By way of summary, seven observations suggest that bawling (rather than weeping) is phylogenetically related to nonhuman separation distress calls:

1. Unlike weeping, bawling is evident from birth.
2. In contrast to weeping, bawling is acoustically much more similar to separation distress calls in a wide range of mammalian species, including nonhuman primates.[28]
3. Separation distress calls in nonhuman primates never involve the production of tears. Tears are ubiquitous in human weeping; some tearing may also accompany bawling, but "tearless" bawling is much more common.

4. Unlike weeping, early infant crying (bawling) is maximum in the *absence* of caregivers—consistent with a separation distress call. By contrast, later child crying (weeping) is maximum in the *presence* of caregivers.

5. Bawling disappears as a human behavior before adolescence; weeping behaviors continue throughout adulthood into old age—suggesting that bawling (but not weeping) is exclusively intended for a caregiver audience.

6. Bawling is typically much louder than weeping—a feature consistent with the fact that separation distress calls favor loud sounds with little or no distinctive visual display. Weeping can occur without any sound, yet it remains easily recognizable because of its distinctive visual features.

7. The punctuated vocalizing characteristic of weeping suggests that weeping more closely resembles the *pant-grunt* of modern apes than ape separation distress calls. This suggests that weeping has a plausible independent evolutionary precursor distinct from separation distress calls.

It bears reminding that although the distress call may have originated as a means of reuniting offspring and caregiver, for most animals, its function has broadened over the eons to include other forms of stress, such as hunger, discomfort, and pain. Human bawling is similarly not restricted to situations of infant/caregiver separation.

As we have noted, unlike bawling, weeping occurs most frequently when a caregiver is present rather than absent. This suggests that weeping serves a different function from bawling. Useful insights into this function may be gained by examining the most extreme form of human crying: the temper tantrum.

The Temper Tantrum

Few human behaviors are as dramatic as the temper tantrum. Temper tantrums are commonly precipitated when a preferred toy, food, or activity is denied or in response to an unwanted caregiver demand (such as getting ready for bed) or prohibitions (such as forbidding playing with a dangerous object).[29] Nearly all temper tantrums involve weeping behaviors, including the shedding of tears.[30] However, temper tantrums also exhibit a range of additional behaviors.

Temper tantrums can involve foot stamping; throwing objects; biting, kicking, or punching caregivers; or collapsing to the floor.[31] Screaming or

Figure 12.1
In tandem with weeping, temper tantrums may also include pouting, where the lips are extended forward. Photo by Tanya Little, by permission.

shrieking is commonplace—although not infant-style bawling;[32] the intensity of these sounds can rival that of a police siren (much to the exasperation of parents). Tantrums sometimes involve verbal threats of self-injury or actual acts of apparent self-injury such as breath-holding and head banging—acts that have a notable capacity for attention-grabbing. In addition to weeping, distinctive facial features often include lowered eyebrows, pouting, and a flushed face (see figure 12.1). The rise in blood pressure associated with a temper tantrum has been known to cause capillaries in a child's cheeks to burst and for the child's eyes to become bloodshot.[33] Temper tantrums are most common in children between eighteen months and four years of age.[34]

In an effort to better understand this jumble of behaviors, Michael Potegal and Richard Davidson of the University of Wisconsin–Madison analyzed hundreds of coded tantrum reports. They found that the component behaviors are consistent with two simultaneous emotions: sadness and anger.[35] Weeping, for example, suggests feelings of grief, whereas behaviors such as striking out, screaming, and lowered eyebrows suggest feelings of anger.[36] In light of this model, several questions arise. Is it really possible to experience more than one emotion concurrently—that is, "mixed" emotions? What accounts for acts or threats of self-injury? And what are we to make of behaviors such as screaming and pouting?

Size Symbolism

In order to better understand the anger component of a temper tantrum, it's useful to review research related to *size symbolism*. In general, apparent size is a common feature of both threat and affiliative/appeasement displays. In making a threat display, animals typically endeavor to make themselves look bigger. For example, a cat will arch its back and make its hair stand on end; a bear may rise up on its hind legs. Conversely, nonthreat, submission, or appeasement can be conveyed by efforts to make an animal look smaller. For example, a submissive dog will crouch close to the ground and tuck in its tail.

Apart from visual aspects of size, the same size symbolism can also be communicated acoustically. One of the best generalizations in the field of acoustics is that large masses or volumes vibrate at a lower frequency than small masses or volumes. Working at the Smithsonian Institution in Washington, ethologist Eugene Morton famously examined vocalizations in a wide range of animal species. Throughout the animal kingdom, Morton found that low-frequency vocalizations tend to be associated with threat or aggression, whereas high-frequency vocalizations tend to be associated with affiliation, appeasement, or deference.[37]

This relationship is also evident in human speech. Harvard University linguist Dwight Bolinger conducted a classic cross-cultural study of speech prosody. Across all the cultures he examined, Bolinger observed that low pitch speech is associated with aggression, whereas high pitch speech is associated with deference or friendliness.[38]

Independent of the pitch or fundamental frequency of the voice, resonant frequencies of vocalizations are also related to the length of the vocal tract. The vocal tract effectively acts as a resonant tube extending from the vocal folds to the tip of the lips.[39] The length of the vocal tract is largely fixed—with the notable exception of the lips, which can be voluntarily extended or retracted. Extending the lips produces a pout that lowers the vocal tract resonance; retracting the lips produces a smile that raises the vocal tract resonance.

University of California, Berkeley, linguist John Ohala noted that size symbolism offers a straightforward account for the pout and smile displays.[40]

If you extend your lips while speaking, the resulting voice will take on a notably aggressive or threatening character. The effect is dramatic and easily demonstrated. Sustain the vowel "oo" with your lips extended far forward. Hold your lips in that position while you speak aloud the remaining words in this paragraph: *Your protruding lips will transform your voice so that it sounds like the cliché Hollywood thug or hooligan. You're now the bad guy. Consistent with size symbolism, pouting alters the voice so as to produce a distinctly bellicose quality.*

Conversely, when the lips are retracted tight against the teeth, the vocal tract length is shortened by about half a centimeter compared with a resting position. This causes an audible upward shift in the tube's fundamental resonance. The result is a sound conveying an appeasing, deferential, or friendly attitude. Retracting the lips is accomplished by flexing the zygomatic muscles used when smiling. The smile doesn't just change the visual appearance of the face; it also has a dramatic effect on the sound of the voice. "Smiling voice" is easily recognized without the need for any visual confirmation. In a telephone conversation, for example, listeners are adept at recognizing whether an unseen speaker is smiling.

Research on the human smile has long focused exclusively on the visual features. For over a century, the smile has been regarded as something of an enigma: smiling tends to expose the teeth, and a prominent display of teeth might suggest a threatening rather than friendly intent. Ohala cogently argued that the smile and pout have an acoustical rather than visual genesis and that their effects originate in sound-size symbolism.[41]

Modern smiling and pouting are visually distinctive displays on their own without the need to make any sound. Indeed, in contrast to weeping, pouting and smiling exhibit no concurrent urge to vocalize. Yet the facial configurations for these displays appear to be simply artifacts of what's necessary to produce the appropriate sound-size symbolic vocalizations. Although it's not necessary to produce any sound, it would be wrong to regard modern pouting and smiling as exclusively visual signals since we are still sensitive to the deferential quality of retracted-lip vocalizations and the bellicose quality of extended-lip vocalizations.

Setting aside the theory, one might wonder whether there is any empirical evidence that pouting is indicative of aggressive intent. In an extensive study of children's emotional behaviors, Ewan Grant at the University of Birmingham found that, apart from raising one's hand in preparing to hit someone, the next behavior most predictive of a physical attack was the child's thrusting their lips forward.[42] Apparently, pouting is not just for show.

Screaming

The temper tantrum reaches its apex when the screaming or shrieking starts. Although not all tantrums involve screaming, most tantrums ultimately reach this odious climax.[43] Screaming is phylogenetically widespread: most mammals include a scream-like alarm in their vocal repertoire. Moreover, from species to species, screams are acoustically nearly identical.[44] Screaming is typically the loudest and most energetic call an animal produces, and that's also the case for human screaming.[45]

Screaming is most commonly evoked by feelings of extreme fear, such as when an animal is caught by a predator. However, screaming arises in many other circumstances[46] and can serve a variety of functions.[47] Among social animals (like humans), most screams arise in response to threats from conspecifics. In most primate species, the effect of screaming is to draw together closely allied individuals to defend one of their members. The effect of screaming on an observer depends on the relationship between the observer and the screamer. If the observer is a close ally, then the scream amounts to a "call to arms." If the observer is not closely allied, then the effect of screaming is to increase attentiveness or vigilance.

In either case, the main effect of screaming is to attract urgent attention. It hardly bears mentioning that screaming is highly salient for human listeners. Peak energy for screaming (whether produced by males or females) lies in the 3,000- to 5,000-Hz frequency range—a range that coincides with the greatest sensitivity in human hearing. Screaming is hard to ignore and is widely regarded as the most distressing and obnoxious of human-generated sounds. Analyses by Swiss researcher Luc Arnal and his colleagues have shown that the acoustic features of screams are closely emulated in commercial sound alarms.[48] The unpleasantness of the scream itself motivates desperate efforts to turn it off. In the presence of a screaming child, everyone within earshot is eager to figure out how to get the screaming to stop.

Pulling Out All the Stops

The temper tantrum highlights an important point related to infant and child crying. Temper tantrums are all about coercing adults into behaving in a manner that the adult is reluctant to do. Characteristic of tantrums is the mixed bag of emotional strategies. Weeping suggests a strategy intended to evoke prosocial compassion in the observing adult. Lowered eyebrows and pouting suggest anger, as though the child is endeavoring to bring about a favorable outcome through threats of aggression. Screaming suggests a fear alarm intended to rally caregivers to come immediately to the child's assistance. When the child and caregiver are closely related, threats or acts of self-injury (breath-holding, head banging, suicide threats) directly imperil the adult's inclusive fitness.[49] The hotchpotch of behaviors evident in the temper tantrum suggests that the child is trying multiple strategies in an effort to coax a reluctant caregiver into some resource provisioning or other action that benefits the child.

The typical adult response to temper tantrums is itself revealing. Unlike plain weeping, adults rarely find temper tantrums emotionally convincing. Adults typically find the behaviors exasperating and are eager simply for the tantrum to end. From an ethological signaling perspective, temper tantrums exhibit a mix of signals that appear contradictory. On the one hand, threatening or aggressive behaviors such as pouting or striking at an adult seem curiously inept. Given the size of the child relative to the adult, it is the rare adult who will feel sufficiently intimidated by the aggression to accede to the child's demands. On the other hand, behaviors such as weeping represent an effort to evoke compassion in the adult. Dropping to the floor or curling into a ball suggests further acts of submission via size symbolism—behaviors that again suggest an effort to evoke compassion. Screaming suggests an all-purpose fear alarm intended to spark prompt assistance from caregivers. Threats or acts of self-injury amount to a form of biological blackmail.

At face value, this multipronged approach seems like a reasonable strategy for shaping a caregiver's behavior to one's goal. Individually, aggressive threat, an expression of alarm, a tearful declaration of misery—each of these might well work in getting an adult to comply with the child's wishes. The problem is that the impact of these individual displays is not additive.

Should one accede to the child's demands because one is afraid of the aggression, should one acquiesce to the child's wishes because one feels compassion, should one come to the child's assistance because the child is in a state of alarm, or should one give in because one's biological fitness is menaced by the child's threats or acts of self-injury? From an adult's perspective, the mixed signaling is incoherent. How can the child expect to evoke compassion in a caregiver while simultaneously threatening the caregiver?

From an adult's perspective, it appears that every effort is being made to compel the adult to produce the desired behavior. It's likely that the child simply does not have enough experience to know which strategy is best. Consequently, the temper tantrum offers a garbled mix of signals that most adults will regard as little more than self-centered coercive behaviors that may seem inauthentic. Is the child sad, angry, or acutely distressed?

It appears that by the age of five, most children will have learned that the jumbled message of the temper tantrum has a limited effectiveness. Over time, the child finds out that simultaneous pouting, weeping, and screaming is unlikely to be successful. Pouting alone may be effective if the adult is motivated to assuage the child's obvious unhappiness. Many children learn that screaming by itself works just fine, especially in public venues where the adult feels socially obliged to quickly bring an end to the ruckus. However, most children discover that weeping by itself is more likely to be effective in evoking compassion, and so successfully able to secure resources from a caregiver.

By the age of twenty-four months or so, most toddlers have sufficient language abilities to communicate verbally their needs or wants. So why is crying necessary? What does crying add that isn't conveyed by demanding a particular toy or refusing to eat one's broccoli? It's important here to state the obvious: crying pushes buttons that language alone cannot reach. It's more difficult for adults to ignore or resist appeals packaged in the form of weeping since weeping is an ethological signal that engages a biologically prepared response. Moreover, it's also important to recognize the specificity of the behavior. A child cannot cross their fingers, shake their head, wiggle their toes, or yank on their earlobes and expect to get the same response. Tears, pouting, and screaming aren't simply arbitrary cultural conventions. The relatively weak response of deaf parents to bawling-evoked flashing lights is not indicative of some sort of parenting deficiency: a flashing light simply fails to evoke the biologically prepared feeling of urgency experienced by hearing the sound.

At the same time, the ability of observers to resist the commonly compelling urges evoked by signals bears witness to the enhanced executive control available to *Homo sapiens* described in chapter 6. People can harden their hearts to human suffering and even respond by perpetrating acts of cruelty that dwarf any sympathy they might feel. Of course, it's much easier to inhibit or suppress our sympathetic feelings when the source of the weeping appears to be trivial. In particular, the incoherent mix of grief and anger evident in the temper tantrum makes it easier for observing adults to dismiss the child's behavior as a blatant act of manipulation and consequently easier to resist.

Mixed Emotions?

The mixture of anger and weeping suggested in figure 12.1 raises a more general question of whether it's possible to feel more than one emotion simultaneously. In the preceding discussion, we have assumed that mixed emotions are possible. So, what is the status of the idea of mixed emotions?

Existing empirical studies suggest that mixed emotions are commonplace.[50] A meta-analysis by Raul Berrios and his colleagues at the University of Sheffield reviewed sixty-three studies and concluded that mixed emotions are "a robust, measurable and non-artifactual experience."[51] In the case of film scenes, for example, research is consistent with the inducing of concurrent positive and negative mixed emotions characteristic of a bittersweet experience.[52] In the case of music, several studies provide experimental evidence of concurrent happy and sad induced emotions.[53]

Costs and Benefits Revisited

In chapter 4, we identified six potential benefits for observers who respond to weeping by offering assistance or terminating aggression. Once again those benefits include kin selection, reciprocal altruism, alliance formation, enhanced social status with respect to the stressed individual, reputation maintenance, and the avoidance of social ostracism or punishment.

In the case of child weeping, the ledger of costs and benefits for the observer differs somewhat from the corresponding ledger when responding to adult weeping. Most child–caregiver interactions involve closely related individuals, so kin selection incentives are foremost. There are excellent reasons for biologically related adults to attend conscientiously to basic survival needs of the child. However, compared with adult weeping, the other benefits in responding altruistically to child weeping are dramatically reduced. A child is unlikely to feel indebted to an adult helper, so favors are generally not reciprocated. Since infants and children already have a low social status in comparison with adults, an enhanced social status with respect to the infant or toddler represents a miniscule benefit for a helping adult. Many weeping episodes arise from rather minor stressors ("Mom didn't let me have any ice cream!"), so except when the child's welfare is truly threatened, adults who are unresponsive to a weeping episode have less to fear that their failure to help will negatively impact their reputation for helping. Apart from providing for basic needs to one's offspring, there is little to be gained by responding to every weeping child's whim.

Apart from the changed ledger of benefits for the observer, we can also consider the changed costs for the weeper. As noted earlier, the principal cost incurred by weeping is the loss of social status. We noted that reduced social status is more costly for high-status individuals of reproductive age. The least cost is incurred by nonreproducing individuals at the bottom of the social ladder, a group whose foremost members include infants and children. Since these individuals incur the least cost associated with weeping, it follows that there are fewer constraints on crying.

Every child discovers that adults have access to many good things that the child didn't know existed. At some point, a child learns of the existence of chocolate, plush toys, ice cream, and any number of novel sources of pleasure. It follows that not all pleas for resource provisioning need have a clear goal. If a child's crying leads a caregiver to offer some resource, then there may be value in urgent crying, even when nothing in particular is needed or requested. Who knows what unforeseen goodie might appear in response to a spirited bout of crying? Insofar as crying can lead to resource provisioning by caregivers, it is possible that some inexplicable episodes of crying might be motivated by such a general or unfocused strategy.

The utility of crying by infants to solicit resources is consistent with a further phenomenon found in infant crying—namely, the tendency for crying to be contagious. The crying of a single infant has long been observed to induce crying among nearby infants.[54] Contagious crying might simply be a response to the unpleasantness of the sound of others crying. However, another possibility is that contagious crying might represent a form of sibling or peer competition. If caregiver attention is given disproportionately to the infant who cries, then a nearby silent infant is apt to receive less. Once again, since infants have little social status, the penalty for crying is small, so there may be considerable value in crying in response to the crying of others.

In the early years, infants and toddlers can tap into adult compassion at very little cost. This window of opportunity closes slightly when toddlers begin earnest social interaction and shuts more fully with sexual maturity when the social costs of weeping have greater consequences for reproductive success. Weeping provides an extraordinary low-cost tool by which enterprising children can loosen the adult grip on resources through a biologically prepared disposition to acts of compassion.

Faking It

Temper tantrums are mostly spontaneous and unconscious. However, some tantrum behaviors suggest voluntary or contrived acts. Temper tantrum researcher Michael Potegal has offered the following apt account:

> Parents report children doing things that suggest they are not necessarily or always caught up in irresistible emotion. Before dropping to the floor, they may look around for a soft place to fall. They may reduce crying when parents leave the room and increase it when they return. Our home videos revealed a little girl who stopped wailing long enough to take a bite of her bread, then resumed where she left off. One particular 3-year-old boy appeared to be a master of tantrum-as-manipulation. In one video in which he was seated with his face buried in the crook of his elbow, the camera caught his surreptitious glance up at his mother. While kneeling on the floor during another tantrum with his back to his mother, he said "You can't see the tears on my face." Her off-camera voice could be heard to acknowledge this (while trying not to laugh).[55]

We will have more to say about voluntary or fake weeping later in chapter 14. We will discover that there are circumstances where obviously feigned or contrived adult weeping can serve an important function that benefits both the observer and the sham weeper.

Finally, we should note that the immediate costs incurred by infants and children from weeping can be so low that persistent crying can lead to the real danger of possible caregiver neglect, abuse, and even infanticide.[56] Such scenarios are consistent with classic parent–offspring resource conflict described by the American evolutionary biologist Robert Trivers.[57]

Reprise

In this chapter, we have expanded on the distinction—first introduced in chapter 1—between bawling and weeping. I have suggested that the two forms of crying manifest unique behaviors, have different evolutionary origins, and represent independent ethological signals that serve different, yet overlapping, functions.

Bawling is the dominant form of crying during the first year of life; bawling is rare in later childhood and virtually nonexistent among adolescents and adults. Weeping is the dominant form of crying for toddlers and older children. Before the age of two, "crying" (i.e., bawling) is maximum in the absence of a caregiver, whereas after the age of two, "crying" (i.e., weeping) is maximum in the presence of a caregiver. These observations reinforce the idea that bawling and weeping serve different functions.

Acoustically, bawling closely resembles the separation distress calls found in other mammals. Deer mothers, for example, respond to human infant bawling in a manner akin to their own species' separation distress call. Since "separation" is characterized by the inability of an infant to see a caregiver, separation distress signals necessarily emphasize sound over vision. In contrast to loud bawling, weeping is readily recognized, even in the absence of any sound. The vocalizations characteristic of bawling makes it a better candidate for a human separation distress call than is the case for weeping. As with other animals, human separation distress calls commonly arise in response to many forms of distress beyond being separated from a caregiver. These include situations of discomfort, hunger, and alarm.

By contrast, in previous chapters, we have noted that weeping invites acts of altruism by evoking prosocial motivations in an observer. If this is the case, then one would expect weeping to arise most frequently when a caregiver is present—rather than absent—and this is indeed what we observe.

Since the cost associated with weeping is principally the loss of social status, and since children have low social status compared with reproductive-aged adults, infants and young children incur very little cost by engaging in weeping. Consequently, many weeping episodes arise in response to events with little impact on survival—such as a misplaced toy or not wanting to go to bed.

The phenomenon of the temper tantrum highlights the fundamental lesson regarding signals. As we've seen, temper tantrums are characterized by mixed displays that raise the question: What do anger, grief, and threats of self-injury share in common? The most plausible answer is that they are all behaviors that might be expected to transform the behavior of an observer (especially someone who is biologically related to the child) so as to respond positively to the child's demand. The peculiar mix of tantrum behaviors (including occasional "fake" or "feigned" behaviors) is less about informing an observer about the feeling state of the child and more about getting the observer to behave in a particular way.

By way of summary, this chapter has offered plausible answers to at least nine questions related to crying among infants and children: Why are there two forms of crying (bawling and weeping) rather than just one? Why is bawling maximum in the *absence* of caregivers, whereas weeping is maximum in the *presence* of caregivers? Why does bawling typically occur without tears? Why is bawling so much louder than weeping? Why is weeping among toddlers likely to also exhibit pouting, as in temper tantrums? Why do temper tantrums sometimes involve acts or threats of self-injury? Why is crying contagious among infants? Why do children cry so much even when they are capable of verbally communicating their needs, and why might infants cry without any apparent reason?[58]

In this chapter, we have focused on sadness-related behaviors among the youngest of humans. In the next chapter, we turn our attention to a distinct sadness-related behavior that is often associated with the oldest of humans—the phenomenon of nostalgia.

13 Nostalgia

An intriguing variant of sadness is nostalgia. The word "nostalgia" was coined in 1688 by the Swiss scholar Johannes Hofer. Hofer created the word by combining the Greek words *nostos* ("return home") and *algia* ("pain"). The direct English equivalent, "homesickness," is a word of even more recent vintage. "Homesickness" didn't become widely used in the English language until about 1750, about a hundred years after Hofer coined the word "nostalgia."

In illustrating the affliction, Hofer described the case of a man from Bern who had traveled to Basel (about 100 kilometers away) in order to study. Shortly after arriving in Basel, the man became ill—what Hofer described as a long-term sadness. Hofer noted that the man's symptoms got continually worse, until the people around him thought that he was in danger of dying. His doctors concluded that the only option remaining was to send him back to his native town of Bern. Hofer continues the story:

> Although the patient was "half dead" by this point, he was placed on a small bed and began the journey back to Bern. No sooner did the journey start than the man began "to draw breath more freely . . . [and] to show a better tranquility of mind." As the traveller approached his hometown, "all the symptoms abated to such a great extent they really relaxed altogether, and he was restored to his whole sane self before he entered Bern."[1]

Although Hofer's term "nostalgia" took some time to catch on among the general public, it was quickly adopted within the medical profession. It wasn't long before "nostalgia" was a common diagnosis. Until the nineteenth century, few people traveled any distance. The most likely reason to spend long periods of time away from home came in the form of conscripted soldiers and press-ganged sailors. Many recruits hadn't traveled more than one or two villages away from home before they were forcibly drafted. For most soldiers, being in the army meant truly being away from home for the first time.

For sailors, long periods of isolation took their toll. Captain James Cook's first great voyage took three years to circumnavigate the globe. The ship's naturalist, Joseph Banks, recorded in his diary how the sailors "were now pretty far gone with the longing for home which the physicians have gone so far as to esteem a disease under the name of Nostalgia." Given the primitive conditions of army and navy life, it's not surprising that the affliction was rampant.[2]

An Affliction of Memory

The use of the word "nostalgia" has changed dramatically since Hofer's time.[3] Modern dictionary definitions emphasize the positive aspects of nostalgia: "a sentimental longing or wistful affection for the past, typically for a period or place with happy personal associations."[4] Although nostalgia entails a certain element of sadness, in modern usage, nostalgia also exhibits a distinctly positive element. The combination of positive and negative feelings results in a mixed emotion commonly characterized as "bittersweet." This seemingly paradoxical mixed emotional quality sets nostalgia apart from both melancholy and grief.

As in the case of melancholy and grief, nostalgia also has a pathological form, and from a modern perspective, it is this form that Hofer was describing in the case of the man from Bern. Although Hofer coined the word "nostalgia," in modern terminology, the condition he described is now deemed "homesickness." Homesickness is not merely an acute form of nostalgia. There are clinical symptoms associated with homesickness, including loss of appetite, not being able to sleep, anhedonia, apathy, sometimes fever, and obsessive thoughts of one's home.[5] The condition is highly stressful.

A key difference between homesickness and nostalgia is that homesickness is purely a negatively valenced state, whereas nostalgia involves a mixed (positive/negative–bittersweet) emotion. Another key difference between homesickness and nostalgia is duration. Episodes of nostalgia tend to be short-lived—typically minutes to hours. One might even encounter some object that triggers a momentary episode of nostalgia that lasts only a few seconds before quickly fading. By contrast, homesickness may last for weeks or as long as one is separated from home.

What is nostalgia? How does nostalgia differ from melancholy or grief? Does nostalgia have a purpose? What is the role of memory in nostalgia?

Is nostalgia a "universal" emotion? How does culture shape nostalgic experience? What accounts for the "bittersweet" feeling? In this chapter, I will argue that nostalgia is distinct from melancholy and grief—although all three share some physiological commonalities, and nostalgia can easily morph into melancholy or grief. I will suggest that nostalgia is evident in every culture and propose that it serves an important biological purpose. We will examine the relationship between nostalgia, autobiographical memory, and reverie. Finally, I will suggest why nostalgia is more enjoyable than either melancholy or grief.

As in our analyses of melancholy and grief, we might begin our discussion of nostalgia by considering its etiology, phenomenology, physiology, and development, as well as its behavioral, sociocultural, and possible evolutionary aspects.

An episode of nostalgia may be triggered by thoughts, external stimuli, or social situations. Internal precipitating causes include thoughts about one's past. External precipitating causes include all kinds of stimuli: an old photograph, the smell of a childhood jacket, a piece of jewelry found in a drawer, or the sound of a particular musical work. In Marcel Proust's famous novel *Remembrance of Things Past*, the taste of a madeleine cake dipped in tea is the starting point leading to an extended bout of nostalgia. Any personally relevant stimulus can serve to bring about a nostalgic episode.

The main phenomenological symptom of nostalgia is *reverie*—a wistful state of reflection about the past.[6] When experiencing nostalgia, we tend to be less attentive to the immediate events around us. Instead, we tend to be transported in time to reminisce about past people, places, objects, events, and situations. As already noted, it's common to describe the phenomenal experience as "bittersweet." Experimental research confirms that nostalgia involves a mixed emotional state in which feelings of sadness and pleasantness are intertwined.[7] In contrast to melancholy or grief, nostalgia is predominantly enjoyable; people frequently seek out the experience of nostalgia. The ratio of reported positive to negative emotions is almost four to one.[8]

As in the case of melancholy, the main physiological symptom of nostalgia is *anergia* or low arousal. Even when nostalgia brings a tear to one's eye, arousal levels tend to remain low, with slow heart rate and shallow respiration. Low arousal is linked to reduced levels of epinephrine, acetylcholine, norepinephrine, and serotonin.

Behaviorally, nostalgia is associated with reduced physical activity. A relaxed or slumped posture is typical. Eyes may lose focus with the person looking at "nothing in particular." Sighing may occur. Less commonly, one or more tears may appear.

Cognitively, nostalgia is associated with thoughts or reminiscences about the past. In fact, pleasant reflections drawn from autobiographical memory might be regarded as the principal defining feature of nostalgia. We will have much more to say about the cognitive aspects of nostalgia later in this chapter.

With regard to development, the principal change is the increasing frequency of nostalgic episodes with advancing age. An eight-year-old might experience a feeling of nostalgia when recalling experiences in a summer camp from the previous year. But generally, children don't often engage in nostalgic reverie. As a person matures into old age, the amount of time spent experiencing nostalgia increases considerably. Nostalgic reveries often focus on memories linked to a particular period of adolescent and early adult life—typically between the ages of roughly fifteen and twenty-five. This period is referred to as the *reminiscence bump*.[9]

Apart from changes due to age, personal circumstances also impact the propensity to engage in nostalgic reverie. Notably, well-off people are less prone to bouts of nostalgia than people whose fortunes are in decline.[10]

Although nostalgia is commonly experienced when alone, it is not uncommon for nostalgic experiences to be shared. In social settings, nostalgia may result in a willingness or eagerness to converse about bygone times. Shared reveries might include relaying personal histories or discussions of shared experiences of past people, places, or events.

In contrast to melancholy and grief, nostalgia can play a notable role in promoting social identity. A sense of social connection can occur for groups of any size. For example, two lovers may experience acute nostalgia from reminiscing about privately shared experiences. Or millions of people can experience nostalgia when viewing a nationally televised event.

Such feelings are commonly used as social motivators. For example, high schools and universities arrange reunion events with the explicit aim that feelings of nostalgia will foster feelings of institutional fealty or loyalty—with positive benefits for fundraising. Feelings of nostalgia have been commonly used to promote nationalist sentiments (for both good and ill).

Cultural expressions of nostalgia are common. Commercial enterprises exist specifically to cater to nostalgic reverie. Examples include "golden oldies" radio formats and products that involve "retro" design elements intended to be evocative of some past. In many cultures, specific occasions are set aside for nostalgic reminiscing. At New Year's Eve, the singing of "Auld Lang Syne" ostensibly encourages reflection about past friendships now lost. In some cultures, specific musical genres aim at evoking feelings of nostalgia, such as Portuguese Fado or Japanese Enka. Events like reunions and anniversaries may be planned in which feelings of nostalgia are expected to feature prominently. In general, there are more social and cultural phenomena aimed at inducing nostalgia than is the case for sadness and grief.

The concept of nostalgia does not appear to be limited to Western culture. A review of nostalgia studies by Frederick Barrett and his colleagues found that nostalgia is evident in every culture in which it has been investigated.[11] Working with an international team of collaborators, Erica Hepper directed a large-scale study of 1,700 participants from eighteen countries (Australia, Cameroon, Chile, China, Ethiopia, Germany, Greece, India, Ireland, Israel, Japan, Netherlands, Poland, Romania, Turkey, Uganda, the United Kingdom, and the United States). Hepper and her colleagues found that people across a range of cultures share strikingly similar conceptions of nostalgia. Whatever you call it, people around the world recognize the experience of reminiscing about their past, focusing on memories associated with positive emotions, yet mixed with an awareness of loss and often involving yearnings or explicit fantasies about recovering elements of that lost past.[12] Finally, apart from modern surveys, descriptions of recognizably nostalgic sentiments can be found in ancient classic Chinese, Indian, and Greek literatures. Nostalgia is a widely shared human experience.

Pleasant Memories

Although people commonly claim that nostalgic reverie is enjoyable, self-reports are not always reliable, and so it's helpful to seek independent corroborating evidence. The financial success of various nostalgia-related industries (such as golden-oldies radio) provides some evidence that people seek out and value nostalgic experiences. However, it would be reassuring to find more direct evidence of nostalgia-induced pleasure.

Such evidence can be found in a lovely series of experiments involving an international collaboration of researchers from China, England, and the Netherlands. In the crucial capstone experiment, Zhou, Wildschut, Sedikides,

Chen, and Vingerhoets made use of a well-known pain-threshold task—the cold pressor test.[13] Participants were instructed to recall either a nostalgic memory or a nonnostalgic memory with emotionally neutral content. Participants were then asked to keep their hands emersed in an icy water bath as long as they could stand it.

Placing your hand in cold water can be pretty unpleasant. When asked to keep your hand in the water as long as possible, the experience quickly morphs into a truly painful experience. Pain tolerance is known to be influenced by several factors, including the release of endorphins. Endorphins have a strong analgesic effect, reducing the sensation of pain. But endorphins are also pleasure molecules.[14] Their presence is symptomatic of the physiological machinery of pleasure.

In their experiment, Xinyue Zhou and her colleagues found that those participants who had engaged in nostalgic reflection kept their hands emersed significantly longer than those who engaged in nonnostalgic reflection.[15] This study offers more direct evidence of the pleasurable effects of nostalgic reverie. Apparently, simply recalling "pleasant" memories can indeed be experienced as pleasurable. For convenience, let's coin a new term that explicitly refers to the pleasurable experience evoked through reminiscence. I propose the term *eumnesia* (yume-NEE-zee-ah). The word is formed from the Greek *eu* (meaning pleasant) and *mnesia* (meaning memory).

Useful Reminiscing

It is possible that nostalgia is a cognitive/emotional by-product or spandrel that serves no useful function. However, as with all common behaviors, it's appropriate to begin by entertaining the possibility that the behavior is indeed functional.[16] So what possible cultural or biological utility might be served by bouts of nostalgia?

Since nostalgia is intertwined with autobiographical memory, we might approach the question of function by first considering the role of memory.[17] A common tendency when thinking about memory is to assume that it is necessarily about the past. From a biological perspective, there is no reason for an organism simply to store information about past events. Memory systems exist, not for the purpose of preserving traces of our personal history but as a way of enhancing the effectiveness of our future behaviors. Information about the past is retained only because it may prove useful later. From a biological perspective, memory is about the future, not the past.[18]

Recall yet again that emotions act as motivational amplifiers—encouraging us to behave in particular ways. If nostalgia serves some useful function, then at least some episodes of nostalgia must ultimately change future behavior. Moreover, if nostalgia has any biological purpose, then the future behaviors provoked by nostalgia must commonly lead to actions that increase the fitness of the person experiencing the nostalgic feelings.

From an evolutionary perspective, I know of no adaptive speculative accounts about nostalgia. Nevertheless, I think that a plausible evolutionary theory can be proposed. A key part of the puzzle is the observation reported in research by Krystine Batcho that nostalgia involves comparative judgments of past versus present states.[19] A second key is the observation by psychologists Mike Nawa and Jerome Platt that less well-off people are more prone to bouts of nostalgia.

In her book *Yesterday's Self*, Andreea Ritivoi chronicles the intense feelings of the one group of people most likely to experience extreme nostalgia—immigrants. In the heyday of U.S. immigration, 30 percent of Polish immigrants returned back to Poland, 46 percent of Greek immigrants returned to Greece, and fully 50 percent of Italian immigrants left the United States to go back to Italy.[20] When we give up a former life, we abandon many good things in the pursuit of something we hope will be better. The immigrant takes a considerable gamble—assuming that on balance, the good will outweigh the bad. However, things don't always work out as hoped.

One might suppose that the decision to stay or return to one's homeland is made following a rational assessment of the pluses and minuses associated with each option. However, research on decision-making suggests that emotions trump dispassionate rationality when it comes to human choice.[21] It is despair or wanderlust or adventure that motivates us to leave and seek our fortune elsewhere. It is nostalgia (or homesickness) that motivates us to return.

Incidentally, many feelings or emotions are linked in oppositional pairs. One feeling motivates us to behave in a particular way, and another feeling motivates us to behave in the opposite way. For example, feelings of hunger will motivate you to eat, whereas feelings of satiation will motivate you to stop eating. If we didn't experience satiation, we might continue eating well beyond what is beneficial to our health. Similarly, feeling sleepy has the predictable consequence of encouraging us to sleep, but after eight or more hours in bed, we will find that the urge to get up and move around becomes irresistible.

My proposal is that *wanderlust* and *nostalgia* are examples of such paired or countervailing feeling states. Both feelings are functional affects that contribute to biological fitness. Feelings of wanderlust (also curiosity or despair) encourage people to traverse mountains, sail across seas, or head for the frontier. Wanderlust encourages us to try new possibilities. Humans are thought to have originated in a small region of Africa. Yet over just a few tens of thousands of years, we ended up spreading over the entire planet. We wouldn't have done that if people didn't regularly experience a strong wanderlust impulse.

Now having itchy feet can have a downside. The adventurous person who climbs over a mountain range, traverses a desert, or sails across an ocean may ultimately find nothing more than desolation. Lots of people discovered new lands that proved to be even more impoverished than the lands they left behind. The search for a better life can easily descend into a desperate struggle for basic survival. Feeling nostalgia, I propose, is the antidote to wanderlust. Feeling nostalgic (or the more pathological feeling of homesickness) encourages us to reconsider past environments, relationships, or circumstances that, if they can be recovered, are better than our current circumstances.

If my proposal is correct, we might ask how nostalgia works. It seems that positive feelings arise from recollections of past circumstances appraised as good. These past images may be juxtaposed against a present appraised as inferior with respect to the subject of the reminiscence. The effectiveness of nostalgia would seem to rely on this combination of *push* and *pull*: a negative feeling that pushes us away and a positive feeling that draws us toward. It is this push–pull partnership, I propose, that accounts for the characteristic "bittersweet" qualia commonly reported when people feel nostalgic. Pleasant memories are enjoyable in their own right. By contrast, melancholy and grief are solely "push" feelings. Melancholy and grief tell us how bad things are—without necessarily positing something good on the horizon that we might aim for. Nostalgia supplements the negative feelings with an image (drawn from memory) of how good things could be. There are undoubtedly elements of our past existence that were less than ideal, but we are nostalgic only for the good things, not the bad things.

Now it would be wrong to think that nostalgia is meant simply to bring us home. Although the word "nostalgia" was originally a literal synonym for homesickness, its modern usage is much broader and better captures the adaptive function I've proposed. Nostalgia can be quite selective. When in the throes of nostalgia, a person may form images of old friends, a former lover, beautiful vistas, past social occasions, a childhood home, or mom's peach pie. As understood today, one can be nostalgic for a beloved pair of shoes. In short, nostalgia includes any sentimental longing or wistful affection for a period, place, object, or event with happy personal associations. This expanded meaning, I would propose, is more biologically appropriate.

Given the specific character of many nostalgic experiences, the nostalgic immigrant need not necessarily give up and return home. Favorite foods can be imported, the internet can deliver media in one's native tongue, and elderly parents can be summoned to join the immigrant's household in the new country. An enterprising individual or immigrant community can make up for many deficiencies in the new land. In each case, the feeling of nostalgia provides the critical impetus. It is the motivational amplifier that brings about changes that improve the quality of one's life.

Of course not all of our desires can be fulfilled. Many aspects of our individual pasts can never be recovered. As a consequence, many nostalgic episodes might seem to be functionally futile; they represent ineffective efforts to help us recover something that is truly gone forever. However, even in these cases, nostalgia can lead us to recognize what is important to us and lead us toward possible surrogates that might, to some extent, replace what has been lost. People do often remarry after the death of a beloved spouse, purchase a house near their (long-gone) childhood home, or find shoes that closely resemble a no longer available treasured pair.

Notice, incidentally, that if the current circumstances are judged as superior to past circumstances, then the individual should be less likely to experience nostalgia. Indeed, as we've noted, well-off people are less prone to bouts of nostalgia. For most people, the relationship between past and present will be mixed: there will be elements of our past that are superior to our present and elements of our present that are superior to our past.

The feelings evoked by autobiographical memory helps us assess whether the current bargain might be improved on. A retired Danish couple enjoys winters in the Spanish Costa del Sol but miss their grandchildren in Copenhagen. In romantic relationships, the good and bad are inextricably bundled together. A woman adores her romantic handyman husband but abhors the couch-potato slob who appears to occupy the same body. All situations involve compromises to various degrees.

Some readers might think that there is no need to posit a distinct emotion of nostalgia when cognitive reflection is all that is necessary for deciding whether to leave or stay. Emphasizing the role of cognition alone, however, fails to recognize that thoughts need to be translated into actions in order for an individual to benefit. There are many examples where someone concludes that doing X would be a good idea, but no action is taken because there is no compulsion to actually do X. Cognition is useless without recruiting some feeling state that ultimately motivates us to act. Antonio Damasio has chronicled cases where certain neurological patients have perfectly intact cognitive and analytic abilities but are unable to act to prevent onerous repercussions because flat affect makes those consequences seem irrelevant or unimportant.[22]

Ultimately, we are all immigrants to greater or lesser degrees. Any change in life produces a contrast between one's former life and one's current life. Changing jobs, moving to a different apartment, attending a different school—all of these sorts of changes give us a chance to appreciate improvements and lament losses. All of us have memories of a past that includes both recoverable and unrecoverable pleasures.

As so often happens with matters of the mind, our immediate subjective experiences are radically different from the long-term biological function: we live in a world of proximal feelings, not distal or ultimate motivations. Phenomenologically, nostalgia is the quintessential experience of backward time travel. We never feel so situated in the past as when we are in the throes of nostalgia. Yet from a biological perspective, nostalgia is one of the ways minds make adaptive use of memory in order to fashion a better future: nostalgia transforms memory into motivation. It leverages the past into desire, and where possible, that desire is ultimately converted into action.

If the purpose of nostalgia is to reshape future behavior, how do we explain why elderly people, approaching the end of life, spend so much time engaged in nostalgic reverie?[23] In the first instance, elderly people typically have more time on their hands than younger working people, so it shouldn't be surprising that idleness might enable more frequent episodes of nostalgia. Moreover, older people have had more life experiences, so they can draw on proportionally more memories, including many more pleasant experiences.

Certainly, the future opportunities for an older person may be dramatically constrained due to limited mobility, poor hearing or vision, or a host of other problems. But that doesn't mean that the nostalgic reveries of the elderly are necessarily nonfunctional. Anyone who has interacted with an elderly parent or grandparent knows that there are still decisions to be made that directly impact the quality of their lives. As the final horizon of life approaches, it remains essential to weigh decisions so that the quality of life is maximized. These decisions include where to live, what celebrations to attend, what possessions to retain, and many other issues. Do you choose to live in a familiar home, closer to the hospital, or closer to one of your children? Planning for the future is not merely about planning years ahead; it's also about planning what to do this afternoon. Emotions inform all of our decision-making right up to the point of death. Nostalgia helps elderly people to sort out what's truly important as the options become fewer.

This is not to suggest that all nostalgic episodes are functional in the biological sense of contributing to fitness. There are pleasures to be had from reminiscing about the past whether or not it leads to tangible improvements in our future lives. That is, eumnesia might be pursued simply for the pleasure it affords. Like the use of contraceptive technology to engage in (nonprocreative) sex, we can take advantage of the proximal positive feelings induced by nostalgia, even if the ultimate biological benefits are bypassed or circumvented.[24]

Nostalgia versus Melancholy

Nostalgia and melancholy are similar in a number of ways. Both states are associated with low physiological arousal, feature a reflective mental state, and evince some degree of negative affect. Both can be reasonably described as types of sadness. This raises the question of whether nostalgia might be regarded as a bona fide emotion distinct from melancholy.

In the first instance, there are at least three important differences between nostalgia and melancholy—those of etiology, phenomenology, and sociocultural expression. With regard to etiology, nostalgia is entirely dependent on autobiographical memory; nostalgia hinges on some sort of cognitive reflection about one's past. Melancholy and grief can also be induced by recalling sad events from one's past. However, nostalgia relies not on recalling past sad events but on recalling past pleasant events. Moreover, both melancholy and grief are commonly evoked by immediate events with little or no involvement of autobiographical memory. The fact that nostalgia taps into past pleasant memories means that the phenomenological experience also differs significantly. Nostalgia evokes a unique subjective experience featuring a blend of positive and negative feelings—the "bittersweet" character that is a classic hallmark of nostalgia. By contrast, melancholy simply feels unpleasant.

An additional difference between nostalgia and melancholic sadness is that human cultures commonly fashion specific events, rituals, or products whose purpose is explicitly intended to evoke feelings of nostalgia. Of course, there are also cultural products intended to evoke melancholic sadness or even grief—as in documentaries whose aim is to induce righteous indignation or spur political action, and tragic arts (to be addressed in chapters 18 and 19) that paradoxically spectators find enjoyable. However, nostalgia-inducing cultural activities or products are typically quite focused on leveraging specific past (pleasant) memories. Altogether, these differences suggest that nostalgia warrants recognition as a unique affect in its own right.

Whereas melancholy exists to make the best of a bad situation, nostalgia exists to make the future more like a preferred past—or, more precisely, some preferred aspect of the past. Both melancholy and nostalgia link cognitive reflection with motivational affect. It is possible that nostalgia also taps into melancholic realism. However, I know of no pertinent research testing whether nostalgia induces more realistic thought.

Finally, we should consider the relationship between nostalgia and weeping. Nostalgic episodes sometimes lead to weeping, often simply some tearing of the eyes. As reviewed in chapter 7, weeping can arise from lots of emotions, including grief, pain, anger, joy, loyalty, piety, laughter, and so on. In the context of nostalgia, the feeling state motivating weeping may remain obscure. If the reminiscence focuses on a past tragic event (such as the death of a loved one), then the memory may offer little pleasure. The bittersweet mixture of sadness and eumnesia that characterizes nostalgia would be absent. In this circumstance, the weeping would be better understood as simple grief. However, if the weeping behavior is accompanied by positive feelings arising from eumnesia, then we might legitimately claim that the weeping is motivated by a feeling of nostalgia—yet another affective state that can be added to our long list of emotions capable of inducing weeping.

Notice, incidentally, that an observer of someone weeping in a nominally nostalgia-inducing situation would have difficulty knowing whether the weeping was due to nostalgia, grief, joy, or some other motivating affect.

Reprise

In this chapter, I have proposed that nostalgia represents a third sadness-related emotion, distinct from melancholy and grief. Although nostalgia entails some feeling of sadness, the experience also involves positive qualities leading to a mixed "bittersweet" character. The sweet component arises from eumnesia—the pleasure to be had from recalling agreeable autobiographical memories.

Like melancholy, nostalgia features cognitive reflection or reverie. I have proposed that nostalgia commonly serves a useful adaptive purpose. By implicitly contrasting a current (inferior) circumstance with a past (superior) circumstance, nostalgia motivates us to return, recover, rekindle, or reconstruct earlier favorable circumstances. Of course there are aspects of one's past that are entirely unrecoverable (such as the death of a loved one). The adaptive function of nostalgia is served only when it leads to objective improvements—including new friendships, replaced objects, or a substitute event—that might at least partially redress truly unrecoverable losses. Nevertheless, nostalgic reverie can sometimes morph into genuine melancholy or grief, whose adaptive benefits we have already discussed. In general, memories are enlisted as motivators that encourage future actions likely to improve a person's lot in life.

By itself, preserving a historical record of past personal events serves no biological purpose. It is only if those memories are enlisted to change future behavior that memory can have any biological value. The biological machinery that enables the existence of long-term autobiographical memory could not have evolved if those memories didn't influence future behaviors in positive ways. In this chapter, I have suggested that nostalgia is one of the ways by which autobiographical memory serves an adaptive role.[25]

In our discussion of nostalgia, we have briefly mentioned various cultural phenomena intended to induce or enhance nostalgic experiences. In the next two chapters, we expand our discussion and consider the role of culture more broadly in emotion-related displays.

If weeping is an ethological signal, then it must be species-wide and therefore truly cross-cultural. However, weeping practices can differ dramatically from culture to culture, and these differences raise challenges for any theory claiming that weeping has a biologically prepared social purpose. In this chapter and the ensuing chapter, we explore the interaction between signals and culture. In this chapter, we briefly address three contrasting examples of culturally defined weeping practices: *public self-injury*, where a mourner engages in various forms of self-inflicted pain; *emotional stoicism*, where cultural norms discourage displays of weeping; and the anthropological curiosity known as the *Welcome of Tears*. In the next chapter we describe how signals and cues are recruited, reshaped, and in some case repurposed in order to serve specific sociocultural functions.

Cultures have their own dynamics, and it would be foolish to expect that naturalistic explanations might somehow account for the extraordinary range of human practices. Nevertheless, we will see that an ethological perspective offers useful insights that can often illuminate many of the unique or baffling cultural practices related to weeping.

The Welcome of Tears

In 1556, Jean de Léry immigrated to a newly formed French colony located on an island near modern-day Rio de Janeiro. Ridiculed for his religious views, de Léry soon abandoned the colony and spent two months living among the Tupinamba people along the Brazilian coast. As he wandered through the forests, de Léry was dumbfounded by the manner in which the Tupinamba greeted him. With each encounter, the Tupinamba would burst into tears and loud weeping. The Tupinamba would simply break down and cry. De Léry was the first European to encounter what would become known as the *Welcome of Tears*.[1]

Thirty years later, the Portuguese Jesuit Fernão Cardim traveled to Brazil, and in 1625, Cardim published a more detailed account of the Welcome of Tears. It turned out that the phenomenon was not limited to native encounters with Europeans or limited to the Tupinamba. In fact, the Welcome of Tears was common throughout the tribes of the Brazilian central plateau region, reaching north to the Amazonian floodplain. In 1928, the Swiss anthropologist Alfred Métraux published the definitive account. Indeed, crying was commonplace not just when greeting strangers but when greeting anyone who was relatively unfamiliar. Wailing, it turned out, was the proper form of greeting for any unusual visitor to a village.[2]

With modern settlement, the practice has declined considerably, but even today, this distinctive greeting can be found among some indigenous peoples of interior Brazil. The Welcome of Tears is one of those delightfully perplexing behaviors that reminds us just how different cultural practices can be. Nevertheless, ethological signaling provides a straightforward account for the Welcome of Tears.

Suppose you are traveling in a land with no central government or formal law enforcement. Encountering a stranger is fraught with danger. One's property might be stolen, or one might be injured or murdered, suffer rape, be enslaved, or even be eaten.[3] When encountering strangers, two broad strategies can be distinguished. One strategy is to impress the stranger through a display of your own power or prowess and indicate a willingness to respond with deadly force if necessary. A second strategy is to offer an olive branch and communicate that you have no aggressive intent and pose no threat. Longstanding research pertaining to conflict suggests that cooperative behavior is typically the more advantageous strategy.[4] In the preponderance of cultures, efforts are made to defuse such stressful situations by reducing the possible perception of threat, such as through smiling, bowing, or open-hand displays (thought to demonstrate the absence of a weapon). As an ethological signal, weeping invites a prosocial response from observers. It amounts to a declaration—"I'm harmless." In especially dangerous situations, weeping is a wholly appropriate way of signaling that your intentions are peaceful or nonthreatening. Incidentally, Métraux noted that wailing was more commonly done by women than by men and that the amount of weeping was related to the size of the encountered party and the mutual unfamiliarity. That is, the display of weeping was proportional to the degree of potential threat.

What has puzzled observers is that the Welcome of Tears seems devoid of any genuine emotion. Having wailed and cried, the greeters then just "turned it off." The whole tearful episode is patently "fake." Such phenomena are commonplace in cultural anthropology and lead to another use of the term *ritualization*. When we think of rituals, what most readily comes to mind are various formal ceremonies like weddings or funerals. However, in cultural anthropology, rituals include a vast collection of patterned behaviors whose forms are stereotypic and socially defined. Examples might include shaking hands when greeting someone or blowing a kiss as a sign of affection. Ritualization, then, is the process by which such patterned behaviors become established in a given culture.

Innate ethological signals sometimes provide the basis for such ritual behaviors. In contrast to spontaneous and unconscious signals, we often observe parallel displays that are contrived and willful. In most cultures, there exist voluntary counterparts that emulate innate spontaneous signals such as smiling, frowning, pouting, disgust, and sneering.

The contrived or ritualized form of weeping used in the Welcome of Tears is, in many ways, superior to actual or spontaneous weeping. First, one cannot always count on involuntary weeping to arise when it would be useful. Moreover, if one's tears were spontaneous "genuine" weeping, it would suggest to the observer that you are truly terrified or genuinely weak or helpless. Between strangers (with no kin connection and little future potential for reciprocal altruism), this response might actually encourage an attack in some circumstances. (Why not take advantage of a stranger who is helpless and deferential?) By contrast, if one engages in contrived or ritualized weeping, it's clear that the behavior is a conscious voluntary act. It suggests that one has chosen to communicate the intent to avoid conflict. Yet, at the same time, the ritualized form says "but I'm not *really* weeping—so look out if you decide to attack anyway; I'm not as helpless as I may appear."

In many ways, what makes the Welcome of Tears a curiosity is its rarity. Unlike social smiling, contrived laughter, mock threat, and so on, many cultures appear not to employ ritualized (i.e., willful and socially recognized) forms of weeping.[5] In the next chapter, we will discuss in more detail how spontaneous and unconscious signals commonly spawn parallel contrived and willful displays that are freighted with cultural significance.

Mourning and Self-Injury

Around the world, traditional mourning practices have included several common behaviors such as cutting or shaving one's hair; smearing ash, clay, or
charcoal on one's face; and dressing in old or black clothing. Tearing of clothing and other acts of resource destruction are also commonly observed. In
several cultures, mourning may be accompanied by acts of self-inflicted pain
such as slapping one's face, pulling one's hair, or beating one's chest. More
extreme forms of self-inflicted pain are also observed. Until the late twenti-
eth century, for example, it was common among the Egyptian Fellahin for
grieving women to slap their faces to the point of drawing blood.[6] Burning
one's skin or cutting oneself is surprisingly common in traditional mourning rituals. Indeed, cutting off or amputating a finger or fingertip is found
in a number of traditional cultural practices such as those of the Blackfoot
Indians, South Pacific Tongans, and the Dani of Papua.[7] Among the Blackfoot
and Dani, finger amputation was done exclusively by women and typically
performed when mourning the death of a spouse.

Western research has tended to regard self-injury and self-mutilation as
a medical pathology. This research has identified a number of reasons why
people engage in acts of nonsuicidal self-injury. These range from congenital
insensitivity to pain, hypersecretion of adrenal hormones, anxiety arising
from intolerable guilt, gender identity dysphoria, and responses to trauma
such as the experiences of child abuse.[8] However, most acts of grief-related
self-injury and self-mutilation around the world can be traced to a vast array
of cultural practices, attitudes, and beliefs, including efforts in folk healing.[9]
Among other motivations, research has established that a major function of
nonsuicidal self-injury is to elicit attention or support from others.[10]

As we've noted, a problem with weeping is that it is not always available
when it might prove useful. Consider the case of a person who is largely
dependent on their spouse for support but whose partner has now died.
Weeping is sure to ensue in the immediate aftermath, and periodic bouts of
weeping are likely to occur for months afterward. But the surviving spouse's
need for assistance is apt to be long term. Relatively brief periods of weeping
may not be sufficient to solicit the perennial and ongoing assistance needed
by someone in such a dire circumstance. For example, weeping is unlikely
to coincide with recurring moments of need, such as daily mealtimes. Social
assistance is likely to be erratic and unreliable.

In the first instance, acts of self-injury, self-mutilation, and resource destruction understandably grab the attention of observers. Similarly, distinctive dress, marking one's face, cutting one's hair, extending one's arms outward, wailing, screaming, and so on are all actions likely to draw attention. Whether or not they are combined with observable weeping, such cultural symbols help to broadcast a person's stressful state. These various behaviors are not easily confused with other functions, and when sustained over time (like the habitual wearing of black clothing or having shaved one's head), they effectively contribute to the *conspicuousness* of the grief state through *persistence* or *redundancy*.[11] When combined with overt weeping, these cultural symbols can be especially effective in evoking observer sympathy.

In general, self-injury is more common among women than men. In particular, it's more common among elderly women beyond their reproductive years. However, in their cross-cultural comparative study of crying, Paul Rosenblatt and colleagues at the University of Minnesota found that sex differences are most evident in those societies that are least egalitarian. In societies where men and women are more equal in status, sex differences in the frequency of self-injury are much reduced.[12] This suggests that the underlying issue relates to social status and resource access, not differences between males and females per se. That is, these acts are more common among those individuals who are most in need.[13] Reinforcing this idea is the observation that self-injury is more common among less educated individuals—people who often have more difficulty securing resources.[14]

Self-injury and resource destruction are cultural practices that incur a considerable cost to the grieving individual. These acts are commonly done in public, increasing their conspicuousness. Most important, self-injury, self-mutilation, tearing of clothing, and other acts of resource destruction are consistent with the handicap principle in signaling theory.[15]

If the aim of weeping in grief situations is to induce prosocial compassion and solicit resource sharing, then many of the most commonly observed mourning-related cultural practices observed around the world are readily understood from an ethological perspective. Many mourning-related cultural practices are consistent with both handicaps that attest to the honesty of the signal and also contribute to the signal's conspicuousness and persistence. The idea that these practices have arisen in order to amplify or embellish an innate signal gains further credence from anthropological observations that such cultural practices are most commonly performed by those members of society with the lowest status and least resources.

In short, many mourning-related practices are readily understood as cultural embellishments that effectively supplement weeping as an ethological signal intended to evoke altruistic assistance from a community of observers. Such behaviors effectively communicate "I really am in trouble here and need ongoing help."

A Stiff Upper Lip

In contrast to activities that make crying more conspicuous, there are also cultural practices that render emotion-related displays like weeping less common or less apparent. For example, the British have long held a reputation for having a reserved emotional manner—the proverbial "stiff upper lip." (The phrase originates as a contrast to a trembling upper lip that might be observed when someone experiences fear.) Charles Darwin regarded stoicism as a national trait. In his 1872 book on emotions, he wrote that "Englishmen rarely cry, except under the pressure of the acutest grief."[16] Commenting on the British penchant for forbearance, the English novelist E. M. Forster famously lamented what he called "the undeveloped hearts" of his fellow countrymen.[17]

For those familiar with the antics of British football fans, the idea that the British hold back their emotions might seem farfetched. So what are we to make of the purported "Keep calm and carry on" national character? Is it really the case that the proper English response to any crisis is a quiet cup of tea?

Incidentally, if there are indeed stable national differences in weeping behaviors, then this raises questions regarding the idea that weeping is an innate emotional disposition that transcends culture. It's possible that purported national differences have some genetic basis. However, as we're about to see, the existing evidence suggests that a genetic account of national differences is highly doubtful.

A place to begin is by asking whether the British are indeed less emotional—more stoic—than people of other nations. Unfortunately, there exists little pertinent research. Some insight can be found, however, in a 2012 Gallup poll entitled "The Emotional State of the World." Gallup conducted a large-scale survey where people in 151 countries were asked about emotions they had experienced the previous day. The results placed the Philippines as home to the most emotional people, with Singapore ranked as having the least emotional people. The United Kingdom was ranked fifty-seventh (just short of the top third) suggesting, if anything, that Brits experience more emotions than the average across the world.[18]

The Gallup survey probed reported feelings rather than emotion-related behaviors, so while it suggests that modern UK citizens experience emotions roughly the same as other inhabitants of the planet, it's not clear whether they are less emotionally demonstrative. Unfortunately, the survey failed to address whether UK citizens are more likely to conceal or mask their feelings compared with others.

On this issue, it's useful to turn to the work of historians. Particularly insightful is the work of Thomas Dixon, who has extensively studied the history of weeping in Britain. In his book, *Weeping Britannia: Portrait of a Nation in Tears*, Dixon chronicled changes in weeping behaviors in Britain over a six-century span.

In late medieval times, public displays of grief in Britain were common and expected. Indeed, failing to show emotion was considered one of the hallmarks of a witch. There were times when British judges regularly wept when announcing harsh sentences, and most condemned criminals wept at their public executions. In literature, poetry, art, and music, overt displays of sentiment were commonplace and considered signs of sophistication and civility. Religious gatherings in Britain were commonly tear-drenched affairs. To modern sensibilities, works such as Henry Mackenzie's *The Man of Feeling* are ludicrously sentimental.[19] In general, the historical research is unequivocal: until the end of the eighteenth century, rather than being the stockade of stoicism, Britain was more a cesspool of sentimentalism.[20]

As detailed in the historical research, the notion of British reserve is rather recent, dating from the end of the eighteenth century.[21] Dixon proposes a compelling theory for changing British attitudes toward the open expression of emotions. He notes that the advent of British stoicism can be traced to the aftermath of the French Revolution. Initially, there was much enthusiasm in Britain for the aims of the revolution and its appeal to the "common feelings of humanity." However, as events in France became increasingly violent, French expressions of passion looked unruly, dangerous, and ultimately barbarous.[22] British enthusiasm waned, and French passion was increasingly viewed with alarm and distain rather than admiration. Dixon notes that

British attitudes became more skeptical regarding public displays of strong emotion, including weeping. According to Dixon, this led to the emergence of the famous British "stiff upper lip."

The remarkable history of changing British attitudes toward open expressions of emotion—especially public weeping—draws attention to the critical influence of social or cultural norms in shaping displays of emotion. The mutability of weeping behaviors within a single culture offers definitive evidence against any genetic explanation for national differences.

It is tempting to conclude that this mutability indicates that emotions are social constructions rather than transcultural or innate dispositions.[23] However, as we already discussed in chapter 6, weeping is susceptible to cognitive influence. When events cause us to become choked up and on the verge of crying, we nevertheless retain some ability to mask, inhibit, or amplify any weeping display.

In this regard, it's telling to consider the earliest known statement— identified by historians—where the English are characterized as having a stoic emotional demeanor.[24] The statement occurs in a volume of *Letters from France* by Helen Maria Williams, dated 1792:

> You will see Frenchmen bathed in tears at a tragedy. An Englishman has quite as much sensibility to a generous or tender sentiment; but he thinks it would be unmanly to weep; and, though half choaked with emotion, he scorns to be overcome, contrives to gain the victory over his feelings, and throws into his countenance as much apathy as he can well wish.[25]

Here Williams explicitly suggests that the English feel just as much sentiment as the French (a suggestion that gains modern credence from the 2012 Gallup poll). Rather, she suggests that the English endeavor to "gain victory" over those feelings. Indeed, she identifies one of the standard cognitive strategies for dampening or evading possible weeping. As we noted in chapter 6, when moved to the edge of tears, one strategy is to contemplate something mundane, such as thinking about what to cook for dinner. Here Williams suggests that the English aim for "as much apathy" as one can muster, even as one is "half choaked with emotion." If one were to accept Williams's interpretation, then English emotional forbearance simply amounts to greater use of executive control in order to mask or inhibit emotion-related displays.

The work of anthropologists and historians provides ample evidence that different cultures manifest different attitudes toward weeping and that the underlying cultural norms can change over time. Pertinent norms include the circumstances in which weeping displays are considered acceptable, sex- and age-related expectations, and other conventions.

Reprise

In chapter 6, we saw that, despite their automaticity, signals are susceptible to some degree of executive control. In that earlier discussion, we emphasized how cost-benefit analyses could encourage someone to modify their behavior in an effort to mask, inhibit, or amplify a signaling behavior. Although such modifications can arise from individual motivations, much executive control can be traced to cultural norms.

There exists a vast range of cultural practices associated with crying. In this chapter, we have considered just three case studies: the Welcome of Tears, mourning and self-injury, and the stoic inhibition of weeping displays. Each of these examples highlights a different form of signal modification. In the case of mourning and self-injury, we see culturally mediated efforts to enhance or amplify a signal. Regarding the British stiff upper lip, we see culturally mediated efforts to suppress or inhibit signaling behavior. And with the Welcome of Tears, we see a culturally defined practice that imitates or emulates a signal.

For self-injury, we've noted that culturally defined forms of public self-injury are consistent with honest signaling theory, hence explaining the apparent paradox that self-injury (and resource destruction) may sometimes enhance biological fitness. Both self-injury and resource destruction illustrate the general ethological principal that signals can be made more compelling to observers when they are recognized as costly to the signaler.[26]

In the case of British stoicism, we saw not just evidence of cultural norms tempering or inhibiting emotion-related displays but also evidence of the malleability of culture and the fact that cultural norms can change over time.

The Welcome of Tears and self-injury grief offer a notable contrast. Self-injury contributes to the perceived authenticity of a nominally automatic signal. In the Welcome of Tears, however, no effort is made to disguise the artificiality of the signal-simulating behavior. Instead, the behavior clearly communicates that the act is voluntary and contrived rather than involuntary and spontaneous. In a sense, the Welcome of Tears says, "I choose to communicate that I'm helpless, but I'm not actually helpless, nor do I feel helpless. I just want to clearly indicate that I harbor no aggressive intent at this time." At the same time, despite its artificiality, the weeping holds the potential to induce prosocial feelings in the observers.

As noted at the beginning of this chapter, cultures can exhibit their own dynamics that transcend any biological rationale. There surely exist cultural practices related to weeping for which there are no valid biological or ethological interpretations. Anthropologists are rightly taught to be skeptical of universalist claims. Indeed, there is a wide range of cultural expressions related to crying—a diverse collection of weeping-related behaviors that suggests caution is warranted. In this chapter, my more modest aim has been to demonstrate that ethological signaling theory offers a unique perspective that can sometimes help illuminate seemingly enigmatic cultural practices.

In the next chapter, we expand our discussion beyond weeping. We will encounter many more examples of ritualization (in the anthropological sense) and consider more broadly how signals and cues are recruited as part of cultural symbol systems.

15 Cultural Cues and Signals

In endeavoring to understand the cultural dimensions of emotion-associated displays, it's useful to expand the discussion beyond signals and consider various human cues as well. In this chapter we will see how cues are often enlisted by cultures to perform signal-like functions and how intentional imitation of signals (like the Welcome of Tears) can be used to tailor the effect on observers.

In human behavior, examples of active cues include scratching, eye blinking, sneezing, coughing, hiccups, yawning, sighing, gasping, clearing one's throat, spitting, chewing, shivering, sweating, snoring, farting, and belching (there are many more).[1] All of these behaviors typically occur as spontaneous involuntary actions. We classify them as cues because they are artifacts of various bodily functions that arise for reasons other than communicating to others. Unlike the rattling of a rattlesnake, behaviors like sneezing and chewing do not appear to be overt acts of communication. Moreover, while actions like scratching, yawning, or hiccups may be noticed and recognized by observers, they tend not to change the behavior of the observer to the benefit of the displayer—one of the defining features of a signal.

Although these cues are typically spontaneous and involuntary, many of these behaviors can be produced or simulated as voluntary actions. For example, nearly everyone can yawn or cough as willful acts. Most of us can voluntarily imitate a plausible sneeze or hiccup. Some talented people are able to belch or burp at will.

Recall that cues have no value for the displayer but commonly benefit observers. Sneezing or coughing might alert an observer to the possibility that the displayer is suffering from an infectious cold. The sight of someone chewing helpfully suggests the presence of food. Any cue that produces a sound (like hiccups or snoring) minimally alerts us to the presence of someone. My cough gives away my presence, and the sound typically allows an observer to infer my precise location. Detecting a person's presence can, in itself, be valuable to an observer.

Ahem

Now there are times when we might want others to be aware of our presence. For example, in a large lecture hall, you might have difficulty getting the instructor to recognize that you have a question you'd like to ask. If raising your hand has proved ineffective, you might supplement your raised hand by deliberately making a sound such as clearing your throat or coughing.

Whether a sound is produced voluntarily or involuntarily, both are capable of attracting an observer's attention. Normally, an involuntary cue has no value for the displayer apart from the corporeal utility of the act itself (chewing, scratching, blinking, and sweating are useful behaviors in themselves). However, if the cue has a predictable effect on an observer, and if that effect is of value to the displayer, then deliberately imitating the cue might be used to change the behavior of the observer to the benefit of the displayer. Intentionally clearing one's throat in order to attract attention is a good example. Such a willfully produced cue begins to resemble a signal insofar as the display is generated in order to benefit the displayer.[2]

Consider another example—the voluntary yawn. Spontaneous or involuntary yawning tends to occur when a person is sleepy. Apart from feeling sleepy as night falls, sleepiness (and so yawning) also tends to arise in situations where there is little to engage a person. Consequently, a yawn may be interpreted as indicating that there is nothing of interest to attract or maintain a person's attention. Accordingly, a person may intentionally emulate a yawn as a way to signify boredom. If my lecture has gone on too long, you might choose to yawn in my presence in a deliberate effort to change my behavior by encouraging me to wrap up my presentation. If I continue to ignore this voluntary cue, you might turn to your friend and yawn conspicuously in order to communicate both your boredom and the cause of your boredom.

Notice that, as a voluntary behavior, a willfully produced cue may or may not be part of everyone's behavioral repertoire. We could well imagine, for example, that throat clearing or coughing may not be used in every culture as a way of intentionally drawing attention to oneself. Moreover, even if the cue is intentionally imitated, the observer may not respond appropriately. A speaker might dismiss a cough as merely a cough or interpret your yawn as merely a spontaneous cue indicating you didn't get enough sleep. In Western culture at least, most people learn to recognize that in certain circumstances, these behaviors have a potential communicative function.

What's important to recognize is that production and recognition of such imitated cues bring the behavior into the domain of culture. When recognized by observers, such voluntary displays effectively define a system of cultural symbols.

Yet another example of a voluntary cue can be found in the burp or belch. Belching commonly happens as a spontaneous involuntary behavior, most often after eating a meal. In some cultures (e.g., Bahrain, China, India), belching is (or has been) regarded as a mark of gastronomic enjoyment, and dinner guests may be expected to belch as a sign of satiation. In these cultures, a person might choose to amplify or enhance the belch in order to increase its conspicuousness. A person of superior alimentary skills might even manufacture a belch de novo. In other cultures, belching may be regarded as crude or unrefined behavior. In these cultures, efforts are made to suppress or mask spontaneous or involuntary belching so as not to offend others. In these cultures, voluntary belching is left to cheeky adolescent males.

Other examples of the cultural recruitment of cues include retching, squinting, and hiccups. For example, in some Western cultures, a single "hic" (emulating the hiccup) may be used as a symbol of drunkenness. In Mandarin Chinese, a single plosive retch (呃, pinyin: è) is used to indicate that one has an upset stomach.[3] When we face possible assault, squinting is potentially useful for protecting one's eyes, so voluntary squinting may be used in some cultures as an indication of cautiousness, wariness, or guarded suspicion.

Kristin Precoda has proposed an interesting speculation regarding head scratching. Spontaneous scratching is a cue, not a signal. We scratch in response to various topical irritants like infestations of lice or fleas. But scratching one's head is often used intentionally—at least in some Western cultures—to indicate that we are perplexed or confused. When baffled, people will sometimes scratch their heads.

As just noted, scratching serves a useful purpose in dislodging unwelcome beasties from much of one's body. Unfortunately, since we can't see the tops of our heads, head lice are a notable nuisance—a problem in virtually all cultures. In order to effectively remove lice from your head, you typically need help from someone else.

Virtually all of our primate relatives engage in mutual grooming. Although humans engage in much less dyadic grooming, co-grooming behaviors can nevertheless be observed in any community in which head lice are a common affliction. At a public meeting on the island of Mwagmwog in Micronesia, I observed women sitting on the ground picking lice out of each other's hair while listening to the chief speak. That is, humans too engage in mutual physical grooming, and until recently, picking head lice from someone else's hair was one of the most common co-grooming behaviors.

Normally, as an involuntary cue, scratching one's head might be recognized by an observant grooming partner as an invitation or suggestion to engage in altruistic grooming. Precoda has speculated that emulating a cue by voluntarily scratching one's head might be interpreted as a general symbolic appeal for help. In Western cultures, such feigned or acted head scratching is limited to cognitive stressors, as when we are puzzled, disoriented, confused, or baffled. Although voluntary head scratching might have originated as a simple cue that commonly encouraged altruistic assistance with physical grooming, it may have become ritualized to serve a more general purpose of appealing for help—including help with cognitive stressors. Hence, head scratching as a cultural symbol indicates that one faces a perplexing problem.[4]

In summary, many cues can be voluntarily emulated and recruited as part of a system of cultural communication. Notice, however, that the intentional imitation of a cue must bootstrap an already existing implication of the spontaneous (involuntary) cue. When recruited as a part of a cultural symbol system, the meanings of a cough, belch, yawn, and so on aren't arbitrary. Throat clearing is used to announce one's presence (rather than, say, boredom), a yawn is used to indicate boredom (rather than, say, satiation), and a belch is used to indicate satiation (rather than, say, bafflement). In this regard, cultural cues depart from a common property of linguistic signs—namely, the arbitrary relationship between the sign and the object it signifies. The behavioral effect or "meaning" of a voluntary (cultural) cue closely parallels the behavioral effect induced by the corresponding involuntary cue. This means that voluntary (cultural) cues are especially easy to learn.

Cultural Signals

In the same way that cues can be willfully imitated and recruited as part of a culture-specific communication system, signals can also be recruited—as long as the signal can be voluntarily emulated or imitated. Like coughing, yawning, and belching, signals like weeping, smiling, and laughing commonly occur as spontaneous involuntary behaviors. However, we can also imitate many or most of these signals. The social smile is a classic example. Although smiling can arise spontaneously, we can also *choose* to smile. Similarly, we can imitate spontaneous laughter, emulate a disgusted face, mimic threat or aggression, reproduce a pout, look as if we are afraid or surprised, and (as in the case of the Welcome of Tears) even pretend to weep. At the same time, some behaviors, like blushing and sweating, are difficult or impossible to emulate voluntarily.

Like voluntary cues, voluntary signals tend to induce behavioral changes in observers that are very close to the changes that would ensue in the case of genuine (spontaneous or involuntary) signals. In the case of the Welcome of Tears, the aim is to induce prosocial proximal feelings in observers and so reduce the likelihood of acts of aggression or violence. However, as we noted in the previous chapter, the obviously fake or feigned version of the weeping conveys an implicit message that the weeping is a contrived intentional act and so the observer should not assume that the displayer is genuinely grief-stricken, helpless, or submissive. Accordingly, feigned weeping can be superior to spontaneous or genuine weeping in tailoring the behavior of the observer to suit the displayer. The same argument applies to other feigned or imitated signals such as the voluntary smile, laugh, disgust, surprise, or threat displays. When observers recognize the willful/contrived origin, the display conveys the supplementary message that the displayer consciously *intends* to make the display.

Ethological Symbols

When the distinction between signals and cues is combined with the distinction between voluntary and involuntary production, we end up with a fourfold classification system rather than the twofold (signal/cue) system: *involuntary signals, involuntary cues, voluntary-imitated signals,* and *voluntary-imitated cues.* In trying to understand the relationship between innate and cultural communication, this will prove to be a useful taxonomy. In the interests of clarity, it's appropriate to consider terminology.

There are several possible ways to label involuntary versions of cues and signals. One possibility is to follow the convention for classifying smiles: the spontaneous/involuntary smile (a.k.a. Duchenne smile[5]) is contrasted with the willful/posed version—referred to as the *social smile*. Following this pattern, one might juxtapose laughter with social laughter, weeping with social weeping, disgust with social disgust, and so on. The problem with the modifier "social" is that it implies that genuine signals are not social. Of course, all signals are social insofar as they exist to be perceived by an observer and transform the ensuing behavior. So the use of the modifier "social" fails to make any meaningful distinction and may lead to confusion.

In the case of cues, the modifier "social" is much more useful. Involuntary belching has no social function. But voluntary belching does. In contrast to signals, in the case of voluntary cues, the modifier "social" is entirely appropriate.

Returning to the case of voluntary signals, one might consider using modifiers like contrived, fake, sham, acted, or mock, but these all convey a negative connotation that's not appropriate. Adjectives like feigned or faux might work. Of the various possibilities, my proposal is to use the modifier "pseudo," as in pseudo-smile, pseudo-laugh, pseudo-weeping, and so on.[6] That is, any voluntary, feigned, imitated, or emulated signal would constitute a *pseudo-signal*.

In my proposed terminology, "pseudo" refers to the willful emulation of a signal, whereas "social" refers to the willful emulation of a cue. Accordingly, we might speak of yawning and social yawning, smiling and pseudo-smiling, belching and social belching, laughter and pseudo-laughter, and so on. Using different adjectives—*social* and *pseudo*—I would suggest, also helps to reinforce the signal/cue distinction. Throughout the remainder of this book, we will make use of this terminology and distinguish *signals*, *pseudo-signals*, *cues*, and *social cues* as needed.

One final proposal regarding terminology. At times, it may be useful to refer collectively to involuntary displays (innate signals and cues) as distinct from voluntary displays (both pseudo-signals and social cues). For this purpose, we might use the collective terms *spontaneous displays* and *posed displays*. Spontaneous displays (signals and cues) are innate, whereas posed displays (pseudo-signals and social cues) are cultural phenomena. Table 15.1 offers a summary of this proposed taxonomy with examples.

Table 15.1

Taxonomy (with examples) of human display behaviors distinguishing spontaneous versus posed and signal versus cue divisions

	Spontaneous Displays (involuntary)	Posed Displays (voluntary)
Signal-based	SIGNAL	PSEUDO-SIGNAL
	Weeping	Pseudo-weeping (e.g., Welcome of Tears)
	Smile	Social smile
	Laugh	Pseudo-laugh
	Pout	Pseudo-pout
	Scream	Pseudo-scream
	. . .	. . .
Cue-based	CUE	SOCIAL CUE
	Cough	Social cough (in some cultures a way of drawing attention)
	Throat-clearing	Social throat-clearing (in some cultures a way of drawing attention)
	Belch	Social belch (in some cultures a way of communicating satisfaction with a meal)
	Yawn	Social yawn (in some cultures a way of indicating boredom)
	Hiccup	Social hic (in some cultures a way of signifying drunkenness)
	Head scratching	Social head scratching (in some cultures a way to indicate puzzlement)
	. . .	. . .

Amplification, Masking, Inhibition, and Emulation

In chapter 6, we identified three forms of executive control: amplification, masking, and inhibition. In this chapter, we have emphasized a fourth form of executive control—namely, emulation. The phenomenon of emulation (imitating a signal or cue) completes, I propose, the list of executive control mechanisms for display behaviors. It's appropriate now to pause and summarize the four ways we willfully recruit or reshape cues and signals.

In the case of signals, first, we can often embellish or *amplify* the signal. We saw examples of signal amplification in the previous chapter where weeping was made more conspicuous by such behaviors as smearing charcoal on one's face, extending one's arms while crying, or wearing black clothing during mourning. In these cases, cultural practices are used to augment the signal by increasing its conspicuousness and duration (what ethologists call *redundancy*).

Second, we can often *mask* a signal, such as covering your face with your hand in order to hide a spontaneous smile or laugh. Other examples of masking include leaving a room so no one can see you weeping or pretending that your tears are due to some eye irritant. In these cases, cultural practices are used to obscure, attenuate, or otherwise make the signal *less* conspicuous to observers.

Third, we can sometimes voluntarily suppress or *inhibit* many signals. In some cases, inhibition can take the form of particular motor actions, such as biting your lower lip in order to sabotage a telltale facial display such as smiling. But most forms of inhibition are cognitive. For example, in order to avoid smiling or laughing, you might intentionally think sad or serious thoughts. Conversely, in order to suppress the urge to cry, you might think of something positive or lighthearted. Where the aim of *masking* is to make the signal less conspicuous, the aim of *inhibition* is to employ cognitive resources (thoughts) so as to avoid producing the signal in the first place.

Fourth, we can voluntarily *emulate* many signals, such as producing a pseudo-smile, rendering a pseudo-laugh, or engaging in the Welcome of Tears. In the first instance, emulating a signal may prove useful when an otherwise valuable signal fails to be generated spontaneously. But, as we have already noted, depending on the circumstance, a posed (fake, feigned, or imitated) signal may have greater utility than the corresponding spontaneous (involuntary) signal since the posed version informs the observer that the displayer has intentionally chosen to make a signal-like display. As we saw in the case of the Welcome of Tears, the obviously fake crying communicates a desire to avoid escalation of possible conflict but simultaneously communicates that the displayer is not actually helpless. Incidentally, observers are generally very good at distinguishing voluntary from involuntary versions of signals.

Of course these same four methods (amplification, masking, inhibition, and emulation) also apply to cues, not just signals. One might choose to enhance or amplify an involuntary cue such as exaggerating a belch or making a naturally

Table 15.2

Examples illustrating four forms of executive control over display behaviors

Mode of modification	Example
Amplification	Thinking sad thoughts in order to extend or intensify a weeping episode.
Masking	Turning away so no one sees the tears in your eyes.
Inhibition	Thinking happy or mundane thoughts to avoid crying.
Emulation	Welcome of Tears.

occurring yawn more conspicuous by extending its duration or through eye contact with an intended observer. Conversely, as in the case of signals, one might also engage in masking or inhibitory behaviors, such as covering one's face to avoid displaying a spontaneous yawn or suppressing a belch in order to avoid culturally defined offensiveness. Finally, one might voluntarily emulate a cue, as when yawning to symbolize boredom, producing a hic to symbolize drunkenness, or belching as a gesture of satiation. These four forms of modification are summarized in table 15.2.

Past experience provides the basis for assessing the likely costs and benefits associated with different forms of executive control. If a spontaneous display led to an unwanted result in the past, then one may see efforts to mask or inhibit the display in similar circumstances. Conversely, if a spontaneous display proved insufficient in the past to produce the preferred result, then one might see efforts to amplify the display. One way to think of these executive control strategies is that they allow the displaying behavior to be tailored to a person's specific circumstance, rather than relying solely on spontaneous evolved behaviors that more crudely anticipate average, typical, or probable consequences. In effect, efforts of cognitive control amount to "behavioral second thoughts" that endeavor to reshape innate tendencies so they are better tailored to the current situation in light of learned past experience.

Of course, the environment in which this learning takes place is one's social environment, and the preeminent constituent of that social environment is one's cultural milieu. Although culture may influence emotion in other ways, the principal way in which culture influences emotion-related displays is through the mechanisms of amplification, masking, inhibition, and emulation.

In the case of emulation, each culture manifests a distinctive repertoire of pseudo-signals and social cues that constitutes a cultural symbol system. Different cultures might recruit and modify involuntary signals and cues in different ways. For example, many more Asian cultures engage in hand covering (masking) of spontaneous smiles than is evident in European cultures.[7] Moreover, it's common to observe sex-related differences in such behaviors, with women more likely to mask a smile than men.

The learned basis for amplification, masking, inhibition, and emulation, however, does not necessarily mean that there aren't strong similarities across cultures. Given the innate origin of signals like weeping and the ubiquity of cues like yawning, we might expect (and indeed observe) considerable overlap between cultures. Some of these symbol systems have long been described under the rubric of "body language." However, an ethological perspective, I propose, provides a better framework for understanding these various behaviors.

Emotion Regulation

Although we have emphasized the use of executive control in shaping communicative displays, we may also use these same control mechanisms for emotion regulation. Feeling states are known to be shaped to some degree by peripheral physiological feedback—such as when smiling sometimes increases feelings of happiness.[8] Even in the absence of an audience, if tears are immanent, we may endeavor to think thoughts that inhibit those tears as a way of subverting or attenuating feelings of sadness. We might intentionally produce a pseudo-laugh as a form of amplification—rendering a mildly positive state somewhat more positive. Similarly, social cues can also sometimes be used for the purpose of emotion regulation, such as intentionally sitting upright or directing one's gaze away from some distressing scene.

Of course it's possible that behaviors intended to regulate one's emotions occur in the presence of observers, in which case it becomes difficult to determine whether the behavior is motivated by regulatory or communicative goals. These different goals can sometimes be at odds with one another. At a funeral, for example, a bereaved spouse may want to suppress weeping in an effort to reduce painful feelings of grief, yet cultural expectations may require or demand a display of weeping. The absence of tears may consequently be interpreted by observers as callous indifference or heartlessness. Here, the individual must navigate between personal emotional regulation and cultural expectations.

Throughout this book we have emphasized that signals are forms of manipulation rather than revelation. Although not reliable, both cues and signals nevertheless provide useful information that can often help an observer decipher the emotional state of the displayer. One might suppose that intentional displays like pseudo-signals and social cues are also necessarily forms of observer manipulation; however, since pseudo-signals and social cues can be used as part of a strategy for emotion regulation, not all posed displays arise as efforts to influence the behavior of observers.

Mockery

In most cultures, pseudo-signals are occasionally used as a form of spite or mockery. Pseudo-signal version of weeping, disgust, or laughter may be used to suggest that a situation does not warrant any genuine weeping, disgust, or laughter. For example, a slightly stressful situation that has moved a mocked individual to tears might be disparaged through deadpan pseudo-weeping. In English, mock weeping might be expressed as "boo, hoo, hoo," with the equivalent Mandarin Chinese conveyed as "wū, wū" (呜呜). The pseudo-weeping of the mocking person indicates that no observer compassion can be expected.

Mocking pseudo-signals are typically marked either through exaggerated caricatures or via deadpan forms. In the case of disgust, an overstated version might entail an exaggerated facial expression and vocalization (eeeeww!).[9] Similarly, a situation that would warrant a polite pseudo-smile might be mocked by producing a perfunctory though highly exaggerated pseudo-smile.

Deadpan versions of pseudo-signals commonly exhibit a slower controlled deliberate emulation, such as mock laughter—as in "haw, haw, haw" or even "hardy, har, har, har." In Mandarin Chinese, spiteful or sarcastic laughter is conveyed as "hē, hē" (呵呵). Both exaggerated or deadpan forms of behavior draw attention to the artifice of acting and, in so doing, make the mocking intention clear.

In all of these various cases, the mocking pseudo-signal communicates that the displayer regards the situation as not warranting the emotion corresponding to the spontaneous signal version of the posed pseudo-signal. For example, mocking pseudo-signal laughter suggests that a "flat" joke lacks any amusement value.

Mockery is commonly used in antagonistic relationships as a form of ridicule, disparagement, insult, or threat. Especially when the mocking individual has a notably higher social rank, the mocking may amount to a form of bullying. However, mocking can also be used among friends as a form of playful taunting, especially between individuals of similar social rank. University of Toronto anthropologist Richard Borshay Lee has chronicled mocking behaviors among the Bushmen of Botswana and Namibia. Among Bushmen, it's common to mock individuals who are notably successful in various ways (such as hunting), as a nominally playful way to curb excessive pride or arrogance.[10]

Mockery is also commonly used in comedy entertainment. It is especially common to mock well-known people of high social station such as politicians. Mockery is regarded by literary theorists as one of the classic forms of satire. In arts and entertainment, it's also found in the form of self-mockery, where a character resorts to extreme pseudo-signals or social cues.[11]

No Signal

In endeavoring to understand human signals, it can be instructive to consider why certain signals *don't* exist in the human behavioral repertoire. From a biological perspective, perhaps the two most important behaviors are eating and engaging in sex. So why don't humans have signals related to hunger or sexual intercourse?

Recall again that signals can evolve only when the interaction is beneficial to both the displayer and the observer. Suppose, for example, that humans had evolved a signal that predictably arises when a person is hungry. Whenever this signal is displayed, observers would be instinctively motivated to offer food to the hungry person. Infant bawling is arguably such a signal—and clearly, feeding one's baby has marked benefits for both infant and kin-related caregivers. However, there is no comparable signal for adult hunger. Why not?

The benefit of a "hunger" signal for the hungry person is clear. Moreover, we would expect all the usual benefits of altruistic behavior to accrue for the person providing the food. A problem arises, however, if the ultimate benefit arising from the altruistic act of giving food to another person is not sufficient to offset the cost of forfeiting the food. As with all animals, for nearly all of human history, food has been a crucial limiting resource that directly impacts survival. The food you give to me is food you cannot consume yourself—a classic zero-sum situation. Given the high value of food, offering food to others has enormous potential for building alliances. So one might suppose that a hunger signal would confer considerable benefits for the altruistic observer. However, since no "hunger" signal has evolved as part of the human behavioral repertoire, the inescapable implication is that its absence suggests a long history of food scarcity.

Similarly, consider a hypothetical signal related to sexual intercourse. A male's erect penis is a cue rather than a signal. But suppose it was a signal that had the effect of evoking in female observers an instinctive motivation to engage in sex with the man. While this would contribute to the man's reproductive success, a woman who fails to choose the best available male as a sex partner would squander the optimum reproductive opportunity. Here, the reproductive interests of the male and female are not in alignment. Although sex with a preferred male enhances the personal fitness for both the man and the woman, this is not the case for a woman should she mate with any arbitrary aroused man. In short, there doesn't exist the consistent win–win situation that would allow a signal to evolve.

Returning to our food/hunger scenario, it bears mentioning that, with the advent of agriculture, humans have often been able to produce an extraordinary abundance of food. If I have more food than I can possibly eat, then there is no longer a zero-sum condition that renders me reluctant to share my food with someone who is hungry. In the absence of an evolved hunger signal, and in situations of food abundance, cultural practices can emerge that provide an opportunity for a mutually beneficial interaction. And indeed, in every observed human culture, the sharing of food is a ubiquitous way of building bonds, cementing friendships, and placating enemies. Notice, however, that in contrast to weeping, the cultural practice of food sharing is initiated by the person possessing the food, not the person in need of the food.

If food abundance had characterized the long-term natural state of human existence, one could well expect that over time, a hunger signal would indeed have evolved since the cost to most observers would be low and the mutual benefits comparatively high. However, biologists note that in the long term, populations typically expand or contract to match the available food resources. Even when food is abundant, that abundance is necessarily temporary until the population/food reaches an equilibrium. The absence of an adult hunger signal provides a tacit testament suggesting that food insecurity has been commonplace throughout human history.

Reprise

In this chapter, we have expanded on our earlier discussion (in chapter 6) regarding executive control. In particular, this chapter has drawn attention to the influence of cultural norms and practices in shaping cognitive control of display behaviors. We have seen that executive control mechanisms are sometimes deployed to modify ethological cues and are not limited to modifying signals.

To the three forms of executive control discussed earlier, we have added a fourth—*emulation*—a behavior that completes our inventory of types of cognitive control of display behaviors. The initialism "AMIE" (French for "friend") provides a useful mnemonic for aiding recall of the four forms of executive control of signals and cues: Amplification, Masking, Inhibition, and Emulation. We are apt to resort to emulation when an otherwise valuable signal or cue fails to arise spontaneously or when it's useful to convey to an observer that the display is willful and intended. We can see the effects of amplification, inhibition, masking, and emulation in common cues (like coughing, yawning, and belching), not just in the case of ethological signals (like weeping or smiling). We have also noted that amplification, inhibition, and emulation are sometimes used for the purposes of personal emotion regulation rather than as a means for shaping the responses of observers.

The intentional or willful production of signal-like or cue-like behaviors suggests that four types of displays can be distinguished—what we have dubbed *signals*, *cues*, *pseudo-signals*, and *social cues*. Social cues and pseudo-signals are cultural conventions that can differ from culture to culture. Nevertheless, social cues and pseudo-signals closely resemble their spontaneous counterparts; consequently, cultures share considerable overlap in how these willful displays are interpreted.

Observers are highly sensitive to whether a cue or signal is voluntary ("posed") or involuntary ("spontaneous"). Feigned or acted displays may be used for a variety of purposes, such as communicating detachment and self-control (as in the Welcome of Tears) or as a form of social aggression (as in mock weeping).

In this chapter, we have also considered why some signals have *failed* to appear in human evolution. Most feeling states do not lead to a distinctive ethological signal. This includes such common feelings as hunger, thirst, feeling sick, loneliness, regret, pride, boredom, and nostalgia. The implication is that most feeling states do not commonly arise in situations where the behaviors of the displayer and observer typically lead to non-zero-sum or win–win interactions. Without such mutual benefits, there are no selection pressures encouraging the emergence of a pertinent ethological signal.

In general, this chapter has drawn attention to the long arm of culture. At the same time, it has provided additional examples where ethological signaling theory offers a useful interpretive lens through which to view cultural practices. Signals and cues provide the raw material for many culturally defined displays. Finally, we might note that the repertoire of social cues and pseudo-signals a person uses effectively marks that person as a member of a given group or community. The posed displays we use amount to cultural symbols that help define our social identity.

In endeavoring to understand the phenomenon of weeping, our account has drawn much inspiration from the field of ethology. The key idea is that many emotion-associated displays can be profitably understood from the perspective of signaling theory. I refer to this perspective as the signaling theory of emotion-associated displays (STEAD). This chapter provides a short summary of STEAD and describes some of its repercussions. But before we present our summary, we need to address one further background topic.

Whence Polygenic Signals

Throughout this book, we've emphasized that a given signal might be evoked by several different emotions. We've noted, for example, that weeping might arise from feelings of grief, joy, pain, patriotism, humor, and so forth. Smiling can ensue from feeling happy, patronizing, socially stressed, and so on.[1] At this point, we might pose the general question: How is it that a given signal can be evoked by so many different motivating emotions?

Consider the following hypothetical scenario. Suppose that in some species, a signal emerges related to thirst and water provisioning. When an individual feels thirsty (W), they tend to issue a signal (X), which induces a proximal feeling in observers (Y) that typically results in behavior (Z) where water is given to the signaler. Notice that the original motivation for generating the signal is that an individual feels thirsty.

Now the feeling of thirst might have been the impetus for the chain of evolutionary events leading to the emergence of X as an ethological signal. However, once signal X exists, it's possible to use that signal simply as a way to achieve outcome Z. If an individual would benefit by evoking behavior

Z, it may not be necessary to experience the motivating state W: the key is simply to produce signal X.

By way of example, suppose that an individual is not currently thirsty but wants water for some other purpose. Perhaps they want to clean some mud-covered food or need water for a thirsty (absent) offspring. Whatever the reasons, the important point is that there might arise other situations where what motivates the signal is a state other than the state that led to the evolution of the signal in the first place. My proposal here is that all signals begin as a solution to a particular problem, but once the signal is in place, it can be used to deal with other possible situations for which the induced observer behavior provides a useful resolution. As long as the new context continues to benefit both individuals, the signal will continue to exist due to positive selection pressures.

In the case of weeping, in chapter 11, I proposed that weeping did indeed originate as a signal inextricably linked to the stress we call grief. The effect of weeping is to induce prosocial feelings in observers, which in turn may lead to a variety of altruistic behaviors or simply result in a positive attitude toward the weeping individual. However, over time, one might encounter situations other than grief where it would be useful to evoke prosocial feelings in observers. Hence, tears of joy, anger, loyalty, and so on described in chapter 7. Tears of laughter offers an especially interesting case. Laughter is a notable component of play behavior but also serves an important bonding function. With the exception of mutual weeping, we never feel so close to someone as when we engage in mutual laughter. In this context, tearful laughter would appear to be the icing on the cake. The tears associated with extreme laughter may well induce an even greater prosocial disposition in the observer.[2]

STEAD Fast

With this additional background, we might now summarize the signal theory of emotion-associated displays. Readers will already be familiar with the STEAD theory since it has been effectively revealed over the course of the preceding chapters. The core of the STEAD theory can be distilled to six main tenets:

1. Several emotion-associated displays—such as bawling, weeping, screaming, smiling, laughing, pouting, disgust, and the threat grimace—are evolved ethological signals whose purpose is to influence the behavior of observers.

2. Signals are displays that typically arise in response to some emotion or feeling state experienced by the signaler. However, most signals can appear in response to more than one motivating emotion.

3. Signals encourage observer behaviors by inducing a proximal emotion in the observer. The same proximal emotion tends to be induced in observers no matter what emotion motivates a given signal. The proximal emotion might not result in some immediate action, but instead produce a change of attitude that may influence an observer's future actions.

4. Although susceptible to executive control in the form of masking, inhibition, or amplification, signals and signal responses are biologically prepared innate tendencies.

5. Signals and signal responses can evolve only if the induced observer behavior typically benefits both the signaler and the observer. Signals initiate interactions that are inherently cooperative.

6. We can emulate signals (pseudo-signals), but observers are typically aware that the emulated signal is willful or contrived, and so sometimes used for deceptive purposes in which the signaler benefits at the expense of the observer.

For interested readers, a more detailed presentation of the STEAD theory can be found in the appendix.

Recognizing Emotions

As noted earlier, the STEAD theory contrasts notably with the widely held view that emotion-associated displays such as smiling or weeping represent communicative acts intended to "express" the displayer's emotional state. There are, I propose, two good reasons why the emotion expression model dominates thinking about emotion-associated displays.

First, the belief in emotion expressions is supported by the fact that we are often correct in deciphering the emotions of others from their display behaviors. We recognize our success in inferring their emotions because we are frequently able to predict the displayer's ensuing behavior. Moreover, our ability to discern the emotions of others is generally greater for signals compared with pseudo-signals, cues, or social cues.

At the same time, much of our success in deciphering someone's emotional state is attributable to unacknowledged contextual information. For example, context is commonly sufficient to resolve whether weeping is motivated by grief, pain, joy, loyalty, or extreme amusement.

Our ability to decipher emotions is also enhanced because some signals may not be polygenic. For example, unlike weeping and smiling, threat signals seem to arise only in situations where the displayer is indeed angry. The disgust signal seems to be linked to just two feeling states—sensory nausea and feelings of moral revulsion.

Altogether, when aware of the context, people are often successful at deciphering the emotional states of others from signal displays even though the selection pressures for signal displays aren't shaped by the goal of communicating one's emotional state to observers. Instead, the selection pressure for signals is the goal of inducing a change of emotional state in observers that (once again) typically results in behaviors that are beneficial to both signaler and observer.

A second reason for the tenacity of the emotion expression model is that predicting the behaviors of others is especially valuable: to the extent that emotions are predictive of future behaviors, we tend to regard deciphering the emotions of the people around us as the paramount project when observing emotion-related displays. We mistakenly conclude that displayers are unconsciously communicating their emotions and that the end goal of display behaviors is emotion recognition. In short, we interpret displays from the biased agenda of the observer. We look at the smiling or weeping person and ask, "What are they feeling?" We forget to ask the more critical question: "What are we (the observers) feeling?"

The claim here is not that signals never inform observers regarding the signaler's emotional state. Instead, the claim is that informing observers of one's emotional state is not the reason why signals exist. Phrases such as "emotional expression" and "facial expression" are regrettably misleading and unhelpful. Less deceptive labels might refer to them as "displays" or "emotion-associated displays."

The STEAD theory provides plausible resolutions to several classic problems that have stymied modern emotion research:

- The STEAD theory explains why several different emotions might lead to the same display.
- As a corollary, in the absence of contextual information, it explains why isolated signal displays are commonly ambiguous for observers endeavoring to decipher the displayer's emotional state.

- As biologically prepared innate dispositions, signaling theory explains why displays commonly arise in the absence of any audience.[3]
- Since perpetually reliable win–win situations are the exception in human interactions, and since signals evolve only under conditions of mutual displayer/observer benefit, the theory explains why the great majority of human emotions are covert and not associated with any characteristic display.
- The theory explains why some emotions—notably anger—may occur either *with* (hot anger) or *without* (cold anger) a discernible display.
- The theory explains why there are commonly both spontaneous and posed versions of most signal displays.
- The signal/cue distinction explains why some efforts to discerns a person's emotional state rely on seemingly obvious displays (such as smiling or weeping), whereas other emotional states are discerned via rather subtle indicators (such as posture, breathing, sweating, gaze direction, or voice quality). For example, the signal/cue distinction explains why grief is much easier to decipher than melancholy.

Repercussions

The signaling theory of emotion-associated displays hold practical lessons for various computational endeavors aimed at emotion recognition. The STEAD theory reinforces the observation that much current research in automated/algorithmic emotion recognition is really aimed at recognizing displays not emotions. Recognizing tear-filled eyes does not mean one has deciphered the feeling of grief.

If one's aim is emotion recognition from photographs, sound recordings, or videos, computer scientists will benefit by attending carefully to context—echoing the strategy responsible for much human success in decrypting the emotional states of others. In addition, if the engineering goal is to create optimum emotion recognition systems, accuracy measures should be compared with the most accurate human observers. That means researchers should be wary of using the responses of young university students as the baseline for measuring accuracy. Existing research suggests that older, more experienced adults are likely to show greater nuance and granularity in recognizing emotions.

The STEAD theory similarly holds implications for therapeutic interventions involving those individuals who have difficulty recognizing emotion-associated displays—such as some autistic spectrum individuals and, more generally, those diagnosed with alexithymia (the inability to recognize emotional states). Pictures of smiling, laughing, or weeping faces are unlikely to provide sufficient training materials unless the pictures are supplemented by major efforts to decipher the contexts in which such displays may arise.

Utterly

Apart from the various spontaneous and posed displays, the principal way by which humans influence the behaviors of others is via language. Instead of posing an angry pseudo-signal, one might simply say "I'm angry."

In linguistics, the effect of an utterance on a listener is referred to as the perlocutionary effect. Saying something can transform the behavior of a listener, such as encouraging certain actions or shaping the listener's attitude toward the speaker. Like pseudo-signals and social cues, language utterances are willful acts. So when someone says "I'm happy," they're conveying not just a statement of their purported feelings but also conveying a willful intention to communicate that claim. Like pseudo-signals and social cues, observers are aware of the willful nature of utterances and so understand they may be motivated by deceptive aims.

As we saw in chapter 15, apart from the effect on observers, pseudo-signals and social cues can be intentionally generated as a form of emotion regulation. We might smile in an effort to cheer ourselves up. Spoken utterances can similarly be used as a means of emotion regulation. For example, swearing (in the absence of any listener) has been shown to increase tolerance for pain.[4] Muttering spiteful insults can amplify feelings of anger. The words we utter can sometimes prove useful in either relieving or amplifying our feelings.[5]

Coda

It bears emphasizing that the STEAD theory should not be construed as an exclusively Machiavellian theory in which human interactions are governed entirely by selfish motives. Signals exist precisely because they typically benefit both signaler and observer. That is, signals are cooperative, not merely manipulative in the pejorative sense. Even in the case of pseudo-signals and social cues, despite the opportunities they offer for self-serving deception, much or most of such voluntary displays are intended to engender cooperative interactions. Similarly, although language too offers endless possibilities for deception, it appears that much or most conversational interactions are mutually beneficial. When someone's speech is perpetually dominated by deceptive motives, we wisely learn to avoid them.

Once again, the goal of this chapter has been to provide a simple summary of the signaling theory of emotion-associated displays. As noted earlier, a more formal and thorough presentation of the STEAD theory can be found in the appendix.

17 Triadic Theory of Immunological Responses to Stress

Having reviewed the main ideas of the STEAD theory, at this point, it's useful to draw together the various lines of discussion and summarize our overarching account of sadness—a story that is rooted in my proposed triadic theory of immunological responses to stress (TTIRS—"tears").

In brief, the TTIRS theory posits three closely related affective states: *sickness*, *melancholy*, and *grief*. These three states offer a comprehensive set of strategies for dealing with a huge range of stressors from insect bites and bruises to bankruptcy and breaking up with a partner. Our review of the research suggests that feeling sick encourages energy conservation that enhances immune system effectiveness by discouraging unnecessary motor behaviors; that the feeling of melancholy enhances cognitive planning through melancholic realism; that the feeling of grief solicits social assistance through the ethological signal of weeping; and that all three affective states are closely linked to the immune system.[1]

A strength of the TTIRS theory is that it offers a plausible overarching narrative that accounts for many observations pertaining to the etiology, physiology, behavior, development, affective experience, and social interaction related to sadness. It also offers some insights into many cultural practices associated with grief, nostalgia, and mourning.

The following numbered summary is intended to clarify the logic of the theory with the potential to better expose gaps and weaknesses. The summary may also provide a guide for identifying components of the theory amenable to empirical testing.

Stress

1. *Types of stress*. We experience many kinds of stress, including physical challenges (such as injury, pathogens, or hunger), cognitive challenges

(such as confusion, frustration, or boredom), and social challenges (such as shame, loneliness, or being bullied).

2. *Stress-induced intentional action.* Many stresses are addressed through general cognitive assessment and planning leading to intention actions such as donning warm clothing, navigating a traffic jam, or seeking companionship.

3. *Specialized responses.* Some stresses are addressed via evolved purpose-specific reactions such as fight/flight/freeze responses to fear, gagging or purging in response to toxic foods, or shivering in response to cold. Among these evolved stress response mechanisms are sickness, melancholy, and grief.

4. *Stress resources.* Sickness, melancholy, and grief recruit corporeal, cognitive, and social resources respectively. That is, sickness draws on corporeal resources that fight infection and repair injury, melancholy draws on cognitive resources that encourage realistic reflection and strategizing, and grief draws on social resources where other people are induced to provide assistance or, if they are the source of the stress, to terminate the stress-inducing behavior.

Sickness and Corporeal Stress

5. *Sickness.* Immune reactions to infection or injury include a combination of peripheral and central responses. Responses include efforts to repair damage, such as local inflammation, and may also include efforts to impede infection such as reduced appetite (anorexia) and fever (pyrexia).

6. *Immune facilitation.* The operation of the immune system is metabolically expensive. The effectiveness of immune responses is enhanced when other metabolic demands are minimized. To this end, an especially useful strategy is to curtail voluntary movement. Feelings of lethargy (anergia), sleepiness (somnolence), pain sensitivity/muscle soreness (hyperalgesia), low mood (malaise), and loss of pleasure (anhedonia) tend to discourage motor activities and so enhance immune effectiveness.

Melancholy and Cognitive Stress

7. *Cognitive response.* Many stressful situations can be resolved—or their negative effects minimized—by careful planning of future behaviors. That is, many stressful circumstances can be successfully addressed through cognitive reflection and planning alone.

8. *Melancholic cognition.* A characteristic feature of melancholy is cognitive reflection.

9. *Melancholic realism.* Compared with a happy or neutral mood, melancholy has been shown to promote more detail-oriented thinking, reduced stereotyping, less judgment bias, greater memory accuracy, reduced gullibility, more patience, greater task perseverance, more social attentiveness and politeness, more accurate assessments of the emotional states of others, a heightened sense of fairness, improved reasoning related to social risks, and a disposition to favor delayed larger rewards over immediate smaller rewards.

10. *Melancholy versus depression.* Depression is a pathology that resembles melancholy. However, cognitively, depression is associated with *rumination* (brooding negative self-assessments, often linked to repeatedly recalling past situations or failures) rather than the more realistic *reflection* that characterizes melancholy.

11. *Sickness versus melancholy.* Melancholy shares many characteristics with sickness, including anergia, somnolence, malaise, and anhedonia—that is, lethargy, sleepiness, low mood, and reduced pleasure. However, melancholy discards many (mostly peripheral) responses associated with sickness, including inflammation, headache, and fever.

Weeping and Social Stress

12. *Social response.* Many stressful situations can be resolved, or their negative effects reduced, by recruiting assistance from friends or bystanders.

13. *Grief.* The most characteristic feature of grief is a display of weeping. Weeping behavior conforms to the definition of an ethological signal.

14. *Mutualism.* Ethological signals evolve only if they typically induce observer behavior that increases the fitness of both the signaling animal and the observing animal.

15. *Benefit of weeping.* The benefit of weeping for the signaling individual is the evoking of prosocial feelings in observers. Such prosocial feelings can lead to acts of altruistic assistance, the termination of aggression, or simply the evoking of a favorable attitude toward the weeper.

16. *Cost of weeping.* Weeping imposes some costs on the weeping individual. Foremost is a reduced social status with respect to the observer. If the observer responds to the weeper by engaging in altruistic acts,

then the weeper also incurs a promissory debt implying an obligation to engage in future reciprocal assistance. This debt may be manifested simply as an expectation of loyalty to the observer.

17. *Different weeping costs.* The social status cost of weeping is greater for reproductive-aged individuals. Since social status has a greater effect on the reproductive success of males, males incur a greater cost for weeping than equivalent-ranked females. A reduction in social status is least costly for infants, children, and individuals of low social status.

18. *Frequency of crying.* The costs of weeping are consistent with observations regarding the frequency of crying. In brief, the likelihood of crying is proportional to the magnitude of the stressor, the capacity of an observer to offer genuine assistance, and inversely proportional to the effect of the loss of social status on reproductive fitness.

Altruism

19. *Ultimate motivations to help.* There are several ways in which altruistic behaviors can enhance the fitness for the person engaged in helping acts. These include gains through kin support (kin selection), the banking of future return favors (reciprocal altruism), alliance formation, increased social status with respect to the person helped, enhanced prosocial reputation among spectators, and the avoidance of social ostracism for failing to assist.

20. *Proximal motivation to help.* The ultimate incentives for helping are opaque to observers. Instead, prosocial behaviors induced in an observer of weeping appear to arise from six proximal motivations: the reducing of personal distress arising from contagious commiseration, the positive feeling of pity or compassion, the positive feeling arising from an enhanced social reputation, the fear of social ostracism for failing to help, anticipation of the private feeling of virtue, and the positive feeling of bonding warmth arising from a social connection with the person being helped.

21. *Mutual weeping.* Observers sometimes respond to a weeping individual by weeping themselves. Insofar as weeping evokes prosocial feelings in observers, both individuals are apt to feel prosocial feelings toward the other person; consequently, mutual weeping often contributes to forming strong bonding relationships.

22. *Automaticity.* As long as a signal-induced interaction commonly benefits both signaler and observer, selection pressures favor automaticity of the

behaviors. In the case of grief, there are biologically prepared tendencies to spontaneously weep under certain circumstances, as well as biologically prepared tendencies for observers to spontaneously experience pro-social feelings toward a weeping individual.

23. *Weeping versus bawling.* Weeping and bawling are distinct signals with different phylogenetic precursors. Bawling is evident at birth and disappears in early childhood. Weeping emerges toward the end of the first year of life and is evident throughout adulthood. Bawling is maximum in the absence of caregivers, consistent with a separation distress call. By contrast, weeping is maximum in the presence of caregivers.

24. *Infant weeping.* Weeping incurs much lower costs for infants and children compared with adults. As a result, there are few constraints limiting the amount of infant or child weeping.

25. *Melancholy versus grief.* Melancholy and grief are different yet complementary states. For any given stressful situation, both melancholy and grief may have value.

26. *Mourning cycle.* For minor stresses, the costs of weeping are likely to outweigh the benefits, and so minor stresses may lead to melancholy without grief. For major stresses, the benefits of weeping are more likely to outweigh the costs. Both grief and melancholy have value, so it is common in stressful situations for inward-directed periods of melancholy to alternate with outward-directed bouts of weeping, producing a "mourning cycle."

Allergic Origins

27. *Embellished allergy.* The physiological foundation for weeping is the allergic response.

28. *Conspicuousness.* Since signals are effective only if they are communicated to observers, signals evolve toward conspicuousness and redundancy or persistence. In weeping, these changes are evident in the compulsion to vocalize and the proliferation of tears.

29. *Compulsion to vocalize.* A useful way to increase conspicuousness is to make a signal multimodal. Apart from sniffling, sneezing, and nose-blowing, the allergic response exhibits no distinctive sonic element. The compulsion to vocalize (which is not part of an allergic response) enhances the communicative effect of tears, contributing to the transformation from an ethological cue to an ethological signal.

30. *Ingressive phonation.* Especially persuasive evidence in support of the compulsion to vocalize is found in ingressive phonation, where the vocal folds are activated even while inhaling—a notably rare behavior. This leads to distinctive vocalized gasping sounds.

31. *Breaking voice and pharyngealization.* As part of the allergic response, the constriction of the pharynx and larynx originally functioned to impede the entry of allergens. With the development of weeping, these vocal tract constrictions afford two characteristic acoustic features: breaking or cracking voice (due to instability between modal and falsetto phonation) and a pharyngealized acoustic resonance. In the evolution from a stress-induced allergy cue to a stress-induced weeping signal, vocal tract constriction qualifies as an *exaptation.*

32. *Psychic tears.* The most conspicuous visual feature of weeping is the production of tears. The appearance of a novel branch of cranial nerve VII specialized for limbic activation of the lacrimal glands testifies to a genetic basis for weeping in humans.

33. *Executive control.* Compared with other animals, humans are endowed with greater executive control. Among humans, otherwise compelling behaviors are susceptible to masking or inhibition. If the social penalty incurred by weeping is assessed as too costly, it is common to hide the signal, engage in cognitive thoughts that reduce the likelihood of overt weeping, or "explain away" weeping as having some other cause (e.g., claims of suffering from an allergy or having an eye irritant).

34. *Responsiveness to weeping.* The tendency of observers to respond prosocially to a weeping individual is similarly susceptible to executive control. Pertinent cognitive factors include the social and biological relationship between observer and weeper, the likelihood of future sustained interaction, the perceived honesty of the signal, the reputation and history of past interactions, the presence of an audience, and the marginal cost to the observer of offering assistance.

Culture

35. *Cultural amplification.* In many cultures, grieving states may be symbolized in various ways, such as the wearing of distinctive clothing, applying charcoal to the face, cutting hair, and other cultural practices and rituals. In addition, grieving states may be associated with behaviors such as hair-pulling or various forms of self-injury. These behaviors

commonly occur in public rather than in private. Many cultural practices associated with mourning are consistent with efforts to amplify, lengthen, or enhance the communicative function of the ethological signal—consistent with the goal of conspicuousness.

36. *Cultural emulation.* The compulsion to weep is not always present in situations where it would otherwise prove useful. Accordingly, in some cultures, weeping may be willfully emulated. Examples include formal "lamenting" such as at funeral rituals and the Welcome of Tears.

37. *Cultural handicaps.* In general, the handicap principle does not apply to ethological signals since signals typically benefit both the signaler and the observer. However, emulating pseudo-signals and cultural elaborations of signals are posed (willful) rather than spontaneous (involuntary) behaviors and so may be used for deceptive purposes. In such circumstances, acts of grief-related self-injury and the destruction of personal property are consistent with the handicap principle.

38. *World of tears.* Although weeping evolved in the context of grief, the inducing of prosocial feelings in observers is useful in many other circumstances. Consequently, weeping is sometimes observed when a weeper experiences other emotions, such as feelings of pain, joy, loyalty, piety, adoration, or extreme laughter.

In addition to offering explanatory accounts of the social and behavioral manifestations of melancholy and grief, the TTIRS theory also offers several speculative claims regarding their physiology and evolution:

Physiological Conjectures

39. *Inflammation.* The immune system's main response to physical stressors is inflammation. Inflammation is caused by a class of endogenous compounds known as proinflammatory cytokines.

40. *Dopamine.* Dopamine facilitates motor movement and also contributes to the experience of pleasure. Proinflammatory cytokines are known to suppress dopamine. Consequently, activation of the immune system can lead to feelings or anergia (lethargy) and anhedonia (loss of pleasure), which discourage motor movement.

41. *Central histamine.* Another proinflammatory cytokine is histamine. In the brain, histamine is implicated in improved cognitive functioning, suggesting that it plays an important role in melancholic realism.

42. *Peripheral histamine.* In the periphery, histamine release leads to allergic symptoms, including watery eyes, nasal congestion, and pharyngeal constriction.

43. *Allergy cue.* Allergic symptoms are moderately apparent to observers. In the evolutionary past, stress-induced allergic symptoms would have amounted to an ethological cue. Since allergies are not infectious, this cue might occasionally encourage approach and assistance from altruistically oriented observers. However, there would have existed no automatic tendency for an observer to respond.

44. *Ritualization.* Due to the mutual benefits for both the stressed individual and the altruistic observer, selection pressures would begin the process of ritualization in which the allergic cue would be transformed into a bona fide ethological signal.

45. *Weeping.* Over time, a stress-induced allergic cue was made more conspicuous by adding a supplementary lacrimal gland nerve that increased the volume of tears and the compulsion to vocalize.

46. *Prosocial ritualization.* A final crucial addition was the appearance of biologically-prepared prosocial feelings induced in observers of weeping.

Figure 17.1 provides a schematic summary of the TTIRS theory, including a sketch of the proposed evolutionary history. Five historical stages are represented: (1) an immune response that is limited to fighting infection and repairing tissue damage, (2) supplemented by motivational changes that enhance immune effectiveness through feelings of anergia and anhedonia, (3) neurochemical changes that promote improved cognitive processing ("melancholic realism") and so expand the range of stressors addressed to include cognitive and social stresses, (4) peripheral histamine release that amplifies visible symptoms of stress and so better encourages possible altruistic assistance, and (5) a full-fledged ethological signal in which a distinctive display (weeping) commonly induces biologically prepared prosocial feelings in observers.

The principal aim of this book has been to explain two nonpathological sadness states—melancholy and grief. In addition, we might quickly summarize our proposed account for a third sadness-related emotion, nostalgia:

Nostalgia

1. *Memory's purpose.* Memory has no biological value unless those memories inspire changes of future behaviors that are likely to enhance fitness. From a biological perspective, memory is about the future, not the past.

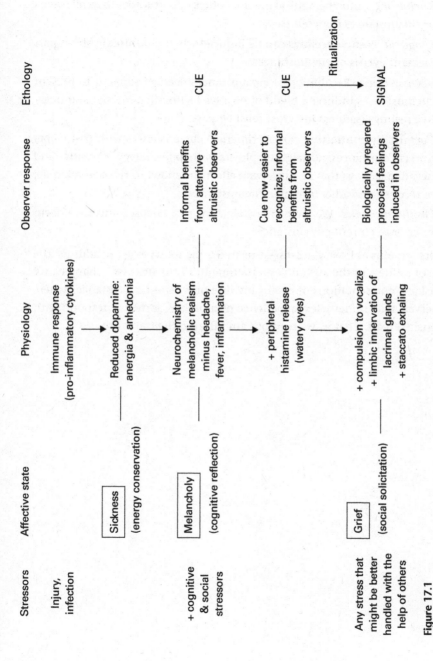

Figure 17.1
Schematic illustration of the triadic theory of immunological responses to stress (TTIRS) theory summarizing a proposed evolutionary history.

2. *Reminiscing*. Autobiographical memory offers opportunities to recall events or circumstances from our past.

3. *Eumnesia*. Positive feelings can be induced when reminiscing about past pleasant events or circumstances.

4. *Bittersweetness*. Recalling past circumstances deemed superior to present circumstances induces a blend of positive feelings (eumnesia) and negative feelings (sadness for a lost relished past).

5. *Contrast*. By contrasting current (inferior) circumstances with past (superior) circumstances, nostalgia employs a push–pull strategy of positive and negative feelings that can motivate efforts intended to recover, rekindle, or reconstruct earlier favorable circumstances.

6. *Positive memories*. We are not nostalgic for past circumstances we deem inferior to current circumstances.

There remains at least one sadness-related topic we have yet to address: the role of sadness in the arts and entertainment. In the next two chapters, we consider representations of ostensibly dire or woeful circumstances where observers may nonetheless experience pleasurable sadness or respond with cherished tears. That is, we turn our attention to the tragic arts.

In industrialized societies, much or perhaps most adult weeping occurs in response to arts and entertainments. Tears are regularly evoked through film, drama, literature, poetry, and other portrayals, both fictional and documentary. Evidently, weeping can be evoked through acts of imagination and representation, apart from real-life situations. Curiously, weeping also frequently occurs in response to nonnarrative instrumental music. In most (though not all) cases, weeping in response to art and entertainment is commonly reported by audience members to be highly pleasurable.

As we have already seen, tears alone are neutral with respect of emotional valence. Tears are commonly induced by positive emotions such as joy. Feelings of joy can certainly arise in response to narrative and dramatic portrayals. However, the tears evoked by arts and entertainment are more commonly linked to scenes characterized by distressing, calamitous, poignant, or heart-rending situations. If grief-related weeping is negatively valenced, why would anyone expose themselves to such anguished literature or drama—let alone seek out such entertainments? In this chapter and the next chapter, we address two questions: What makes tragic arts so appealing? And how can tragic arts induce pleasurable weeping?

The paradox of "enjoying negative emotions" was famously addressed in Aristotle's classic work, *Poetics* (fourth century BCE). Aristotle was particularly struck by the popular appeal of tragic drama. How, he wondered, could portrayals of misery and misfortune possibly be enjoyable for audiences? Two drama-induced emotions that Aristotle found especially puzzling are the feelings of fear and pity (*eleos*).[1]

The problem raised by Aristotle has inspired philosophical discussion and debate spanning more than two millennia. Classic literature pertaining to the topic includes writings by Dubos, Hume, Diderot, Kant, Burke, Schopenhauer, and Nietzsche—among others. In recent decades, the "problem of negative emotions in the arts" has kindled a veritable industry within aesthetic philosophy.[2] French philosopher Carole Talon-Hugon has conveniently characterized the paradox via the following informal syllogism:

(1) Negative emotions are psychologically unpleasant.

(2) Art can cause us to experience negative emotions.

(3) We find pleasure in this experience.[3]

Clearly, at least one of the three statements in this syllogism must be wrong. Philosophers have proposed many ideas and theories, but there is certainly no consensus regarding a solution to the paradox. Unfortunately, with only a few exceptions, the existing philosophical literature rarely engages the pertinent physiological, psychological, or ethological research related to emotions such as grief or melancholy.[4] In this chapter and the ensuing chapter 19, we will see that attending to ethological and evolutionary considerations helps to clarify the apparent paradox and leads to plausible resolutions of at least some of the main issues. In this chapter, we address a few of the more popular ideas, many of which turn out to be minor issues or simply wrong. The ensuing chapter will offer a more compelling account.

Story-Induced Emotions

A useful place to start is by considering how it's possible that stories or narratives would have *any* emotional impact at all on audiences. When watching a movie or a theatrical production, reading a novel, or listening to a storyteller, it is patently obvious that we are engaging a depiction, portrayal, or representation of events, not the events themselves. We are cognitively aware of the artifice, yet the story nevertheless has the capacity to move us. If emotions are motivational amplifiers that encourage us to behave in ways that typically enhance inclusive fitness, then why would "mere" narratives (contrived representations of potentially fake or unreal situations) have any effect at all on us? Like an animal encountering its own image in a mirror, isn't our best response to simply ignore the illusion? One might expect that selection pressures would, over time, make us less emotionally responsive to stories. Surely, our responses to the world would be more finely tuned if we acted only on firsthand information we know to be factual.

It helps if we separately address nominally fictional from nonfictional stories. Let's begin by considering the case of nominally factual (nonfictional or documentary) stories.

Oxford anthropologist Robin Dunbar has drawn attention to the importance of gossip in human evolution. Language serves many functions, but Dunbar has suggested that one of the earliest and most valuable functions is that language allows us to exchange information about other people.[5] We learn a lot by observing the behaviors of others. We learn who is stronger or weaker, who is allied with whom, who is trustworthy, and a host of other socially useful snippets of information. Much information can be gathered firsthand through our own observations. But exchanging information through gossip (or modern variants of gossip such as film documentaries or trusted news sources) allows us to tap into a much larger network of observers. In other words, there can be considerable value in secondhand information. Relying exclusively on our own firsthand observations can be unduly restrictive.

Research on memory and learning has long recognized that lessons are better retained when they induce some emotion. For example, we are more likely to retain the memory that Alice was betrayed by Bill if that information is accompanied by a feeling of disgust or revulsion. The benefit of nominally factual gossip is enhanced if we allow the gossip to stir our emotions.

Gossip also allows for the possibility of disinformation, so acts of gossip themselves become behaviors by which we may, over time, understand whose stories are trustworthy or deceptive. When people engage in repeated interactions over long periods of time, a reputation for honesty typically proves more valuable than the short-term gains arising from deceit. Research has shown that even a single act of duplicity can have long-term detrimental impacts for the person who engages in cheating.[6] Like any behavior, gossip itself provides opportunities to learn who can be trusted.

The risks of misinformation notwithstanding, attending to secondhand information can be very useful, and that information might be better retained in memory if we become emotionally engaged. In light of the potential value, we might also expect that a good bit of gossip would be experienced as enjoyable. Indeed, several scholars have suggested that one source for the appreciation of stories is the pleasure of learning, or the appeal of useful informative.[7]

The above account might explain why it makes sense for us to experience emotional reactions to secondhand tales of factual events, but what about stories that are entirely made up? One might suppose that factual stories have a greater capacity for evoking emotions than fictional stories. Although this idea is intuitively appealing, it is not supported by the empirical research. Whether a story is presented as fictional or factual has no effect on induced emotion. Research by Jeffrey Strange and Cynthia Leung at Columbia University and Melanie Green and Timothy Brock at the Ohio State University suggests that the key to induced emotion is the degree to which the spectator is transported into the portrayed world.[8] When equally engaged in a factual or fictional narrative, it turns out that the experienced emotions are equally intense.[9] So why do people respond emotionally to narratives that they know to be fake or fictional?

One might suppose that if fictional narratives have little, no, or negative survival value, then selection pressures would favor cognitive inhibition of emotional engagement. That is, cognitive awareness of the fictional status of a narrative would ultimately suppress or attenuate its emotional impact and so minimize the potential effect of the story on future behavior. The fact that this doesn't occur suggests that fictional narratives (at least the ones we find sufficiently interesting to attend to) might yet be functionally useful.

A number of scholars have suggested that fictional narratives often offer opportunities to learn useful life lessons. There are plenty of utilitarian scenarios one might propose. For example, stories often construct hypothetical situations that help us imagine how we might act in comparable circumstances. A fictional story about encountering a wild animal might encourage us to be more vigilant when traveling through a forest. Or a story about personal abandonment might alert us to the importance of maintaining close friendships. In *The Adventures of Tom Sawyer*, Mark Twain relays a story illustrating the value of "reverse psychology." Tom has been assigned the chore of whitewashing Aunt Polly's fence, but Tom has no desire to ruin his Saturday by painting fences. He hatches an unusual strategy; he begins painting the fence slowly and calmly as though savoring the pleasure of the assignment. In doing so, he manages to convince his unsuspecting pals of the desirability of this task and so induces his friends to do the job for him.

Even the most fantastic hypothetical scenarios often provide cautionary tales or useful tips that stimulate us into considering how we might respond in similar situations. When Odysseus was torn between the desire to hear the seductive sounds of the Sirens and knowing that the Siren calls were dangerous, he came up with the clever solution of ordering his crew to block their ears with beeswax while having himself bound to the mast of his ship—allowing him to listen to their beautiful compelling songs while avoiding the lure that would otherwise lead the ship to its doom. Imagining how we might react in different situations can have future benefits.

Another possibility is that fictional narratives have little or no fitness value but offer a nonadaptive form of pleasure seeking. Like refined sugar, pornography, or heroin, fictional narratives may simply take advantage of a pleasure channel that evolved for other important purposes. For example, it may be that tolerance of nonadaptive fictional narrative is a price we pay for maintaining psychological mechanisms that allow us to learn valuable factual information through secondhand channels. The point is, even when we cognitively recognize the fake, fictional, or imaginative origin of a narrative, we may nevertheless be incapable of turning off the emotional responses to these stories since these same responses must be available for secondhand gossip that is indeed valuable.[10]

Finally, we should note that cognitive assessment is not wholly ineffective in tempering the emotional impact of fictional narratives. This is most evident in horror films where more than one audience member has survived the feeling of terror by resorting to the mental mantra, "It's just a movie; this is just a movie." The horror film genre testifies to both our ability and our *inability* to control our emotional responses. Evidently, cognitive awareness of the fictional status of a story can temper but not completely disable our emotional responses.

The Pleasures of Rubbernecking

The capacity for narratives (both factual and fictional) to evoke emotions still leaves open the question of pleasure. How is it possible for positive feelings to arise from information we might regard as tragic, disgusting, or gruesome? Whether information is firsthand or secondhand, the motivation for information gathering is most evident in the feeling of *curiosity*. The allure of inquisitiveness is evident in the *Republic*, where Plato relays the following story:

> Leontius, the son of Aglaion, coming up one day from Piraeus, under the north wall on the outside, observed some dead bodies lying on the ground at the place of execution. He felt the desire to see them, and also dread and abhorrence of them; for a time he struggled and covered his eyes, but at length the desire got the better

of him; and forcing them open, he ran up to the dead bodies, saying, Look, ye wretches, take your fill of the fair sight.[11]

Plato's story highlights conflicting motives—producing mixed emotions. At least three motives seem evident. At some level, Leontius is clearly attracted to the sight. He wants to examine the dead bodies. We might label this motive "curiosity." At the same time, he anticipates the repugnant nature of the scene. We might label this second feeling disgust—or, more precisely, the anticipation of disgust, namely, "dread." Finally, Leontius appears uncomfortable with his own morbid curiosity, perhaps believing that it is unbecoming and that a moral person would simply pass by without gawking. We might label this third, more speculative, feeling "moral revulsion." In the end, his curiosity prevails: his ultimate behavior is to *look*. At the same time, he voices a negative self-assessment of his behavior. Despite his misgivings, he is unable to suppress his curiosity.

Perhaps the best modern equivalent is evident in rubbernecking, where motorists slow down when passing an automobile accident. We may even feel disappointed when a police officer impatiently waves us through, giving us no opportunity to view the mayhem. This hunger for information is not restricted to *Homo sapiens*. Drawn to my window by the sound of two cats engaging in a noisy altercation, I am struck by the sight of the many neighborhood cats whose small faces are similarly glued to various windows throughout the apartment complex. There are, of course, excellent fitness reasons for the neighborhood cats to want to learn who wins and who loses in any feline brawl.

In his book reviewing the science of curiosity, Mario Livio notes that curiosity has been the main driver of scientific inquiry throughout history. Immanuel Kant spoke of the "appetite for knowledge," and in many languages, people resort to similar gustatory metaphors: we hunger for recent news and have a thirst for knowledge. Information has clear utility, so we should not be surprised that, like seeking and consuming food, we are motivated to actively seek information and take pleasure in its consumption.

Research on curiosity suggests that there are at least two forms: epistemic curiosity and perceptual curiosity.[12] Epistemic curiosity pertains to cognitive knowledge, such as learning that Jim and Carol are brother and sister. Perceptual curiosity pertains to perceptual puzzles, such as deciphering an unusual smell or resolving an initially confusing photograph.

Neuroimaging studies suggest that both epistemic and perceptual curiosity are able to evoke pleasure. For example, Caltech researcher Min Jeong Kang and her colleagues scanned the brains of volunteers while participants were presented with forty simple knowledge questions. For example, one of the questions was, "What is the name of the galaxy that the earth is part of?" (Answer: Milky Way.) Participants guessed the answer to each question and rated how confident they were about their answers. They also indicated how curious they were to learn the correct answer. When presented with the correct answer, Kang and her colleagues observed increased activity in brain regions associated with the anticipation of reward; they also observed that the degree of activation was positively correlated with the earlier ratings of expressed curiosity. The results suggest that the pleasure we derive from information is proportional to our curiosity.[13]

With respect to *perceptual* curiosity, a pertinent study has been conducted by Marieke Jepma and her colleagues at Leiden University in the Netherlands. They scanned the brains of volunteers while participants tried to identify the objects in intentionally blurred photographs. Degraded visual information was found to activate brain regions known to be associated with unpleasant experiences. When the visual information was restored, reward-associated circuits were activated. In other words, perceptual curiosity resembles brain activity associated with hunger or deprivation, whereas providing information resembles brain activity associated with the pleasure of consuming food.[14]

Neither the Kang or Jepma experiments resolve the underlying mechanisms of curiosity. Nor do they offer definitive evidence that the evoked pleasure actually originates in information anticipation and consumption. Research concerning curiosity is in its infancy, and the phenomenon appears to be rather complex.[15] Nevertheless, there is little doubt that people and nonhuman animals crave information. Information gathering is valued even when the content of that information is negative. No one should be surprised to learn that people are happy to learn the latest juicy gossip.

How does research on curiosity relate to the paradox of tragedy? McGill University philosopher David Davies has drawn attention to curiosity as a contributing motivation for audiences to expose themselves to the tragic arts. If curiosity encourages information gathering, and if information gathering evokes a positively valenced affect, then engaging tragedy is likely to induce at least some element of pleasure whether the content is positive or negative. Like Leontius, it's often hard for us to resist our own curiosity, even when that curiosity warrants the adjective "morbid."[16]

By way of summary, we have noted that there is value to secondhand information and that secondhand information is best retained in memory when an emotional response is induced. We've seen empirical evidence that emotions tend to be evoked even if the information is known to be fake or fictional. We have also noted that people are motivated to seek out information, that people take pleasure in anticipating and consuming information, and that the pleasure of information is independent of the valence of the content: that is, we are eager to receive information even about disturbing or gruesome events.

This raises the reflexive question of why anyone might read this book. Unless you have been assigned this book by a dictatorial teacher, your principal motivation for reading is likely curiosity. As the author of this book, it's more than a little disconcerting to realize that the research implies that the pleasure of reading this book is likely to be the same, whether readers consider my text grounded in scientific research or wholly made up.

Although curiosity is a potent motivator and might explain much of the appeal of tragic arts, it still fails to address how audience members might enjoy feelings of sadness evoked by many tragic arts.

The Paradox of Happiness

A common, though naive, theory is that happy stories make people feel happy and sad stories make people feel sad. If feeling happy is more pleasurable than feeling sad, then we ought to prefer happy stories over sad stories. In fact, happiness is a surprisingly hard sell in the arts and entertainment, especially in drama and literature. Most happy stories are boring, rarely evoke happiness in audiences, and offer little enjoyment.

The eighteenth-century Scottish philosopher David Hume described theatrical scenes portraying happiness as "unbearable." Most of us would agree. Can you imagine a film portraying hours of happiness that viewers would truly enjoy? Suppose we watch a film that is filled with positive heartwarming scenes from a person's life. It begins with childhood images of the person playing happily with other children. There are scenes of high school and college graduations, amusing parties, pleasing vacations at the beach, and touching wedding ceremonies. The film continues with scenes showing uplifting acts of kindness, segments showing gratifying accomplishments, celebrations for a highly successful career, and public praise of achievements, culminating with smiling images of the protagonist's own children embarking on similarly happy lives. Who on earth would want to watch such a tedious production?

There is, of course, a positive answer: grandparents. There exist millions of such cheery videos, and many of them are among the most prized possession of parents and grandparents. What other people find boring, bland, and insipid, parents and grandparents will savor watching on repeated occasions.

The contrasting responses are readily understood if we consider them from an evolutionary perspective. Pleasure is the perennial companion of those behaviors that commonly enhance inclusive fitness. There are excellent biological reasons why grandparents might relish the successes of their children and grandchildren. If the video were of someone else's offspring, one would expect the delight to be considerably reduced. And indeed, we find retirement homes in which residents take pleasure in relaying the achievements of their offspring to their friends, while simultaneously complaining of intense boredom when hearing of the successes of their friend's offspring.

Notice that our enjoyment of "happily ever after" appears to be correlated with both kinship and affiliative proximity. We take greatest pleasure in the happiness of our own offspring, then perhaps more distant relatives. We also take some pleasure in the happiness of close friends, members of our social group, and perhaps people we regard as belonging to broader groups with which we identify (neighborhood, subculture, nation).[17] In short, the delight of watching "happily ever after" appears to be directly proportional to how that happiness attests to an enhanced inclusive fitness for the viewer—including both kin-related and social-affiliative groups.

The difficulty of portraying happiness is well recognized by professional writers and producers. The plots for many successful stories center on overcoming some adversity, often ending with "and they all lived happily ever after." Although "happily ever after" may be a pleasing place for a story to end, writers and producers recognize that it provides a pitiful plot for a sequel.

By way of summary, I am suggesting that what Aristotle regarded as the problem of tragedy—and what later philosophers expanded and renamed the paradox of negative emotion in the arts—has an overlooked companion: the paradox of happiness in the arts. The paradox is simply that spectators find most depictions of happiness to be uninteresting and devoid of enjoyment.

This apparent paradox can be plausibly resolved by examining it from an evolutionary psychological perspective. When observing portrayals of happiness, whether we experience pleasure or indifferent boredom depends on whether the situation can be interpreted as representing or implying an increase in our own inclusive fitness. As we will see shortly, the apparent paradox of tragedy can be similarly approached by attending to questions of inclusive fitness.

Before we explore our ethologically/evolutionary-inspired account, it's appropriate to return to our earlier discussion of *catharsis*—Aristotle's long-admired notion, where artistic portrayals of negative affect might cause a purging of negative emotions in the observer.

Catharsis

Unquestionably, Aristotle's theory of catharsis has been the most popular theoretical account of the tragic arts. Aristotle's proposal was primarily motivated by his reaction to Plato's earlier claim (in the *Republic*) that poetry (the arts generally) can lead audience members to become overemotional or hysterical. Plato regarded uncontrolled emotions as dangerous and consequently held a dim view of the more passionate or melodramatic arts. Aristotle, by contrast, argued that such feelings can be healthy. For Aristotle, poetry, drama, and music are ultimately good because the emotions unleashed provide a healthy outlet for purging negative feelings.

In the case of weeping, recall from chapter 4 that there is little evidence that crying, by itself, has any beneficial effect, apart from the social support induced in observers and the capacity of crying to suppress the hostile behavior of others. The evidence that weeping by itself has any cathartic effect is weak to nonexistent.

Well, perhaps weeping is an exception. One can appreciate tragic arts without ever being brought to tears. Unfortunately, even if the theory of catharsis were true, there is a more serious problem with catharsis as an explanation for resolving the paradox of negative emotions in the arts and entertainment.

The theory of catharsis is an *ultimate* account, not a *proximal* account. Something can be beneficial to one's health without being enjoyable. The problem of tragic arts is to explain not why viewing tragic arts might be good for us but why we *enjoy* them. The purported health-conferring effects of catharsis fail to provide the necessary proximal motivation for exposing ourselves to tragic portrayals. There must be more immediate feelings of pleasure or delight; otherwise, no one would bother.

This same argument applies to sundry variants of catharsis. For example, apart from weeping, we might ask whether (nontearful) melancholic sadness has any cathartic effect. We might suppose that an episode of melancholy could help a person when "working through" some problem—taking advantage of the cognitive benefits of melancholic realism. Again, the snag is that although melancholic realism may confer important mental benefits, we have no evidence that melancholy itself is enjoyable. Consequently, the ultimate benefits of melancholic realism fail to explain the proximal allure of tragic arts.

Righteous Indignation

So where lies the enjoyment in tragedy? Under what circumstances are stressful states enjoyable for audiences? As we learned in our discussion of the paradox of happiness, it helps to pay close attention to point of view and to consider in detail the repercussions for inclusive fitness of different scenarios. Not all tragedies are the same. The tragedies that befall our competitors or enemies may actually increase our inclusive fitness. Tragedies that decrease our inclusive fitness include those of family, friends, and allies, as well as our own personal misfortunes.

In the arts and entertainment, considerable satisfaction can accompany the misfortunes of a villain. An instructive example is offered in the second installment of *The Terminator* film franchise (*Judgment Day*). The film ends with two deaths—the death of a villainous T-1000 robot (played by Robert Patrick) and the intentional death by suicide of the heroic T-800 robot (played by Arnold Schwarzenegger). Both robots die in the same charming manner: dissolved in a vat of molten metal. Yet the emotional effects on the audience are utterly contrasting—the death of the villain evokes feelings of righteous indignation and celebration, whereas the death of the hero elicits feelings of sadness, compassion, and admiration. Feelings of compassion are amplified in this scene by the tearful display of the teenage John Connor character (played by Edward Furlong), who is understandably distressed by the loss of his T-800 defender and companion. That is, in addition to our feelings of loss for Schwarzenegger's heroic character, viewers may also feel direct compassion for the Connor character.

What does it mean for a character to be a hero as opposed to a villain? In artistic and entertainment depictions, it's essential for the author to establish a positive relationship between a protagonist and the audience. In order for tragedy or misfortune to evoke an emotional response in viewers, the character must be tailored so as to be recognizable as either friend or foe. Whether the portrayal is fictional or documentary, we must feel that the protagonist is someone whose actions benefit us; the protagonist is, or would be, a worthy companion or valued ally. Violent action films are often replete with dozens of incidental deaths of fleeting characters ("collateral damage") that evoke little response in the audience. In order to feel strong emotions (either positive or negative), the audience must be vested in a character as either ally or adversary. Said another way, tragedy itself has an undefined valence. As with the "happy videos" discussed earlier, it is how the tragedy is interpreted with respect to our own inclusive fitness that colors the tragic events as either positive or negative.

Finally, we need to consider the proximal feelings evoked when a perceived foe gets their comeuppance. Here, I propose, that the salient emotion is a feeling of righteous indignation. Throughout history, people have clamored for and experienced considerable satisfaction when witnessing the punishment of those deemed morally wicked. Economists have long noted that people will pay to see a bad person punished.[18]

These feelings are not limited to nonfictional events. A few years ago, I attended a commercial showing of the film *Wind River* directed by Taylor Sheridan. In this film, the villains are especially callous and heinous. At the end of the film, the story's protagonist—a Fish and Wildlife Service agent, Cory Lambert—manages to prevail over the villains. As the bad guys meet their end, one by one, I was struck by the spontaneous applause of the cinema audience. There was a palpable atmosphere of joy as each of the bad guys met his demise. Not all tragedies include a final reckoning for the bad guys. But for those that do, at least one of the potential pleasures for spectators can be found in the positive experience of righteous indignation.

Reprise

We are motivated to attend to information through the positively valenced feeling of *curiosity*. Curiosity is not limited to information with positive content. In fact, curiosity is greatest for information whose content is stressful, threatening, or gruesome. Secondhand information (such as gossip) allows us to extend our network of observations beyond our own eyes and ears. The lessons of secondhand information are best retained in memory when they evoke an emotional response. Although fictional information may be less valuable than factual information, it appears that fictional narratives are as capable of inducing spectator emotion as factual secondhand information.

Despite the widespread popularity of the idea of catharsis, there is scant evidence that weeping confers any therapeutic or health benefits apart from the social assistance induced in observers. Moreover, even if exposure to tragic arts were shown to be healthy for us, the health benefits alone would not be sufficient for people to want to engage the tragic arts; the motivation for engaging tragic arts must be proximal, not merely ultimate.

What later philosophers expanded and renamed the paradox of negative emotion in the arts has an overlooked companion: the paradox of happiness in the arts. In the case of happiness, we saw that the apparent paradox can be plausibly resolved by examining specific scenarios from an evolutionary psychological perspective. When observing portrayals of happiness, whether we experience pleasure or indifference depends on whether the situation represents or presages an increase or decrease in our own inclusive fitness.

In this chapter, we have drawn attention to two conditions in which portrayals of tragedy or distress might lead to positive experiences for spectators: the simple pleasure of information gathering and the pleasurable feeling of righteous indignation when the tragedy befalls a perceived enemy or villain. However, these two conditions play a relatively minor role in the enjoyment of tragic arts. In the next chapter, we consider a more compelling mechanism by which tragic entertainments might evoke pleasure for audiences.

19 Tragic Arts (II)

In addressing the paradox of negative emotions in the arts, it's helpful to distinguish three closely related questions: First, how might tragedy be enjoyable? Second, how might tragedy lead to weeping behaviors among audience members? And finally, how can we enjoy watching a tragedy that moves us to tears? Addressing these questions separately is important: not all tragedies are enjoyable, not all enjoyable tragedies move us to tears, not all tears are evoked by tragedies, and not all tears are enjoyable. Given the complexity, it's helpful to proceed step by step in our analysis.

In the previous chapter, we discussed two relatively minor considerations related to the pleasure of tragic arts: the pleasures related to information gathering or curiosity and the pleasures experienced when bad guys get their comeuppance—the pleasure of *righteous indignation*. A problem with these accounts is that they are limited to narrative arts where it's possible to convey concrete information and where characters can be clearly depicted as either good or bad. When we consider nonnarrative arts, such as abstract art or music, these accounts are less helpful in understanding the enjoyment of portrayals of negative emotion. A particularly challenging art is to be found in sad instrumental music. Devoid of narrative lyrics, instrumental music is nevertheless quite capable of portraying or conveying sadness, of bringing listeners to tears, and of inducing tearful feelings of sublime pleasure.[1]

In this chapter, much, though not all, of our discussion will focus on the thorny case of sad instrumental music. The phenomenon of sad music has attracted considerable empirical research over the past three decades, and we will see that research on sad music enjoyment offers a plausible account for much of the enjoyment of tragic portrayals in general—both narrative and nonnarrative.[2]

As an appropriate starting point, we might briefly consider what it is that makes some music sound sad.

Wretched Refrains

The phenomenon of sad music is rich, elaborate, and captivating. The unabridged story I will leave for another occasion; however, for our purposes here, we can at least summarize the main features. In general, music-related affect involves cultural norms, learned associations, emulation of acoustic cues and signals, and other elements.[3]

Much music conveys particular affects by emulating characteristics of the human voice. For example, "happy" musical passages emulate many of the acoustic characteristics of a happy voice.[4] This acoustic parallel can be referred to as the *voice–music homolog*. In the case of sadness, it's important to distinguish the sounds of melancholy from the sounds of grief. When in a state of melancholy, recall that people speak more slowly, quieter, and with a lower overall pitch, a more restricted or monotone pitch movement, a more mumbled articulation, and a darker sound. By contrast, a grief voice tends to be louder (sometimes wailing) and higher in pitch (especially falsetto), and it involves punctuated vocalized exhaling, ingressive phonation (audible gasping), a pharyngealized or tense timbre, and the highly characteristic sound of a cracking or breaking voice (i.e., abrupt pitch movements, typically from very high to moderately high).

These features are echoed in two different types of "sad" musical passages. "Melancholic" passages have been shown to exhibit slower overall tempos, quieter dynamic levels, lower overall pitch height, small melodic intervals, legato (rather than detached) articulation, and darker timbres.[5] In the case of darker timbres, for example, instruments that exhibit lower spectral content are judged by listeners to be sadder sounding. Hence, the guitar is perceived to sound "sadder" compared with the banjo, the cello compared with the viola, the viola compared with the violin, and the English horn compared with the oboe.[6]

The difference between grief-like and melancholy-like musical passages is readily apparent in Samuel Barber's famous *Adagio* for strings—an iconic work that was voted in an international BBC poll to be the saddest of all musical works. Many people will recognize this as a work played at Princess Diana's funeral or in Hollywood movies such as Oliver Stone's *Platoon* and

David Lynch's *Elephant Man*. Barber's *Adagio* opens with a passage that is slow, with a quiet dynamic, low overall pitch, and small pitch movements. However, for most listeners, the most moving passage occurs later in the work where the strings play using a loud dynamic level, very high in pitch, with an intense texture. That is, the work exhibits both melancholy-like and grief-like passages. In effect, Barber's *Adagio* emulates the mourning cycle, alternating between the two cardinal forms of sadness.

Working in my lab at the Ohio State University, Lindsay Warrenburg identified hundreds of nominally "sad" musical passages that are readily classified as "grief-like" or "melancholy-like." She has also shown that much of the past confusion in sad music research can be traced to the failure of researchers to distinguish melancholy-like or grief-like passage used as experimental stimuli. It's common for researchers to simply describe their musical stimuli as "sad."[7]

There are other factors involved in conveying sadness through musical sounds, including enculturated features (such as different scales or modes), tragic or touching lyrics, and linking sadness with learned associations such as the use of specific instruments or certain musical gestures (such as the playing of "Taps" at a military funeral). Research also implicates personality traits (like neuroticism) that shape the likelihood that a given listener will experience a passage as sad sounding. However, the broad outlines of "sad sounds" can be traced to a straightforward voice–music homolog for melancholy and grief.

Sources of Pleasure

If the sounds that characterize sad music resemble the vocal sounds produced by sad people, then it seems reasonable to suppose that listeners might react similarly to both. In chapter 4, we identified six proximal feelings that tend to arise when we witness distress in others. These feelings are intended to motivate behaviors that benefit the distressed individual but ultimately benefit the altruist as well. Like all emotions, these feelings are valenced—positive or negative—as a way of inducing us to act appropriately. It makes sense, then, to consider whether one or more of these proximal feelings might account for the pleasure often experienced in response to tragic arts.

One proximal motivation is the *reduced commiseration* that occurs when our efforts to alleviate a stressed person's suffering simultaneously reduce our own vicarious distress. However, even in the case of narrative arts, like film, literature, or storytelling, no one actually suffers; nor can audience members do anything to temper a stressed character's pain. In the case of sad instrumental music, there doesn't appear to be any agent or persona whose affliction we might help to alleviate.

A second proximal motivation is the positive feeling of *compassion*—a vivid feeling of pity, care, concern, tenderheartedness, kindness, or benevolence for a person experiencing distress. This positive feeling might well arise in narrative arts, but the case of sad instrumental music is again problematic. If nominally sad music is able to evoke compassion, an obvious question is, "Compassion for whom?" In the case of vocal music, sad lyrics make it theoretically possible to empathize with the singer or some character conveyed in the lyrics. However, listeners have long reported that sadness can be compellingly conveyed (and enjoyed) by purely instrumental music—as in the case of Barber's *Adagio*. It is nominally sad instrumental music that most clearly raises the conundrum of agency: Who might be the object of our purported feeling of compassion? Do listeners feel compassion for the trombone or the electric guitar? Do we imagine that the instrumentalist is in a state of grief? Do we somehow feel pity for the absent composer? New Zealand philosopher Stephen Davies has been explicit in articulating these misgivings. Davies argues that music is not an agent, and so by definition, music is incapable of "being sad."[8]

Two further altruistic motivations relate to feelings of social reputation—both the allure of social approval for helping and the fear of social ostracism for failing to help. In the tragic arts, however, there is no actual need for us to respond by trying to be helpful. When watching Shakespeare's *Romeo and Juliet*, we do not applaud the spectator who rushes onto the stage in order to snatch the dagger from Juliet's trembling hand. Nor is it the case that, when watching *Les Misérables*, we are haunted by thoughts that others will regard us as morally delinquent for failing to bring food to toss at the apparently starving cast members.

A similar problem attends the fifth proximal motivation encouraging prosocial behaviors—namely, the anticipation of the private feeling of *virtue* or enhanced self-worth that arises following acts of charity. Since audience members don't engage in altruistic acts, there is no way to feel virtuous for having acted charitably.

Finally, a sixth proximal motivation is the positive feeling of *bonding warmth*—arising from the prospect of expanding or bolstering one's network of friends or companions to include the individual benefiting from our largesse. Here again, the problem is that spectators don't engage in charitable acts toward any character, nor is a character likely to feel friendship toward an unseen audience member. Moreover, in the case of instrumental music, once again there appears to be no sentient agent with whom we might bond.

At this point, the proximal feelings typically induced when we observe individuals in distress seem unlikely sources of pleasure that might account for the enjoyment of sad or tragic arts. Later in this chapter, however, we will return and reconsider these proximal emotional responses.

Individual Differences

A helpful starting point begins with the recognition that there are individual differences in how people respond to tragic arts. Not all arts and entertainment experiences are enjoyable. Not everyone enjoys watching horror films or coping with emotionally overwrought dramas. People can find certain works disturbing, upsetting, offensive, or repellent. When viewing a tragedy, for example, feelings of distress or commiseration might move a spectator to genuine grief, marked by weeping and accompanied by distinctly unpleasant feelings. Indeed, people will leave a movie theater in mid-showing, stop reading a book, or turn off the music when moved to the point of acute melancholy or grief. Behaviors that terminate the experience suggest that the work is not enjoyable for the spectator.

In the specific case of music, individual differences are especially pronounced. Several surveys have confirmed that listeners are roughly equally divided between those who like and those who dislike sad music. This roughly 50/50 proportion has been observed whether surveys poll music students,[9] the general online population,[10] or even taxi drivers from a variety of non-Western cultures.[11] When asked about nominally sad music, the reported level of enjoyment spans the complete range from intense enjoyment to intense dislike.[12] In surveys we've conducted, roughly 10 percent of listeners claim that nominally sad music is the music they most enjoy.

From a research perspective, this range of responses represents a happy opportunity to conduct studies that aim to identify why listener experiences differ. Note that the contrasting responses of people to the same music suggest that listeners' enjoyment of sad music is shaped by differences in past experience, dispositional differences (e.g., personality), or a combination of the two.[13] Recent research points to the importance of empathy.

Empathy

Empathy is commonly defined as the capacity to understand or feel what another person is experiencing. Research suggests that empathy involves both affective and cognitive components.[14] The affective component entails an involuntary emotional reaction induced in response to the observed emotions of others.[15] The cognitive component entails an intellectualized recognition or a dispassionate understanding of the emotional states of others—without the observer necessarily experiencing any emotion.[16]

The experience of commiseration offers an example of affective empathy. Recall from chapter 4 our description of Tania Singer's experiment where she showed that inflicting obvious pain on a person in the presence of a romantic partner caused brain regions associated with pain to be activated in the scanned brains of the partner not receiving the pain.

When we observe other people, our own emotional responses fall into two broad categories. As in the case of commiseration, one category consists of emotions that echo or mirror the observed or inferred emotion. For example, when observing a nominally angry person, we might ourselves also feel some degree of anger. Such emotional reactions are generally referred to as *contagious* emotions. A second category consists of emotional responses that differ from what we observe or infer in the other person. For example, when observing a nominally angry person, we might feel some degree of fear. This second category of emotional reactions might be referred to as *repercussive* emotions.

When observing or inferring the emotions of others, it's common to experience both contagious and repercussive emotions simultaneously. For example, when observing a sad person, it's common for observers to feel both compassion as well as some element of sadness. In our discussion of ethological signals, we have so far focused exclusively on repercussive emotions and have paid little attention to contagious emotions. Why, we might ask, do we experience contagious emotions? What's the purpose of feeling sad when we observe someone weeping or feeling happy when we observe someone in joyous celebration?

Emotional contagion may or may not serve any useful purpose. For example, it has been argued that emotional contagion may simply be a nonfunctional by-product of associative learning.[17] However, at least three theories have been proposed positing possible benefits of emotional contagion.[18] One potential benefit is that emotional contagion may enhance emotion recognition by observers. That is, vicarious feelings alert the observer to the emotional state of the displayer.[19] Even if the intensity of the contagious emotion is rather subdued, the vicarious mirroring of a displayed feeling may still prove useful for discerning someone's emotional state.

In other circumstances, widespread emotional contagion might prove beneficial insofar as it facilitates coordinated group action. For example, a sports team is likely to perform better when all of its members share the same emotion.[20]

In yet other circumstances, emotional contagion might serve to enhance social bonding. For example, close romantic partners may become even closer when one member fully shares the joy or distress of the other member.[21]

In the case of sadness, we call the contagious emotion "commiseration." Earlier we described Thomas Hobbes's observation that giving sixpence to a beggar allayed his own pain. Since Hobbes gave money, it's useful to pause and consider the effect of giving different monetary amounts. For example, what if Hobbes had given the beggar just a half penny?

Contagious sadness (commiseration) is important, I propose, in allowing us to assess the effectiveness of our altruistic actions. Suppose, for example, that we encounter an injured person lying along the side of the road in obvious pain. The person is unlikely to be helped if we offer to loan our phone so that the person might call a friend or a taxi. Giving the injured person $100 might represent a real sacrifice on our part. But neither of these actions is likely to impact the injured person's pain. To the extent that we are able to *share* a person's pain, commiseration tells us that our actions are wholly inadequate. I propose that the contagious feeling of commiseration effectively provides a sort of gauge or meter that helps us tailor our altruistic actions so as to best help the person in need. As our efforts to help someone bear fruit, it is the reduction in our own commiserative stress that tells us we have successfully reduced the helped individual's misery. If Hobbes had given the beggar the minimum amount of money, it's doubtful that his own "painn" would have been as effectively reduced.

Commiseration is only one component of empathy. Working at the University of Texas at Austin, Mark Davis conducted seminal research investigating how people differ with respect to the experience of empathy. Davis's influential research suggests that trait empathy can be characterized by four stable factors, referred to as empathic concern, personal distress, perspective taking, and fantasy.[22]

Empathic concern is the disposition to feel concern, sympathy, or compassion for another person experiencing some stress or misfortune. We might summarize this facet of empathy via the statement "I feel pity for you." *Personal distress* is the disposition to mirror or echo feelings of personal anxiety or unease when witnessing stress or tension in others ("I feel your pain"). *Perspective taking* is the cognitive tendency to spontaneously adopt the psychological point of view of others ("I understand where you're coming from"). Finally, *fantasy* is ostensibly the cognitive ability to be absorbed and imagine the feeling state for a fictional character, such as portrayed in literature, drama, or films ("I can imagine that situation"). Together, *empathic concern* and *personal distress* represent the affective or emotional aspect of empathy, whereas *perspective taking* and *fantasy* represent the cognitive aspect of empathy. Davis went on to develop a validated survey (known as the Interpersonal Reactivity Index—IRI) that measures these four empathy traits for any given individual.[23]

With regard to the affective components of empathy, empathic concern and personal distress exhibit notably contrasting valences. With her colleagues, Tania Singer has shown that compassion (what Davis called empathic concern) and commiseration (what Davis called personal distress) activate different neural circuits: compassion is positively valenced, whereas commiseration is negatively valenced.[24]

Now people differ in the degree to which they experience empathy. For example, there are well-documented sex differences, with females self-reporting experiencing empathy more readily than males.[25] Apart from the sex differences, there are stable personality or trait differences: some people are simply more empathetic than others. As you might expect, people who are emotionally moved by sad or tragic arts tend to score higher on overall measures of empathy.[26] However, high empathy alone does not necessarily mean that a person will enjoy sad or tragic arts. It depends on the mix of the four component facets of empathy.

Who Likes Sad Music?

So who are these people who like listening to sad music? Working in Finland, Jonna Vuoskoski and her colleagues exposed listeners to a variety of musical excerpts and asked participants to rate how much they liked each excerpt.[27] Participants also completed the IRI empathy survey. As expected, listeners who scored high on empathy were most affected by sad music—as evident in both self-reports as well as in physiological measures.[28] However, the high-empathy listeners who most enjoyed the sad musical excerpts exhibited elevated scores for only two of the four facets of empathy.

With regard to the cognitive facets of empathy, listeners who most enjoyed the sad music scored high on empathetic *fantasy* but not *perspective taking*. Said another way, sad-music lovers are people who generally experienced "I can imagine that" but have typical or ordinary scores for experiencing "I understand where this is coming from."

With regard to the emotional or affective facets of empathy, those listeners who most enjoyed sad music scored high on *empathetic concern* but not on *personal distress*. Said another way, sad-music lovers experienced "I feel pity" but scored moderately on "I feel your pain." That is, sad-music lovers tended to experience a high level of *compassion* but a nominal level of *commiseration*.

This same pattern of empathy factors has been replicated in several subsequent studies, including studies conducted in Finland, Japan, and Austria.[29] Overall, these results might be summarized as follows: those listeners who most enjoy sad music tend to experience high levels compassion and have a well-developed imaginative ability to recognize the experience of a fictional character or persona. But they do not themselves generally experience greater personal distress, nor are they especially sensitive to the perspective of another. In general, sad-music lovers generally experience "I feel pity" and "I can imagine that," but they do not show any enhanced emotional contagion by which they might vicariously share in the portrayed pain or sadness ("I feel your pain").[30,31]

Pleasurable Compassion Theory

As we've just learned, the research shows that people who enjoy listening to sad music score high on compassion with nominal levels of commiseration. Since compassion is positively valenced and commiseration is negatively valenced, it makes sense that enjoyment is more likely when the balance is tipped in the direction of feeling greater compassion.

In collaboration with Jonna Vuoskoski, we have dubbed this the *pleasurable compassion theory*.[32] Like any theory, ours raises a number of questions: First, if compassion is positively valenced, then wouldn't the most enjoyable trag-edies be real-world tragedies? As David Hume noted, if we enjoy sad or tragic displays, then a visit to a hospital ward should be highly pleasurable.[33] Two further questions arise from the specific case of sad music. Who, one might ask, does the listener feel compassion for? As we noted at the beginning of this chapter, in the case of instrumental music, it's not clear that there is any sad or stressed agent who would warrant a listener's feeling of compassion. Finally, if compassion is the predominant emotion evoked in sad-music lov-ers, why don't listeners commonly describe their musical experience as evok-ing feelings of pity or compassion? In fact, I'm not aware of a single study of music-induced emotion that points to compassion, sympathy, or pity as significant reported musically evoked affects. Each of these three questions is addressed in turn below.

Fiction and Fantasy

In attempting to resolve the paradox of tragedy, a number of philosophers have emphasized the "fictional" aspect of artistic portrayals. The idea is that because spectators recognize that the tragedy is not real, genuine sadness is not evoked in viewers and so audience members do not experience true nega-tive feelings. As we noted in the previous chapter, the empirical research does not support this intuitively appealing idea. Narrative research has shown that whether a story is presented as fictional or factual has no effect on induced emotion.[34]

Nevertheless, recognizing the artificial nature of the art or entertainment itself remains important because spectators are absolved of the need to offer altruistic assistance. In this regard, portrayals of tragedies (either fictional or documentary) have an advantage over real (actual, unfolding) tragedies, since viewers can experience the positive feelings of compassion without feeling any guilt for failing to engage in compassionate acts. Nor is there any cost to the observer, such as the need to share resources with a stressed protagonist or the erosion of social reputation due to inaction. Visiting a cinema is not the same as visiting a homeless shelter. As members of a cinema audience, we sit in silence, entirely untroubled by our failure to help. We feel pity or sympathy without being burdened by the guilt of inaction. In short, Hume's critique does not apply because, in contrast to actual tragedies, the compas-sion evoked in artistic or entertainment contexts is cost-free. The spectator does not have to part with precious resources, worry about their reputation, fear social ostracism, or wrestle with feelings of guilt.

Apart from being cost-free, there are other reasons why portrayals of tragedy (both fictional and documentary) might be more pleasurable than actual tragedies. First, as noted in chapter 4, when a tragedy unfolds in our immediate presence, the positive feelings arising from compassion tend to be overshadowed by our cognitive fixation on the precipitating stressful events or circumstances. If your house were engulfed in flames and you were crying, it would be wholly inappropriate for me to thank you for the opportunity to console you. In the presence of genuine suffering, relishing the feeling of compassion would be properly interpreted by observers as self-centered, insensitive, and morally offensive. By contrast, when the suffering is depicted rather than real, no such moral lapse occurs. Leaving a theater after watching a tragic drama, it would be socially acceptable to declare a feeling of enjoyment. Leaving a funeral parlor after comforting friends, it would be entirely socially unacceptable (and personally disturbing) to acknowledge such positive feelings.

Second, fictional or documentary accounts are often crafted in more affecting ways than occurs in naked reality. There is skill and craft in arts and entertainments that counts for something. That is to say, the pleasures of a well-crafted tragedy need not be limited to the positive feelings associated with the evoking of compassion. There exist many other pleasure-inducing facets in an artwork.[35]

Compassion for Whom?

The claim that the enjoyment of sad music arises principally from the positive feeling of compassion flies in the face of the problem of agency described at the beginning of this chapter. Indeed, in instrumental music, who is this purported creature for whom listeners might feel pity or sympathy? When viewed from a logical perspective, we ought not to feel any socially oriented emotion toward an entity that is not an agent capable of feeling pain or distress. Recall Stephen Davies's argument that there can be no such thing as "sad music" because music is incapable of feeling sad.

Such arguments, however, presume that humans are strictly rational beings and that human emotions are driven predominantly or solely by cognitive/logical assessments. Classic research in the fields of affective neuroscience and behavioral economics suggests this assumption is incorrect.[36] Similar agency-related paradoxes are evident in other domains, such as pornography, where printed photographs or illuminated computer screens can cause sexual arousal in observers despite the complete absence of any human presence that would warrant feelings or behaviors whose ultimate purpose is pair-bonding or procreation. More generally, humans have been shown to have an involuntary tendency to "mind-read" nonhuman and virtual objects. For example, viewers readily assign intentions and emotions to animations of abstract geometric objects such as moving squares and triangles.[37] Empirical research has also shown this to be the case with music, demonstrating that listeners do in fact have a propensity to hear music as if it were mimicking or representing the actions and intentions of a (virtual) person.[38]

Moreover, in experiments conducted by Ronald Friedman and Christa Taylor at the University of Albany, they found that whether listeners believe some music to be machine-composed or human-composed has no measurable impact on listener emotional responses.[39] In this regard, indifference to the agency status for music parallels the indifference between fictional and nonfictional narratives for story-induced emotions.

The issue of agency leads conveniently to our third question: Why don't sad-music lovers commonly describe their experiences as evoking a feeling of compassion? As already mentioned, I am not aware of a single study of music-induced emotion that points to *compassion*, *sympathy*, or *pity* as commonly reported musically evoked emotions.

Confabulation

The most commonly reported emotions induced by music have been chronicled in a major study conducted by Marcel Zentner, Klaus Scherer, and Didier Grandjean at the University of Geneva.[40] As one might expect, sad-music listeners commonly report feeling sad. However, much more common descriptions are the feeling of *being touched* or the feeling of *tenderness*.[41]

Dictionary definitions for "tenderness" include such synonyms as *gentleness* and *kindness*—words strongly suggestive of a prosocial disposition. That is, tenderness implies positive feelings toward some agent or persona. Similarly, dictionary definitions for "touched" include "tugging at someone's heartstrings." The phrase "feeling touched" seems very close to the concept of compassion.

Another related emotional concept that is frequently encountered in arts-related literature is the concept of *being moved*.[42] The concept is commonly discussed in the aesthetics literature, but it has also received recent scientific attention.[43] "Being moved" has figured prominently in research conducted by Winfried Menninghaus and his colleagues at the Max Planck Institute for Empirical Aesthetics.[44] They conceptualize "being moved" as a mixed but predominantly positive emotion that plays an important role in the enjoyment of tragic art. Importantly, Menninghaus has emphasized the empirical evidence linking "being moved" with prosocial bonding behaviors.[45] At the same time, he has argued that this prosocial function is largely opaque to spectators. With regard to art-elicited instances of "being moved" and prosocial motivations, Menninghaus has written,

> Classical treatises on being moved by art barely ever explicitly speak of such [prosocial] norms. In fact, art-works that explicitly propagate such prosocial norms and self-ideals are often, if not mostly, bad art. We therefore suggest that it may be important for the poetics of being moved that the prosocial implications of this feeling largely escape a conscious representation and are only brought to the fore by scientific analysis.[46]

Although prosocial compassion is a biologically prepared response to sadness signals and cues, I think that listeners unconsciously recognize that "compassion," "pity," and "sympathy" just don't feel like appropriate descriptors for feelings induced by music. Casting around for descriptive labels that don't imply the existence of a sad or grief-stricken persona, we resort to terms like "tenderness," "feeling touched," "feeling moved," and their various synonyms. Such descriptors allow us to sidestep the clash between our involuntary emotional responses and our cognitive rationalizations of what we are feeling—a clash highlighted in the philosophical writings of Stephen Davies.

Research has long established that, when faced with unaccountable feelings, minds are adept at confabulating a story in an effort to make sense of the situation.[47] For listeners uncomfortable with the logical implications, terms like "tenderness," being "touched," or "moved" offer convenient alternative descriptors. We gloss compassion, pity, or sympathy without requiring or implying a human agent or animate target suggested by that emotion. We'll see some further vivid examples of such emotion-related confabulations in our final chapter ("The Meaning of Sadness").

Feeling Sad

This still leaves the question of the feeling of sadness: when listening to nominally sad music, listeners commonly describe their feelings as simply one of sadness. This description is common, both among sad-music lovers and those listeners who dislike sad music.

Curious about the role of "being moved" in sad-music listening, Jonna Vuoskoski and Tuomas Eerola carried out an experiment in which they endeavored to trace the causal connections between sadness, being moved, and enjoyment.[48]

In the first instance, they observed that increased feelings of sadness are correlated with increased enjoyment: felt sadness appears to contribute to the enjoyment of sadness-inducing music. However, a key aspect of their study was a mediation analysis—a statistical procedure which showed that felt sadness contributes to the enjoyment of sadness-inducing music by intensifying feelings of being moved. That is to say, "being moved" was pivotal in order for listeners to enjoy nominally sad music.[49] If "being moved" and "compassion" are synonymous, then the mediation analysis can be interpreted as indicating that the causal train is not sadness evokes pleasure, but sadness evokes compassionate feelings, which in turn evokes pleasure.

Notice that this latter causal scenario is consistent with an interpretation in which the feeling of sadness represents a *contagious* emotion whose purpose is to facilitate deciphering the emotion perceived in others and that deciphering this emotion induces the appropriate *repercussive* emotion (being moved or compassion), which is positively valenced.

Evidently, sad-music listeners commonly experience a state of mixed emotions involving both sadness and being moved or compassion. This suggests that whether a listener likes or dislikes sad music, the experience of sadness is a cognitively salient emotion and so tends to commonly appear in introspective reports of how listeners feel.

In general, it bears noting that people have trouble describing precisely what they feel when listening to music. Moreover, the problem is compounded by the often rapid changes characteristic of music; the mood or tenor of many musical passages can change several times within a few seconds.[50]

Tear-Jerking

There remains the question of spectator tears. What is it about tragic arts that can bring tears to one's eyes or a lump to one's throat? In particular, how can some tearful experiences be enjoyable?

At this point, attentive readers ought to be able to answer these questions for themselves. We have already established that one should not equate displays with their motivating emotions. Weeping is an emotion-related display, not itself an emotion. Tears can be induced by both negative and positive feelings. Whatever causes a person to become teary-eyed, we cannot assume the tears to be necessarily symptomatic of negative affect.

It is possible that the tears induced in a spectator are indeed tears of grief and so negatively valenced. For example, in a theatrical portrayal, we may feel so attached to a protagonist that her demise leaves us in a state of acute distress. Negatively valenced tears can also arise as part of a commiserative response: that is, tears might be shed as part of a contagious mirror response as when we empathize with a grief-stricken protagonist. As we noted earlier, not all arts or entertainment produce pleasurable experiences.

Of course, contagious tears can also arise in response to joyous situations or portrayals. For example, in the case of music, positively valenced tears can be induced by patriotic music or music that evokes strong feelings of nostalgia. It is not just sad music that may bring listeners to tears.

In the case of tragic arts—including sad instrumental music—whether or not we find the experience enjoyable appears to be linked to whether positive feelings such as compassion outweigh negative feelings such as commiseration.[51] As we've seen, our experiences are shaped by individual differences, especially differences in trait empathy. So exposure to the same tragic artwork can induce different responses in different spectators.

By way of summary, even if a tearful listener reports feeling sad, one should not assume that the tears are symptomatic of grief or that the experience of the listener is exclusively negatively valenced. For tearful listeners who report enjoying the music, the likelihood is that their experience of (contagious) sadness is accompanied by high levels of unacknowledged (repercussive) compassion—glossed as "being moved" or "touched"—and that the display of tears (like many of forms of enjoyable weeping) cannot be assumed to indicate a negatively valenced experience.

Reprise

At the beginning of chapter 18, we relayed philosopher Carole Talon-Hugon's informal syllogism encapsulating the apparent paradox of the enjoyment of negative emotions in the arts:

(1) Negative emotions are psychologically unpleasant.

(2) Art can cause us to experience negative emotions.

(3) We find pleasure in this experience.

In light of the scientific research, we can see that the flaw in the logic lies in an incomplete second premise. We might rewrite the logic as follows:

(1) Negative emotions are psychologically unpleasant.

(2) Art can cause us to experience negative emotions. However, tragic portrayals in art can induce a variety of contagious and repercussive emotions in observers. These might include negative emotions (such as disgust, fear, or commiseration) and positive emotions (such as righteous indignation, curiosity, compassion, or appreciation of craft). When attending to a tragic art, the magnitude of the various emotions experienced by a given observer depends on personality traits such as various facets of empathy and on cognitive interpretations—shaped by unconscious concerns of inclusive fitness—such as whether a protagonist is regarded as friend or foe.

(3) Whether a person takes pleasure in a tragic art depends on the balance of induced positive and negative emotions.

As we have noted, unlike real-world tragedies, artistic representations or portrayals of stressful situations liberate the observer from costly obligations to engage in altruistic assistance and absolve the observer from feeling guilty or having to worry about social ostracism for inactivity. Consequently, artistic representations of tragedy typically make it easier for spectators to recognize or acknowledge feelings of pleasure or enjoyment.

Especially in the case of instrumental music, the absence of an obvious stressed character or agent discourages listeners from describing their feeling using socially loaded terms such as pity or sympathy. Consequently, synonyms like "feeling tender," "being touched," or "being moved" are common characterizations. Self-report descriptions are broadly consistent with induced prosocial feelings—such as feelings of kindness, affection, devotion, tugging on one's heartstrings, or feelings associated with suspending aggression—such as feeling peaceful or gentle.

With regard to audience weeping, art can evoke tears in many ways, some positively valenced, others negatively valenced. When positive feelings predominate—such as through feelings of compassion—being moved to tears can be one of the most sublime experiences that art can achieve.

In earlier chapters, we highlighted the difficulties when trying to decipher someone's emotional state. In the previous chapter, we saw that we can encounter similar difficulties when trying to describe our own emotional responses to various tragic arts. We can experience various symptoms and wonder why they've appeared and what they might suggest.[1] In effect, we wrestle with the question: what do our feelings mean?

The research literature is filled with experimental reports documenting how people commonly misconstrue or misattribute their emotional states. A classic experiment in this regard is one carried out by Donald Dutton and Arthur Aron from the University of British Columbia.[2] Male university students were brought to the picturesque Capilano canyon and randomly assigned to cross the canyon via one of two bridges: either a solid concrete bridge or a wood and cable suspension footbridge that wobbles and sways disconcertingly. Having traversed to the other side, participants were then shown a photograph and asked to tell a story related to the picture. As explained to the participants, the ostensible purpose of the research was to study the effect of the natural environment on creativity. The experimenter was an attractive female graduate student.

At the end of the experiment, the female experimenter offered the male participants her telephone number and said they were welcome to call her if they had any questions about the experiment. Dutton and Aron were curious to determine whether the participating men would telephone the female experimenter.

Those participants who had walked across the footbridge were much more likely to contact the experimenter—and ask her for a date—than those participants who had walked across the nearby concrete bridge. In other words, the men who had walked across the vertigo-inducing suspension bridge were more likely to misconstrue their woozy feelings as suggesting they were romantically attracted to the female experimenter.[3]

Such misattributions are not limited to young men encountering an attractive woman. Emotion researcher Lisa Feldman Barrett tells a parallel personal story:

> Back when I was in graduate school, a guy in my psychology program asked me out on a date. I didn't know him very well and was reluctant to go because, honestly, I wasn't particularly attracted to him, but I had been cooped up too long in the lab that day, so I agreed. As we sat together in a coffee shop, to my surprise, I felt my face flush several times as we spoke. My stomach fluttered and I started having trouble concentrating. Okay, I realized, I was wrong. I am clearly attracted to him. We parted an hour later—after I agreed to go out with him again—and I headed home, intrigued. I walked into my apartment, dropped my keys on the floor, threw up, and spent the next seven days in bed with the flu.[4]

Recognizing our emotional states is linked, at least in part, to perceptions of our bodily state—a form of perception referred to as *interoception*. Interoception attends to a range of visceral, proprioceptive, somatosensory, and chemosensory conditions. Some aspects of interoception are accessible to conscious awareness such as when we are aware of our labored breathing, sweating, heart palpitations, or various "gut feelings." But mostly, interoception entails an unconscious monitoring and representation by the brain of our corporeal milieu, such as tracking blood sugar levels.

In endeavoring to recognize our emotions, interoceptive states are typically ambiguous. Most internal states admit multiple possible interpretations. For example, when a person sweats, it could mean that one is experiencing fear or has a fever, or it could be a hot flash of menopause or simply the result of warm weather. The feeling state doesn't come with a preattached label like fear, fever, hot flash, or warm weather.

As in the case of recognizing the emotional states of others, recognizing our own emotions depends a lot on context. If I'm about to appear before a large audience, I'm more likely to interpret my sweaty hands as suggesting stage fright rather than, say, a fever.[5]

The evidence suggests that, over time, we learn to decipher our emotions by combining contextual information with those interoceptive sensations that are accessible to awareness. Consider, for example, the feeling of hunger. A baby periodically experiences stomach muscle contractions triggered by high concentrations of the hormone ghrelin—an unpleasant sensation. However, the infant has no idea of the source of the unpleasantness. Over time, babies learn that feeding will cause the negative sensation to go away. Later yet, a toddler will learn to attach a linguistic label ("hungry") to the particular interoception.

In a sense, perceived bodily states are akin to learned ethological cues: our sensations provide observers (ourselves) with clues that over time we learn to interpret with varying degrees of success. In the same way that we interpret the emotional states of others, when it comes to our own emotions, we must decipher or decode (rather than simply "recognize") what we ourselves feel. In the same way that we should be skeptical of the idea that our bodies express or convey our emotions to others, we should be skeptical of the idea that our bodily responses are meant to tell us what emotion we're feeling.

Labels and Narratives

When interpreting our feeling states, we commonly rely on narratives whose purpose is to explain or account for our feelings. These narratives are typically distilled to simple labels such as loneliness, love, and hunger. Through these labels, we endeavor to give meaning to our experiences.

Semioticians distinguish two aspects of meaning: denotative and connotative.[6] Denotative meaning pertains to the relationship between a sign and an object, as when we define "tears" as "a watery substance produced by the eyes." Connotative meaning is a much broader concept—one form of which is a type of causal story, as when we say "his tears mean their relationship must be on the rocks." These causal narratives sometimes include moral interpretations such as ascribing responsibility, praise, or blame.

If we must learn to interpret our feeling states, then we might expect that some people are better at learning appropriate interpretations than others. And indeed, there is ample evidence that people can differ widely in their "emotional intelligence."[7] An important aspect of emotional intelligence is a person's ability to discriminate or resolve subtly different emotions. Lisa Barrett likens the situation to differences in color labeling. Some people rely on a single color category ("blue"), whereas others discriminate many subcategories (azure, cobalt, ultramarine, etc.). In the same way, some people may rely on a single emotion term ("anger"), whereas others distinguish various shades of anger, such as irritation, scorn, or vengeance.[8] Barrett offers the helpful term "emotion granularity" to refer to the degree of refinement in a person's emotional vocabulary. Research on emotional development in children suggests that emotional granularity tends to become more refined over time. For example, concepts like anger, sadness, and fear typically aren't distinguished from each other until around the age of three.[9]

So what use is an emotional label or story? Why, one might ask, do we bother to engage in labeling and causal narratives related to different affective states?

Causal stories offer at least three potential benefits. First, if we correctly understand the origins of some feeling state, this knowledge often gives us some control over events: when we understand where a feeling comes from, that knowledge often allows us to sidestep situations likely to lead to adverse outcomes or expedite situations that lead to favorable outcomes. Second, even if we are powerless to control a situation, good narratives may help us anticipate the inevitable: a good narrative helps us prepare for the future. Third, good narratives tell us who the villains and heroes are; we know who or what to condemn or celebrate. Especially in situations where we suffer, good narratives can give meaning to our suffering.

The value of interpretive narratives is highlighted in situations where we have no narrative at all. For me, a haunting example is the death of Marilyn, a pianist and wife of a former professional colleague. Marilyn suffered for many years before her death. At one point, she abandoned playing the piano because she found the sounds too painful. Yet medical science (including psychiatry) was never able to identify her condition—even after an autopsy following her early death. The autopsy merely showed that her brain was chock-full of endorphins—endogenous molecules that are commonly released in an effort to ameliorate pain. The absence of a suitable narrative can leave us floundering in a sea of meaninglessness. One might be tempted to say that Marilyn suffered without meaning.

Feeling-related narratives can be constructed from our personal experience, but most of the narratives we use to characterize our feelings we absorb from our surrounding culture. To the extent that we share a common humanity, people share many feeling states in similar situations, and consequently, different cultures often construct similar narratives. At the same time, cultures often diverge, producing different labels and narratives for seemingly identical states, such as the Tahitian state of "fatigue" used for situations that a Western-trained therapist might describe as either melancholy or depression.

However, one doesn't need to compare different cultures in order to see diverging interpretations. The effects of different narratives can be seen by looking at historical changes within a single culture. In her work on English Renaissance conceptions of sadness, historian Erin Sullivan chronicled two contrasting interpretations of sadness prevalent in England in the sixteenth and seventeenth centuries. One prominent conception dominated contemporary medical thinking. For medical practitioners, melancholy was regarded as a consequence of an excess of black bile (the literal origin of the word "melancholy"). This medical view regarded sadness as the result of a humoral imbalance that, if prolonged over time, could produce a variety of physiological and psychological disorders. This interpretation is captured in the contemporary slogan: "Sorrowes to men diseases bring."[10]

By contrast, among sixteenth- and seventeenth-century English Protestant theologians, sadness was regarded as symptomatic of piety. Weeping during a religious conversion was an auspicious sign. Moreover, a devotee who was prone to persistent sorrow testified to the humbling of the spirit and a sensitivity to the sinfulness of the world: "Godly sorrow causeth (in us) repentance unto salvation."[11]

In summary, for English medical practitioners of the period, all forms of sadness were regarded as pathological, whereas their spiritual compatriots viewed sadness as potentially symptomatic of divine grace. Depending on which narrative was adopted, a chronically "sad" person could feel either medically imperiled or heartened by the prospect of imminent salvation.

As another example, consider the sickness known as the Black Death. During the height of the Black Death in Europe (1346–1353), millions of people experienced the depths of fear, despair, and grief. How does one interpret this horrific scourge and the feelings it evoked?

At the time, many believed that the pandemic was a punishment delivered by God for their sins. Others regarded its origin as a conspiracy enacted by Jews who, among other accusations, were thought to have poisoned wells. In 1349, the Jewish communities in Cologne and Mainz were destroyed and some 2,000 Jews were massacred in Strasbourg. By 1351, over 200 Jewish communities had been attacked. Across Europe, people with skin diseases were murdered, including lepers, and even people suffering from acne. Today, we have a different perspective.

The benefits of explanatory narratives are maximized only if the narrative has merit—that is, where the narrative captures some reliable relationship in the world. In the case of the Black Death, our modern interpretation is that it was a high-mortality infection caused by the bacterium *Yersinia pestis*. Our modern narrative allows us to respond to such a situation in a manner that better serves human well-being. Indeed, the bacterium that caused the Black Death continues to circulate to this day, but it hardly kills anyone because we rely on different cultural narratives about its origin and how to deal with it. A person today who is infected with the plague is less likely to blame Jews, consider it a punishment for the sinfulness of the world, or regard it as testifying to their own moral depravity and more likely to appreciate that a simple antibiotic will bring about a cure. Many of our modern narratives are not merely different. They are better insofar as they improve personal well-being, reduce unnecessary anxiety, and better protect the innocent.

Sad Stories

This book has presented a series of sadness-related explanatory narratives. Specifically, I've presented narratives for three distinct states: grief, melancholy, and nostalgia. Coincidentally, I've also offered interpretations of depression and sickness. The narratives presented in this book are products of a particular time and place. Science is a perpetual work in progress, not a completed project. So one shouldn't assume that my stories are necessarily the right ones. However, to the extent that science can be a progressive enterprise, over time, ongoing research is likely to help us better understand our experiences.

Improved narratives ultimately lead us to more accurate self-conceptions. Suppose I have a feeling state that might be described as consistent with either melancholic sadness or depression. My symptoms might include feelings of lethargy and malaise or unhappiness. I could interpret my state as signifying many possibilities:

For example, I could conclude that I am feeling melancholic and that this state arose because I misplaced my wallet. When interpreted as melancholy, I might conclude that I am simply the unlucky victim of a temporary misfortune and that my bad feeling is likely to pass in the near future. Alternatively, I could interpret my state as merely fatigue, suggesting that I have been working too hard. When interpreted as fatigue, I might conclude that perhaps a good night's rest will prove sufficient.

If my family has a history of depression, I might regard my sadness as symptomatic of a depressive pathology that has no particular significance apart from implying an unfortunate genetic pedigree. I might conclude that I ought to seek psychiatric intervention, including possible prescription medication. If I conclude that I am afflicted by nostalgia, then I might consider changes in lifestyle such as moving closer to friends, purchasing uplifting memorabilia, or other lifestyle changes that might improve the quality of my life.

If I am experiencing spontaneous weeping in circumstances that would normally warrant no emotional response, there is a good chance that I am suffering from pseudobulbar affect. Such a condition would recommend not being tested as a possible witch but instead being scanned for evidence of a stroke.

It helps to know the likely source of one's feelings. If the cause of my malaise is that I've misplaced my wallet, a regimen of Prozac won't help. However, at the same time, it must be acknowledged that empirically weak narratives can sometimes have a palliative value. A sixteenth-century individual suffering from what we would deem chronic depression, without the aid of modern psychotherapy or medications, might well have been better off believing that their relentless sadness originated in their deep awareness of the sinfulness of the world and that their lamentable condition potentially marked them as a fortunate recipient of divine grace.

How we interpret our feelings can dramatically affect both our long-term prognosis as well as our phenomenological experience. As we have noted, when our interpretations are based on an accurate understanding of the world, they help us to better understand ourselves and have the capacity to increase our well-being—not to mention the welfare of those around us.

Cognitive Mediation

In understanding our emotional states, words provide us with useful labels on which to hang our sensations and attribute meanings. Words often facilitate forming concepts and, in the case of emotions, can assist in increasing emotional granularity. However, one must be wary of placing too much emphasis on words. The absence of a denotative emotional term in some language does not necessarily mean that people who speak only that language don't experience the associated emotion. There is no English equivalent of the German *Schadenfreude*, but that doesn't mean English speakers don't sometimes take pleasure in the misfortunes of people they dislike. Similarly, the Japanese *age-otori* has no English equivalent, but English speakers will have little difficulty identifying with the feeling of irritation and disappoint-ment that comes from looking worse after a haircut.[12]

Most historians are aware of the dangers of placing too much empha-sis on words. Just because the word "nostalgia" wasn't coined until 1688 doesn't mean that people didn't experience nostalgia before the seventeenth century. Similarly, in the psychology of emotion, it's important not to pre-sume that one's ability to experience a particular feeling state depends on one's vocabulary.

That's not to say that concepts or narratives don't shape our overall feel-ings. Indeed, they do. For an ambitious businessperson, a feeling labeled "envy" might spur them to work more assiduously. However, for a devout Catholic, an experience deemed "envy" might spur them to be more vigilant to suppress such sinful feelings—since, in Catholic theology, envy is one of the seven deadly sins.

When we attach a label to a feeling state, it does draw the feeling into a particular narrative. Because of the cultural origin of most narratives, it's tempting to regard the experience as a psychological or social construction. The word "construction" is something of an exaggeration because our narra-tives are constrained by the specific behaviors a feeling commonly induces. Fleeing, for example, might arise from fear, shame, or an urgent call of nature. But it's harder to accommodate the fleeing behavior in narratives like nostalgia or loneliness. Nothing is created from scratch, as implied by the word "construction." We assemble much of our mental world, but not just as we please.

At the same time, the narratives we use to characterize our feeling states are not merely interpretations. The word "interpretation" suggests that the experience itself is fixed and that we are merely construing that experience in a particular way. The word "interpretation" fails to accommodate the fact that how we think of an experience can profoundly shape the experience itself. In short, it's an exaggeration to say emotions are cognitively constructed, and it's an understatement to say that emotions are cognitively interpreted.

A better term for characterizing the effect of cognitive labels and narra-tives is, I propose, *cognitive mediation*. What we think influences our emo-tions, while our emotions simultaneously influence what and how we think. But these relationships are not boundlessly malleable.

By way of summary, we can distinguish at least four links between cogni-tion and emotion—between thoughts and feelings. First, emotions can be evoked simply from thought alone. For example, as noted earlier, a moment of alarm might be induced simply by asking whether you forgot your mother's birthday. Second, cognition plays a central role in executive control. In earlier chapters, we explored how we may choose to amplify, inhibit, mask, or emulate various emotional displays depending on our assessment of the situation. Executive control also shapes how we react to the signals produced by others. Third, our moods can influence how and what we think. Such influences are often regarded as incidental artifacts—inadvertent biases. However, as we've seen in the case of melancholy, the main function of this saddened mood appears to be precisely one of shaping more prudent or realistic thought. Indeed, according to the theory proposed in this book, if melancholic sadness fails to change how we think, it has failed to achieve its biological purpose. Finally, in this chapter, we've broached the slippery issue of meaning. One source of meaning is to be found in the stories we tell ourselves about our experiences. Cultural narratives can shape how we feel about our feelings, and in this chapter we've noted how culturally embedded causal accounts can sometimes help us better cope with a particular situation.

Ultimately, from a biological perspective, what counts is not what we think but how we behave. Of course, thinking is important insofar as it influences our behaviors. Nevertheless, it bears reminding that much feeling-induced behavior arises without us giving much thought. Opening a food container that has spent far too long in my refrigerator, a smell-induced feeling of disgust can be evoked without the involvement of any thought or cognitive assessment. Moreover, cultures don't always provide labels or narratives for a given feeling state. There are many reasons why we may find tears welling up in our eyes or experience a choked-up feeling. This might include witnessing an act of kindness, observing an infant's smile, watching a sunset, or hearing a kitten's mews. We are easily at a loss for words and commonly resort to vague descriptions, such as "being touched" or "being moved."

So why, one might ask, do we have such trouble deciphering many of our emotions? Once again, feeling states exist to change our behaviors, not to inform us what they signify: they tell us what to do, not what they mean.[13] The opaque character of our feeling states very likely reflects the evolved origin of many or most emotions. Nonhuman animals probably experience vivid feeling states like fear and hunger. For most (perhaps all) animals, there is likely little conscious awareness or thought involved in how they respond: the feelings are there to motivate normally adaptive actions, not to convey to the animal some concept or meaning.[14] Emotions can still function in appropriate ways, even for the person who has no language abilities—and so no emotion-related narratives or labels.

When our emotion-related narratives are grounded in reality, how we conceive of our emotions can be enormously beneficial in helping us deal more effectively with an emotion-inducing situation. But the thoughts themselves rarely constitute the emotion itself. The secondary role of cognition should not be surprising since the ability to think about emotions is a biological latecomer. Our ancestors surely experienced emotions long before we had language.

Proximal Living

We live in a physical world of intensely competing chemical patterns that are the ultimate foundation for evolution by natural selection. Yet our subjective experience has us living in a world of people and objects, social networks, and ineffable feelings. The two worlds are intimately connected; indeed, the second world is an astonishing emergent property of the first. But as conscious beings, the world we experience is the latter, not the former.

Although our feelings are broadly shaped by biological imperatives, our emotional experiences are confined to the realm of proximal feelings. As noted earlier, while love may simply be Nature's way of encouraging pair-bonding and procreation, the feeling of love is no less profound an experience despite its prosaic origin and function.

When we weep, our subjective experience is not one of manipulating others to our benefit. Although people do, from time to time, shed overtly manipulative "crocodile" tears, our tearful experiences are typically associated with proximal feelings such as being moved or touched, deep feelings of affection or joy, feelings of connection, or profound feelings of loss arising from an unhappy human condition. The tears we shed do not feel like shrewd tricks for commandeering resources from hapless bystanders.

Similarly, when we observe others weeping, our phenomenological experience is not one of gleefully helping others, confidence that they are incurring a great debt to us that may be repaid later, or that our social reputation will receive a welcome boost. Instead, our experience is one of expressing genuine compassion and sympathy, often accompanied by powerful feelings of social connection. The unbidden automaticity of these feelings contributes to the sense of sincerity, intimacy, and vulnerability—transforming melancholy and grief into intense human experiences infused with meaning and value.

Sociobiological theories such as those offered in this book have a long history of poor reception by the general public. It is not simply that self-centered efforts to maximize biological fitness are offensive when viewed from common moral standards. The accounts themselves simply do not accord with our subjective experiences. In our lived social world, sickness is about weakness and vulnerability, nostalgia is about a lost cherished past, melancholy is about worry and regret, and grief is about helplessness and compassion.

The Cooperative Species

Throughout this book, I have characterized signals as forms of manipulation rather than information. Unfortunately, the word "manipulation" holds unduly negative connotations. Signals evolve only when they benefit both the signaler *and* the observer, so the "manipulation" typically benefits both parties. Said another way, all signals are ultimately cooperative. Even if a signal isn't overtly prosocial (as in the case of a threat display), the interaction is mutually advantageous. Indeed, it is the criterion of mutual benefit that allows a signal to evolve in the first place. This explains why the most sociable animal species are those with the richest repertoire of signals.

As a species, *Homo sapiens* has been extraordinarily successful. It has often been noted that our success is attributable to our impressive capacity for cooperation. Harvard biologist Edward Wilson spent a lifetime studying social insects. He never ceased to be amazed by the eusocial character of various ant, wasp, and termite species. In his book, *Genesis: The Deep Origin of Societies*, Wilson drew attention to the fact that *Homo sapiens* exhibit the most eusocial tendencies of all primate species. We are cooperative to an extent that is not merely rare among primates but exceptionally rare among animal species in general. In fact, few other vertebrate species come close to the cooperative social disposition evident in our species.[15]

In the end, signals testify to our remarkable cooperative disposition. Weeping would not exist if we weren't ready to offer help to those in need. We are so disposed to help others, we respond not merely when others effectively request our help (through weeping), but we seek out opportunities to help in the absence of overt requests—as in our hypersensitivity to the wholly ambiguous melancholic cues of others.

Our species is named *Homo sapiens*, a designation intended to draw attention to our intelligence. We are, to be sure, a smart lot. But our success is mostly due to ratcheting that intelligence through cooperation. A more appropriate designation for our species might well be *Homo cooperans*.[16]

Ultimately, we best understand signaling behaviors when we view them from a cooperative perspective. This applies not merely to weeping but to surprise, fear, laughter, smiling, and all the other human signals. Even displays of disgust, screaming, and threat exist as behaviors that ultimately benefit both observers and displayers. Altogether, they form a remarkable palette of behaviors that have contributed enormously to our success as a species. It is a striking and inspiring story, part of which I hope to have conveyed in this book.

Acknowledgments

Few things are more valuable for an author than receiving good advice. Especially in interdisciplinary endeavors, it's essential to seek the guidance of knowledgeable people. As readers will appreciate, the subject matter for this book draws on many fields—psychology, physiology, neuroscience, evolutionary biology, animal behavior, medicine, emotion research, social psychology, and cultural anthropology—all in aid of understanding sadness. In the realm of psychology alone, I have benefited from the critical feedback from scholars working in cognitive psychology, developmental psychology, social psychology, and clinical psychology.

My most sincere thanks to friends and colleagues who read earlier drafts and offered helpful suggestions and encouragement: Tuomas Eerola, Joe Forgas, Sandra Garrido, John Hoey, Heather Lench, Susan Lingle, Da Lin, Tanya Little, Michael Potegal, Kristin Precoda, Lindsey Reymore, Emery Schubert, Daniel Shanahan, Ofer Tchernichovski, Michael Trimble, and the late Sandra Trehub.

Above all, I am most indebted to three colleagues who spent many hours thoroughly engaging my text and offering extensive feedback. My sincere thanks to Swiss-Norwegian psychologist Rolf Reber, to Italian philosopher of emotion Federico Lauria, and to Dutch crying expert Ad Vingerhoets. My heartfelt gratitude to Rolf, Federico, and Ad for their careful readings and valuable counsel.

Postscript: Why Is Emotion Theorizing Such a Mess?

In the preamble, I noted that research on human emotions has been in turmoil for more than a century. Actually, it's not so much the experimental research that has been the source of chaos as the *theorizing* about emotions. In her recent critical history of the field of emotion research, philosopher Ruth Leys quite correctly notes that "there is no consensus regarding the science of emotion's most basic assumptions."[1] In this postscript, I offer not a theory of emotions but a theory of why emotion theorizing is so hard and why there is so little agreement about the most basic concepts or principles. I also offer a modest proposal for future research efforts.

In attempting to understand emotions, it's helpful, I would suggest, to consider human organs (like stomachs and gallbladders). Over the centuries, anatomists identified and described the many organs of the body, such as the liver, heart, and pancreas. Accompanying the descriptive enterprise, many efforts were made to define "organ." Despite the various efforts, there is no widely accepted definition. Indeed, most biologists have concluded that it's impossible to define "organ."[2]

Anatomists and physiologists have also made various efforts to *categorize* different types of organs. One could classify organs according to their size, color, or texture, but such groupings aren't helpful. We'd prefer categories that reflect similar functions or similar mechanisms. Today, there is only one classification scheme that is accepted by anatomists—the distinction between so-called solid organs (like kidneys and livers) and hollow organs (like intestines and bladders). That's it. That's the one accepted classification scheme anatomists use for organs. The sad truth is that there really isn't a good "theory" of human or animal organs.

The source of the problem is that the organs are so very different from one another. Each organ serves a unique purpose or set of functions. What the heart does is very different from what the pancreas does. Spleens, ovaries, lungs, and brains are just so very different from one another. Are the spinal cord, pineal gland, prostate, and tonsils organs? What do the various organs share in common?

All organs are made of cells, although that's also true of toenails. Organs exhibit relatively well-defined boundaries or borders, but these can also be observed in cancerous tumors. Organs all receive circulating blood, but that's also the case with fat tissue and hemorrhoids.

When it comes to *function*, there is only one single function all organs share: they all contribute to keeping an individual alive. However, their contributions are all unique. Some organs secrete stuff, some extract stuff, some move stuff around, and some process information. Most organs do some combination of things.

The point is, the body consists of an assortment of complicated parts. It's hard to identify something they all share or even to group them into categories that provide some useful classification.

I propose that this is also the case with what are generally referred to as "emotions." When we try to identify the shared common features that define emotions, we face the same problem: we are trying to impose some logical scheme onto a bunch of things that are strikingly diverse. That's not to say that there aren't some features that are shared by some emotions. The problems arise when attempting to find features shared by all emotions—something that might warrant the use of a single collective term like "emotions."

As I noted in the preamble, if one defines "emotion" in a particular way, then it's possible to exclude all the embarrassing states that don't conform to one's intuitions. As readers of this book will recognize, my own inclination is to cast a wide net that includes any distinctive feeling state. As already mentioned, the reason why I prefer such an inclusive perspective comes from my research experience in the field of music-induced emotion. Most of the distinctive feeling states that listeners commonly experience when listening to music simply don't appear on any usual list of human emotions. Of course, "feeling state" already amounts to a definition. In this regard, I accept a common "folk" view—what Paul Griffiths calls the "feeling theory of emotion"—a perspective that links emotions to subjective feeling states.[3]

The ostensible problem with emphasizing "feelings" is that it includes many states that don't conform to the modern usage of the English word "emotion." Feelings embrace all sorts of experiences, from feeling playful, sleepy, or sexually aroused to boredom, confusion, and the pleasure of some moment of insight. So what about other ways of characterizing emotions?

In the case of Ekman's basic emotions, an important shared property is the *visibility* of a shared facial display. Joy, anger, fear, surprise, disgust, and grief are associated with distinctive facial configurations. However, in the feeling theory of emotion, plenty of states span the complete range of "visibility." At one extreme, we have feelings of grief and disgust that may be easily observed. At the other extreme, we have feelings of hunger and nostalgia that are entirely invisible. In between, we have states like sleepiness and melancholy that are partially observable.

Well, what about the role of *cognitive* involvement? At one extreme, we indeed have feelings that are purely cognitive, such as curiosity, confusion, and pride. At the other extreme, however, we have feelings that are entirely noncognitive, like thirst, itchiness, and smell-induced disgust.

Another way emotions differ is whether associated displaying behaviors are *volitional*. At one extreme, we have states like blushing that are impossible for us to produce willfully. At the other extreme, we have no difficulty producing a voluntary smile or frown. In between these extremes, there are displays that are partially voluntary. For example, we can emulate nearly all aspects of weeping with the crucial exception of tears—which cannot be willfully generated.

Emotions also differ with regard to their *social* effect. At one extreme, we have displays like weeping that predictably induce feelings (such as prosocial compassion) in most or nearly all social observers. At the other extreme, we have displays like blushing that appear to have no specific ability to induce a feeling state in observers.

Even what we regard as a single emotion can sometimes exhibit quite different manifestations. Such *variability* is evident in the two classic forms of anger: "hot anger" (associated with a threat displays) and "cold anger" (felt but not expressed or evident).

Emotions certainly differ in their strength or *intensity*. Some feeling states (like pain, anger, disappointment, or ecstasy) can be highly vivid. Other states (like suspicion or hope) may be experienced as rather subdued feelings. Some (like amusement) can span the complete range of intensities—from muted giggling to hysterical guffaws.

Emotions also differ with regard to their *duration*. Some emotional states (like disgust) are typically short-lived, sometimes lasting only a couple of seconds. Other emotional states (like affection) typically extend over a much longer period.

Most emotions have a recognizable *function*. But not all emotions. For example, one of the most commonly induced strong emotions when listening to music is the experience of *frisson*, where the listener experiences pleasurable chills or shivers. It's not at all clear what the function of music-induced frisson is, if any.

Feeling states also differ considerably in their *mechanism*. As noted in the preamble, some feeling states (like feeling cold) involve their own special-purpose system of sensory neurons. Others, like satiation, are known to be linked to specific hormones. An emotion like nostalgia is intimately linked to memory. And fear is linked to particular brain circuits.

Emotions also differ with respect to their *contagiousness*. Yawning (a display associated with feeling sleepy) is highly contagious. Joy, anger, and nausea, a little less so. Feelings of sadness are somewhat less contagious. And feelings of loneliness and jealousy exhibit little contagiousness.

Some emotions act as immediate *action motivators*, as when the feeling of hatred increases the likelihood of striking out. But other emotions motivate changes in how we think—such as the feeling of loneliness, suspicion—or melancholy. These latter emotions can ultimately influence behavior, but not directly.

Some emotions have clear *phylogenetic* connections. For example, we have feelings like fear and itchiness that we share with other animals and that probably rely on similar or identical physiological foundations. At the other extreme, we have feelings like shame, which appears to be unique to humans.

Finally, some emotions are *culture* specific, such as the Tahitian notion of *haumani* and the Portuguese feeling of *saudade*. Other examples include the Tagalog notion of *gigil* (the irresistible urge to pinch or squeeze someone because they are cute or cherished). Moreover, cultures sometimes differ with respect to emotional susceptibility or *sensitivity*. For example, in some cultures, people more easily take offense or feel insulted than in other cultures.

So "emotions," at least when defined within the "feeling theory of emotion," differ in their physiological mechanisms; their intensity and duration; their contagiousness; whether they are private or public; whether they are observable or covert; the degree to which an associated display is generated voluntarily or involuntarily; whether or not a display influences the behavior of observers; whether they have some phylogenetic analog in other animals; their degree of universality, cultural specificity, or cultural sensitivity; and whether they serve some function and, if so, whether that function is behavioral or cognitive.

I propose that the reason why emotion theorizing is in such disarray is that the category "emotion"—like the biological category "organ"—turns out to be a potpourri of diverse states. Each emotion has a unique origin and purpose that makes it quite unlike the other emotions. Apart from these phenomena being associated with distinctive subjective feeling states (which in this discussion they admittedly share only by virtue of definition), there is little else that all of these various states share. Like the concept of "organ," I expect that researchers will never ultimately come to agree on a definition of "emotion." My point is that the very diversity of feeling states makes it virtually impossible to generalize.

Conclusion

I propose that the way out of the "emotion theory mess" is to pay closer attention to the individual emotions themselves. We need more detailed descriptions of individual feeling states like feeling cold, tasting something disgusting, getting angry with someone, experiencing something really cute, or understanding why a musical chord might send shivers up the spine of a listener. Like physiologists trying to make sense of various organs, it helps to focus one's research on the specific operation of individual parts. Nor should we restrict ourselves to whiz-bang feeling states. Human behavior is shaped by many more subdued feelings (like the faint feeling of irritation from static posture discussed in the preamble).

In the process of analyzing individual feeling states, we are sure to ultimately recognize various features that are shared by several "emotions." My hope is that my STEAD theory represents an example of such limited encapsulated theorizing that draws attention to commonalities that can indeed be observed in a number of emotional states but in the full knowledge that these commonalities are not shared by all emotions.

My modest proposal here is simply to suggest that researchers abandon the task of defining emotions, recognize the huge diversity of human feeling states, and aim to produce comprehensive bio-psycho-sociocultural accounts of individual feeling states or feeling-related behaviors. I think the future of emotion research is rosy as long as we give up trying to defend one narrow conception of emotions against other equally narrow conceptions of emotion.

Appendix: The STEAD Theory

A conceptual foundation underlying this book has been my signal theory of emotion-associated displays (STEAD). The following numbered summary gathers together the main theoretical claims and assumptions. As in our summary of the TTIRS theory in chapter 17, the aim is to clarify the logic, help expose weaknesses, and provide a guide for possible future empirical tests.

For knowledgeable readers, the STEAD theory might be regarded as a mixture of Paul Ekman's neurocultural theory, Alan Fridlund's behavioral ecological view, and Lisa Feldman Barrett's psychological constructivism. Like neurocultural theory, I conceive of most emotion-related displays as innate adaptive phenomena that are shaped by executive control whose influence also reflects sociocultural conventions. Like the behavioral ecological view, I am inspired by the work of Krebs and Dawkins, who cogently argued that animal communication is motivated principally by manipulating observer behavior rather than informing observers. Like psychological constructivism, I regard language and cultural scripts as playing a major role in mediating our affective experiences.

Unlike neurocultural theory, although displays are motivated by displayer emotions, I consider most displays as conveying no particular emotion, that the concept of "basic" emotions conflates emotions and displays, and that most purported basic emotion displays are, in fact, ethological signals. Unlike the behavioral ecological view, I am inspired by ethological signaling theory beyond the work of Krebs and Dawkins. Specifically, I regard the signal/cue distinction as pivotal and that signals are largely immune to deception since they evolve and are sustained because they typically benefit both displayer and observer. Also, unlike Fridlund's behavioral ecological view, I regard feeling states as critically important since the main function of emotion is to motivate behavior.[1] Unlike psychological constructivism, I believe many emotions are indeed "natural kinds," that language is not necessary in order to experience emotions, and that nonhuman animals and infants are perfectly capable of experiencing emotions.

The logic underlying the STEAD theory is presented below.

1. *Ultimate versus proximal causation.* In understanding behavior, it is essential to distinguish ultimate from proximal causation. For example, the ultimate goal of food consumption is to provide metabolic energy; however, the proximal motivation to eat is found in the combination of push and pull feelings—namely, feelings of hunger and the gustatory pleasure of eating.

2. *Proximal motivations.* Proximal motivations are commonly manifested as vivid feeling states, variously described as affects, emotions, sentiments, passions, drives, feelings, or moods. We are more likely to engage in certain behaviors in the presence of such feelings than in their absence. For example, we are more likely to help someone in the presence of a feeling of compassion than when compassion is absent. The word "passion" helpfully captures the joint ideas of both a distinctive feeling state and the impulse to engage in particular activities.

3. *Emotion.* For convenience here, we will use the word "emotion" to broadly include any feeling state that motivates behavior.

4. *Consequential emotions.* Emotions are most intense in those circumstances where the motivated behaviors have the greatest impact on fitness. This includes feelings related to safety, injury, illness, food, procreation, parenting, friendship, betrayal, and social status.

5. *Emotional valence.* Emotions are phenomenologically positive or negative. Positive emotions are states that a person wants to repeat or continue, whereas negative emotions are states that a person wants to avoid or end. In animal behavior, positive emotions are associated with advance or approach, whereas negative emotions are associated with avoidance or withdrawal.

6. *Future-oriented conditioning.* In addition to the role of shaping immediate behavior, emotions can also serve a long-term conditioning function, teaching us to avoid or seek out certain circumstances in the future. When experiencing pain, for example, a negative feeling state encourages us to remedy the immediate situation but also provides

a long-term lesson that sensitizes us to avoid similar future potential pain-inducing circumstances.

7. *Emotion duration.* Emotions can be short-lived (such as a momentary feeling of disgust) or long-lasting (such as a sustained or recurring feelings of loneliness). Different time spans are echoed in language, such as distinguishing brief *emotions*, daylong *moods*, and lifelong changes of *temperament.* The duration of an affective state is typically tailored so as to encourage those behavioral changes best able to enhance overall fitness. For example, dealing with a disgusting smell can typically be done more quickly than dealing with loneliness. Long-term changes of temperament often arise from traumatic experiences that may make a person perpetually vigilant or wary of certain situations, or perpetually attracted to various behaviors or substances.

8. *Emotional intensity.* Emotions also differ in their intensity. Some emotions (like pain or anger) can be overpowering and compelling. Other emotions (like melancholy or suspicion) can be subdued or muted in intensity. As in the case of duration, the intensity of an affective state is typically tailored so as to encourage those behavioral changes best able to enhance overall fitness.

9. *Polar emotions.* Emotions sometimes form oppositional pairs. For example, feelings of hunger encourage us to eat, whereas feelings of satiation encourage us to stop eating. Feelings of wanderlust encourage us to be adventurous, such as when seeking a better life; feelings of nostalgia encourage us to return when that better life proves elusive.

10. *Anticipating behaviors.* The behaviors of other people commonly impact our own well-being. Anticipating the behaviors of others can help us minimize harmful consequences or maximize possible opportunities.

11. *Unpredictable behaviors.* Our actions are often less effective when others are able to anticipate them. For example, if someone is able to anticipate our anger-motivated attack, that person would benefit by running away or by striking first.

12. *Predictable behaviors.* Sometimes it is useful for others to be able to predict our behaviors. Such circumstances can arise only in situations of mutual benefit, where the response of the observer increases the fitness of both individuals. An example would be collaborative hunting efforts.

13. *Recognizing emotions.* Since many behaviors are motivated by emotions, one of the best ways to anticipate someone's behavior is by correctly recognizing or inferring their emotional state. As a corollary, our behaviors are less predictable if others are unable to recognize or infer *our* emotional states.

14. *Inferring emotions.* There are two ways we might recognize the emotional states of others. First, it is possible that someone explicitly communicates their emotion to us. Second, we might successfully infer a person's emotional state through sensitive observation and deduction.

15. *Displays.* Theoretically, apart from telling someone how you feel, one might suppose that a person's emotional state might be communicated through some distinctive "display." Ostensibly, examples of possible candidate displays might include smiling, laughing, sneering, yawning, rolling one's eyes, sticking out one's tongue, and so on. One might expect that some displays are culture specific, whereas others can be observed cross-culturally. Some displays might be willful, whereas others are spontaneous.

16. *Inconspicuous emotions.* Most emotions or feeling states have no associated display—the emotions are invisible or covert. Examples might include regret, pride, loneliness, cold anger, nostalgia, wanderlust, suspicion, indignation, awe, boredom, contentment, compassion, and melancholy, as well as such basic feelings as hunger and thirst. Some states might be inferred by attending to subtle clues such as recognizing that someone is hot (by observing sweat) or cold (by observing shivering).

17. *Context.* Inconspicuous emotions can often be inferred by attending to the situation or behavioral context. For example, seeing someone running away from a bear, we might infer that they are experiencing fear; seeing someone running toward their recently returned child, we might infer that they are experiencing joy. Running by itself tells us little about a person's emotional state.

18. *Inducing observer emotion.* Apart from anticipating the behaviors of others, it is especially valuable when we can induce others to behave in ways that are beneficial to us. Since emotions motivate behaviors, a useful way to encourage others to behave in a particular way is by inducing in them an appropriate emotion.

19. *Resistance and acceptance of influence.* We should resist efforts by others to influence our emotions if the resulting emotion motivates us to

behave in a way that is detrimental to us. Conversely, we should welcome efforts by others to influence our emotions if the resulting emotion motivates us to behave in a way that is beneficial to us.

20. *Inducing behavioral change.* There is no point in explicitly communicating one's emotion if it doesn't ultimately influence the observer's behavior. Of course, the desired behavior might not occur immediately. Instead, the communication might lead to a change of attitude that could influence an observer's future behavior.

21. *Minimum criteria for emotion display.* Explicit communication of one's emotion would be expected to occur only if it commonly leads to observer behavior that benefits the displayer.

22. *Mutual benefit of induced behavior.* Accordingly, overt emotion-related displays would be expected to arise only when they induce an emotional state in the observer that typically leads to observer behavior that benefits both individuals.

23. *Reliable transactions.* Since such interactions benefit both individuals, selection pressures will tend to shape the emotion-related display so it is communicated clearly and shape the observer's emotional response so it becomes reliable. The display is typically more conspicuous when it involves more than one sensory modality, so displays are not limited to visual features. However, a given display may exclude sounds, smells, or visual elements depending on their potential for attracting predators.

24. *Signal.* Over time, such mutually beneficial interactions can be made more efficient and reliable by rendering some or all of the component behaviors innate or instinctive. At this point, we may refer to the display using the ethological term—*signal*.

25. *Emotion-motivated signals.* Signals frequently arise in response to an emotion experienced by the signaler. The emotion originates in some threat or opportunity that is best addressed by recruiting the cooperation of an observer.

26. *Without audience.* To the extent that a display is spontaneous or innate, a display can arise in the absence of an observer.

27. *Plausible human signals.* Examples of likely human signals include smiling, pouting, bawling, weeping, laughing, screaming, disgust, sneering, threat, blushing, and surprise.

28. *Signal stability.* If a signal was commonly used to deceive an observer—that is, to cause the observer to behave in a way that is detrimental to

the observer—then selection pressures would, over time, lead to the extinction of any innate responding tendency. Consequently, the very existence of a signal testifies to its general stability and the fact that it is rarely used for deceptive gain.

29. *Signal recruitment.* If a signal is able to induce a predictable emotion in an observer, then it should be possible to recruit that signal to induce that emotion in situations other than the one in which the signal originally evolved. As long as the new situation continues to benefit both individuals, the signal will continue to exist due to positive selection pressures. For example, although weeping might have evolved in the context of grief, the pertinent prosocial emotion can be induced in an observer by weeping in other circumstances, such as in tears of loyalty or affection. Most mature signals reach this polygenic state.

30. *Signal–emotion dissociation.* Over time, a mature signal is likely to become divorced from the emotion that originally engendered it. At this point, the signal no longer allows the observer to reliably recognize which emotion led to the signal. Those displays that conform to an ethological signal cannot be presumed to communicate one specific emotional state (as supposed in point 15 above).

31. *Emotion recognition: Signaler versus observer.* Although a signal might arise from several possible emotions, as noted, the emotion induced in the observer tends to remain stable. Consequently, it follows that the emotion that is most reliably predicted by a mature signal is the induced emotion in the observer, not the motivating emotion of the signaler.

32. *Altruistic benefits.* Helping someone may incur costs in terms of time and resources. However, helping others can benefit the helper in at least six ways: via kin selection, where assisting one's closest relatives can facilitate propagating shared genes; through reciprocal altruism, where the person helped incurs an implied debt to potentially assist the altruist in the future; by enhanced social status for the altruist with respect to the person helped; through enhancing one's social reputation (among immediate spectators and to a larger audience via gossip); by avoiding social ostracism for failing to act; and by building a social alliance with the helped individual.

33. *Proximal altruistic motivations.* Altruistic assistance can be directed at both people who are in an obviously stressful state, as well as people

who are not immediately troubled or in dire need. Altruistic acts can be motivated six proximal feelings: (1) anticipating the feeling of virtue or enhanced self-worth, (2) anticipating the feeling of bonding warmth—connecting with the person being assisted, (3) the allure of enhanced social reputation (for those who may observe or hear about our altruistic acts), (4) the positive feeling of compassion, (5) anticipation of the feeling of reduced commiseration, and (6) fear of social ostracism for failing to act.

34. *Observer-initiated transactions.* Since altruistic behaviors often benefit the altruist, one need not simply wait for requests for assistance in order to act. If an opportunity to be helpful presents itself, there may be benefits to engage in helping behaviors even in the absence of an overt signal. Indeed, one ought to be vigilant and seek out opportunities to help others when the benefits outweigh the costs. Consequently, most altruistic acts do not arise in response to a signal such as weeping or bawling.

35. *Handicapped signals.* In general, signals benefit both displayer and observer so there is no need to ensure the honesty of the signal through a handicap. Nevertheless, a signal may be made more conspicuous or the effect of a signal may be amplified by associated handicap efforts. An example of a handicap would be an act of self-injury in conjunction with grief-induced weeping.

36. *Cues.* Apart from overt signaling behavior, information is commonly conveyed as a by-product of other behaviors such as sweating or breathing. We may refer to such inadvertent information using the pertinent ethological term—*cue*. Examples of human cues include walking, running, sleeping, scratching, eye blinking, squinting, gaze direction (orienting), sneezing, coughing, hiccups, yawning, sighing, gasping, clearing one's throat, spitting, chewing, shivering, snoring, and belching (there are many more). Cues are not intentional communicative acts, but they may nevertheless be informative—as when someone chewing indicates the availability of food, someone sleeping suggests a safe environment, or someone's cough alerts the observer to the presence of the coughing individual. There exist many more cues than signals; signals arise only in mutually beneficial conditions, and those conditions may be infrequent.

37. *Emotional cues.* Observers can sometimes use cues to infer a person's emotional state, as when shivering or trembling might suggest fear, or when sneezing or coughing might suggest illness.

38. *Ritualization.* All signals evolve from ethological cues. Cues normally benefit only the observer; cues don't benefit the displayer. However, circumstances can change so that a given cue benefits the displayer as well as the observer. Joint benefits provide the selection pressure that may ultimately transform the cue into a signal. This transformation typically involves making the signal more conspicuous and stereotypic. But the main transformation is rendering the display and response behaviors more automatic or instinctive.

39. *Innate signals.* Most human social behaviors are volitional. Signals are a notable exception. Signals involve innate tendencies to generate the signal and innate tendencies to respond to the signal. We tend to involuntarily scream when threatened, weep when experiencing grief, and exhibit a surprise facial display when astonished or shocked. Similarly, we tend to involuntarily feel alarm when someone else screams, feel compassion when seeing someone cry, and follow the gaze of someone who shows surprise.

40. *Signal development.* Innate tendencies may not be exclusively genetic in origin but may develop in conjunction with learning.

41. *Executive control.* Despite the automaticity, compared with other animals, humans can draw on greater cognitive resources that allow us to shape instinctive behaviors. Our capacity for executive control is evident both when generating signals and when responding to signals.

42. *Cooperative versus competitive assessment.* The role of executive control in responding to a signal is most evident in our assessment of our social relationship with the signaler. Although it may be to our benefit to assist friends and family, it may be to our detriment if we assist competitors or enemies. For example, it may be beneficial to suppress any prosocial feelings induced by observing a weeping adversary. Since signals evolve because they confer mutual benefits to the displayer and the observer, what must evolve in tandem with, and as part of the signal response, is the assessment of whether the other individual is a friend/collaborator or a foe/competitor. Strangers form a third category. We may be wary of strangers, but strangers may also represent opportunities to expand our

network of allies or partners. Accordingly, we should be more willing to trust (i.e., collaborate with) strangers than enemies. Such assessments should be an omnipresent facet of our automatic response when reacting to a signal.

43. *Costs and benefits.* The role of executive control in generating signals is notably evident in the influence of cost-benefit assessments. We endeavor to *mask* or *inhibit* a display when the cost of a spontaneous display is assessed as greater than the benefit. And we may resort to *amplification* or *emulation* when the consequences of the display are assessed as especially beneficial.

44. *Executive control: Masking.* Executive control and the limits of executive control are evident in efforts to mask a signal, such as by hiding one's face, seeking isolation, or pretending that the display has some nonemotional origin (e.g., implying that the appearance of tears is the result of something caught in your eye).

45. *Executive control: Inhibition.* Executive control is also evident in various strategies intended to suppress or inhibit a signal. For example, the impulse to laugh might be suppressed by thinking of something serious; the impulse to weep might be inhibited by thinking of something mundane.

46. *Executive control: Amplification.* Executive control is further evident in various efforts to draw attention to the signal, such as moving into a public arena where one is more likely to be observed, thinking thoughts that may prolong the signal duration, or supplementing the signal with conspicuous culturally normative displays. In the case of grief-induced weeping, examples of cultural amplification may include wearing black clothing, shaving one's head, or engaging in acts of self-injury.

47. *Executive limits: Emulation.* The limits of executive control are notably evident in our limited ability to willfully generate signals that observers assess as "authentic" (that is, spontaneous, involuntary, not deliberate). At times when a signal would be useful (but not forthcoming), we may resort to imitating or emulating the signal. Examples of such *pseudo-signals* include social smiling, social laughter, feigned surprise, imitated disgust, and pretend weeping.

48. *Interpreting pseudo-signals.* Observers are generally adept at distinguishing authentic (spontaneous) signals from contrived (posed) pseudo-signals.

49. *Social cues*. Apart from emulating signals, it is also possible to emulate *cues*. Depending on the culture, people might voluntarily imitate yawning to symbolize boredom, produce a "hic" sound to symbolize drunkenness, belch as a symbol of satiation, and so on.

50. *Cultural symbol systems*. As voluntary rather than innate behaviors, pseudo-signals and social cues are susceptible to variation in use and interpretation. The repertoire of pseudo-signals and social cues employed in some community contributes to a cultural symbol system that helps to define a cultural group.

51. *Four display types*. As described above, displays can be usefully categorized into one of four types: *signals*, *pseudo-signals*, *cues*, and *social cues*. Signals and cues are involuntary *spontaneous* displays; pseudo-signals and social cues are willful *posed* displays that mimic their spontaneous counterparts.

52. *Intention*. In general, pseudo-signals communicate the signaler's aim to transform the observer's behavior in the same way that the corresponding spontaneous signal would. However, since observers are typically adept at distinguishing pseudo-signals from spontaneous signals, pseudo-signals simultaneously communicate that the signal is intentional and that the signaler retains some control. For example, pseudo-weeping (as in the Welcome of Tears) might communicate that the displayer is eager to induce a prosocial response in the observer without necessarily communicating helplessness or fear.

53. *Prosocial signals*. To the extent that signals benefit both the displayer and observer, all signals are prosocial. Even a display like threat/aggression exists only because it typically benefits both the displayer and the observer. It has been observed that those species that have the largest repertoire of signals are also the most social species.

54. *Manipulation rather than information*. Insofar as mature signals are not linked to one specific emotion, signals should not be regarded as efforts to reveal one's emotional state to others. The purpose of signals is to manipulate the behavior of observers; signals are tools for initiating a mutually beneficial social transaction. Pseudo-signals and social cues also aim to manipulate the behavior of the observer but may additionally be deployed to intentionally convey the displayer's emotional state.

55. *Compound displays*. It is possible to produce more than one display simultaneously, as in the combination of weeping, screaming, and pouting/threat often observed in temper tantrums.

56. *Mixed emotions.* Research suggests that it is possible to experience more than one emotion concurrently. An example is again found in the temper tantrum, where the motivating affects may be a combination of anger and grief.

57. *Common associations.* Although a signal may be evoked by many different emotions, some motivating emotions are more likely than others. Hence, we tend to associate weeping primarily with grief and smiling with happiness. Although these associations are statistically reasonable guesses by the observer, they are not reliable.

58. *Misconstrued signals.* Because a given signal is more commonly associated with a particular emotional state, observers (and many emotion researchers) wrongly conclude that the purpose of the display is to communicate or reveal the displayer's emotional state.

59. *Observer wisdom.* With age and experience, people are more reluctant to associate a given display with a single motivating emotion. We learn, for example, that not all smiles denote happiness and that not all tears denote grief.

60. *Deciphering emotions.* It is signals that are conspicuous, not emotions. When an observer correctly infers the underlying emotion that led to a signal, it is not because the displayer has issued an unambiguous message. Instead, correct inference of the motivating emotion is due to the observer's attentive piecing together of contextual and other information. In other words, emotions are not expressed; they are deciphered. Concepts like "facial expression" and "emotional expression" are not helpful since they perpetuate the incorrect idea that the purpose of a display is to convey, disclose, or reveal our emotions to others.

61. *Emotion expression model.* Once it is understood that behaviors like frowning, disgust, threat/aggression, weeping, laughing, and smiling are signals rather than emotional expressions, it follows that probably all emotions are covert. Like the invisible feelings of loneliness or affection, feelings of grief or joy are not directly communicated despite their nominal association with weeping or smiling. The emotion expression model traditionally assumed by most theories of emotion is biologically implausible and has impeded progress in understanding emotion-associated displays.

Glossary of Terms

accommodation: A type of learning in which new information that conflicts with expectation leads a person to alter their existing schema or create a new schema. Accommodation is more likely to occur when in a melancholic mood compared with a happy mood. One of a pair of concepts proposed by Jean Piaget: compare *assimilation*. See also *melancholic realism*.

acetylcholine: A *neurotransmitter* evident both *centrally* and *peripherally*. In the peripheral nervous system, low acetylcholine is associated with poor muscle tone and slow muscle reactivity. Low acetylcholine makes us feel slow, lethargic, and sleepy.

adenosine triphosphate (ATP): An organic compound that is the universal biological energy source. ATP is critical for such energy-consuming functions as muscle contraction and nerve conduction. In conjunction with histamine, ATP is released by mast cells.

altruism: The disposition to behave in a way that benefits someone else. See also *bonding warmth, commiseration, compassion, empathy, virtuous*.

AMIE: A mnemonic for aiding recall of four forms of executive control of signals and cues: *Amplification, Masking, Inhibition,* and *Emulation*.

amplification: A type of executive control in which a displayer endeavors to make the signal more conspicuous such as a mourner wearing distinctive clothing, appearing in public, marking one's face with charcoal, or engaging in acts of self-injury. See also *AMIE, executive control, masking, inhibition, emulation*.

anergia: A feeling of low energy or lethargy. See also *melancholy, sickness*.

anhedonia: A feeling state associated with a reduction or loss of pleasure. See also *melancholy, sickness*.

anorexia: A loss of appetite. See also *melancholy, sickness*.

assimilation: A type of learning in which new information that conflicts with expectation is simply absorbed within a person's existing schema without altering the schema or creating a new schema. Compared with a melancholic mood, assimilation is more likely to occur when in a happy or positive mood. One of a pair of concepts proposed by Jean Piaget: compare *accommodation*.

automaticity: For signals, the tendency for display behaviors and signal-induced observer behaviors to arise from innate or instinctive dispositions. See also *stereotypy*.

bawling: A type of crying characterized by a series of extended vocalized exhales: *waaaah, waaaah, waaaah . . .* , where each cry coincides with a full exhale. Bawling is prominent in the first year of life, occurs rarely in toddlers, and is virtually never observed in older children and adults. In this book, bawling is claimed to be an ethological *signal*, functionally distinct from weeping, and whose phylogenetic origin is the *separation distress call*. Compare *sobbing, weeping*. See also *crying*.

being moved: Also "being touched." An amorphous feeling of being stirred, warmed, or affected, including feelings of compassion, tenderheartedness, gentleness, peacefulness, being impressed, warmed, upset, or distressed. Often associated with the globus sensation or feeling choked up. See also *kama muta, bonding warmth*. See *choked up, globus sensation*.

blood–brain barrier: The unique property of blood vessel membranes that allow only some molecules or compounds to pass from the blood into the brain or vice versa. A barrier that chemically distinguishes two realms, commonly referred to using the adjectives *central* (brain) and *peripheral* (body).

bonding warmth: The positively valenced feeling that rewards and motivates an individual to form social bonds or friendships, such as via charitable or altruistic acts. See also *being moved, kama muta*.

breaking voice: A distinctive sound associated with weeping, produced by a transition from modal (normal) to falsetto (high) voice or vice versa. A unique vocal "cracking" sound that arises due to phonological instability caused by muscle constriction of the voice box. See *weeping*. See also *creaky voice, falsetto voice, gasping, ingressive phonation, pharyngealization, sobbing*.

broaden-and-build theory: The tendency, when in a positive mood, for people to take a broad "big picture" approach. When problem-solving, a positive mood commonly leads to greater flexibility, creativity, and a wider field of attention. Complex sources of information tend to be simplified and so improve task efficiency. Compare *melancholic realism*. See also *accommodation*.

central: Pertaining to the brain. Contrasts with *peripheral*. See also *blood–brain barrier*.

choked up: See *globus sensation*.

choking up: In this book, inconspicuous symptoms associated with weeping; symptoms that are difficult or impossible for observers to notice such as the globus sensation (constricted pharynx), the feeling of pending or incipient tears, or a slight downward movement of the corners of the mouth. Often precedes a conspicuous full-blown "wailing and waterworks" weeping behavior. May be associated with grief, "being moved," and other emotions. Compare *weeping*. See also *crying, bawling, globus sensation, grief, being moved, kama muta.*

cold anger: Anger that is felt but not manifested in any aggression or threat display. Compare *hot anger.*

colic: Sustained infant crying (*bawling* rather than *weeping*) that arises for no apparent reason.

commiseration: A vicarious feeling of stress or pain induced by observing another person suffering stress or pain. See *contagious emotion, empathy, prosocial.* Compare *compassion.*

compassion: A feeling of sympathy, pity, or concern for the suffering or misfortune of others. See *empathic concern, prosocial.* Compare *commiseration, empathy.*

conspecific: Belonging to the same species.

conspicuousness: For signals, the tendency for display behaviors to be easily perceived by observers. Conspicuous displays often involve more than one sensory modality and are often sustained ("persistent") or repeated over time. See also *redundant, signal.*

contagious emotion: Any emotion that is induced by observing the emotion of another person, where the induced emotion is the same as the emotion of the observed person. For example, where a person experiences vicarious happiness in response to observing the happiness of someone else. See *commiseration.* Contrasts with *repercussive emotion.*

cortisol: A hormone released in response to stress. Its presence activates antistress and anti-inflammatory responses.

creaky voice: A way of vocalizing (phonating) that resembles a "creaking" sound, such as the sound of a creaking door. Produced when the vocal folds are tightly compressed by drawing together the arytenoid cartilages. The compression causes the vocal folds to become slack and so vibrate at a very low frequency. Commonly evident when a weeping individual speaks. Also known as vocal fry or laryngealization. See *weeping.* See also *breaking voice, falsetto voice, gasping, ingressive phonation, pharyngealization, sobbing.*

crying: For the purposes of this book, a general term that includes *bawling* and *weeping.*

cue: As classically defined in ethology: a feature of the world that can be used by an animal as a guide to future action and that benefits only the observing animal. Compare *signal.*

deception: Any display behavior that causes an observer to respond in a way that unexpectedly proves detrimental to the observer while benefiting the displayer.

depression: A sadness pathology characterized by malaise, anhedonia, lethargy, sleep disruption, and feelings of hopelessness, sometimes leading to suicidal thoughts. Depressed individuals tend to dwell on negative thoughts, especially *rumination* regarding past or recent failures. Compare *melancholy*. See also *sadness*.

depressive realism: In this book, a deprecated term. See *melancholic realism*.

display: Any behavior that an observer would recognize as distinctive, including both ethological *cues* and *signals*. Examples include smiling, pouting, sneezing, scratching, sticking out one's tongue, crouching, or screaming.

Duchenne smile: An intense form of smiling characterized by squinting—as evident in the appearance of "crow's feet" in the vicinity of the left and right temples (referred to as the Duchenne marker). Paul Ekman proposed that the presence or absence of the Duchenne marker distinguishes "genuine" or spontaneous involuntary smiles from social (willful, posed) smiles. However, non-Duchenne smiles can also arise as spontaneous involuntary displays, and many individuals are capable of voluntarily producing a full Duchenne smile. See Fridlund (1994, pp. 116–117) for a critique. See *posed display, signal, social smile*.

emotion: For the purposes of this book: 1. Compelling phenomenological and brain states (such as anger, fear, compassion, loneliness, or hunger) whose purpose is to motivate behaviors that are commonly beneficial. 2. Any short-lived feeling state lasting seconds to minutes. Commonly contrasted with *mood* and *temperament*.

emotion-associated display: In this book, the preferred term for behaviors such as smiling, weeping, laughing, and screaming. A term used in preference to emotional "expression," since a given display may be motivated by many possible emotions. See *emotional expression model, emotional expression, polygenic signals*.

emotion expression model: The widespread and long-standing idea that we convey, disclose, or reveal our emotions to others through emotional displays (such as facial expressions) and that observers typically recognize the emotion conveyed or transmitted by the displayer. For example, that weeping is an expression of grief, which is actively communicated to observers who recognize that the person is feeling grief. Or that smiling reveals to observers that the smiling person is happy. A theory that is challenged by Alan Fridlund, as well as in this book. Compare *signal theory of emotion-associated displays (STEAD)*. See also *emotional expression, facial expression*.

emotional expression: In this book, a deprecated term since it wrongly implies that the purpose of a display is to convey, disclose, or reveal the displayer's emotional state. The preferred term is *emotion-associated display*. See also *emotion expression model, facial expression, signal*.

empathic concern: The disposition to feel concern, sympathy, or compassion for another person experiencing some stress or misfortune; one of four personality facets in Mark Davis's Interpersonal Reactivity Index whose purpose is to characterize empathetic traits. See *compassion, prosocial*. Compare *commiseration, empathy*.

empathy: The capacity to understand or feel what another person is experiencing; the capacity to place oneself in another person's situation. See *compassion, commiseration*.

emulation: The willful or voluntary imitation of an innate *cue* or *signal*, such as coughing in order to attract attention, mock weeping, or the social (willful) smile. See *AMIE, executive control*.

endogenous compound: Any chemical that is produced by the body itself, such as histamine, acetylcholine, serotonin, cortisol, and interferon-α. Contrasts with *exogenous compounds* (external chemicals that are ingested, injected, inhaled, or applied to the skin).

ethology: The study of animal behavior. See also *cue, signal*.

eumnesia: (yume-NEE-zee-ah) The positive feeling state associated with recalling pleasant past memories. Formed from the Greek *eu* (meaning pleasant) and *mnesia* (meaning memory). Characteristic of *nostalgia*.

exaptation: The use or recruitment of a biological structure or behavior for a purpose other than that for which it initially evolved. In this book, an example of an exaptation is the recruitment of the allergic response as the foundation for weeping.

executive control: The cognitive ability to inhibit or modify otherwise innate or involuntary behaviors, as well as the ability to voluntarily mimic or feign an innate behavior that is otherwise not forthcoming. See also *AMIE, amplification, emulation, inhibition, masking*.

exogenous compound: Any chemical originating outside the body. A compound that is ingested, injected, inhaled, or applied to the skin. Contrasts with *endogenous compound*.

facial display: In this book, the preferred term (rather than *facial expression*) for distinctive facial behaviors such as smiling, frowning, pouting, disgust, and sneering. See also *emotional expression, emotion expression model, cue, signal*.

facial expression: In this book, a deprecated term since it wrongly implies that the purpose of a facial display is to convey ("express") a person's emotional state. The preferred term is *facial display*. See also *emotional expression, emotion expression model, signal*.

falsetto voice: An especially high voice range or register, not commonly used when speaking. Roughly an octave higher than normal or modal voice. See *weeping*. See also *breaking voice, creaky voice, falsetto voice, gasping, ingressive phonation, pharyngealization, sobbing*. Compare *modal voice*.

fitness: In the strict sense, the reproductive success of an individual as indexed by that individual's contribution to the ensuing generation's gene pool. In this book, the word "fitness" is used in the probabilistic sense advocated by Maynard Smith. Specifically, fitness is a property not of an individual, or an individual behavior, but of a class of individuals with a propensity to behave in a particular way that is likely to increase their contribution to the ensuing generation's gene pool.

gasping: An unvoiced wheezing sound produced while inhaling. Most noticeable when the vocal tract is constricted as commonly occurs when weeping. Compare with *ingressive phonation*. See *weeping*. See also *breaking voice, creaky voice, falsetto voice, pharyngealization, sobbing*.

globus sensation: The technical term for the experience of being "choked up"; a sensation that arises due to a tightening of the muscles of the pharynx. Commonly associated with *grief, being moved*, and other intense emotions. See also *breaking voice, choking up, sobbing*.

grief: An acute sadness state commonly induced in reaction to major distressing events such as the death of a loved one or the failure of a valued romantic relationship. A state characterized by overt weeping or by less conspicuous sensations, including the sensation of imminent tears or the globus sensation (feeling choked up)—states that are best regarded as preludes, harbingers, or arrested precursors of weeping. Although weeping, imminent tears, and the globus sensation can be induced by other emotions, grief can be distinguished because it is always linked to the negatively valenced subjective experience of acute sadness. Contrast with *melancholy*. See also *choking up, globus sensation, mourning cycle, sadness, triadic theory of immunological responses to stress, weeping*.

handicap principle: The idea, proposed by Amotz Zahavi, that in order to minimize deception or bluffing, signals should be costly for the signaling animal and that the observing animal should be able to readily recognize the incurred cost. Ongoing research, however, has shown that not all signals involve or require a handicap. See *signal*.

histamine: An *endogenous* proinflammatory *cytokine* and *neurotransmitter*. In the *periphery*, most histamine is released from mast cells. Peripheral histamine plays a major role in the allergic response, producing symptoms such as nasal congestion and watery eyes. In the *central* nervous system, histamine is found in various tissues, notably in the hypothalamus. Central histamine is linked to several cognitive functions associated with improved rational thought. See also *melancholic realism*.

hot anger: Anger that leads to an aggression or threat display. Compare *cold anger*.

hyperalgesia: Increased pain sensitivity, often evident in the form of muscle soreness. A *peripheral* phenomenon associated with the release of prostaglandin E. See also *prostaglandins*.

hypermelancholy: In this book, a conjectured biologically defunct state in human evolution; an archaic stress state representing an intermediate stage or bridge between melancholy and grief. An intense form of melancholy that is conjectured to have included activation of peripheral histamine producing an allergic response (a cue) that made it easier for observers to recognize a person's stressed state. See also *protomelancholy*.

illness: In this book, a conjectured ancient stress response (not observable in humans) characterized by inflammation, *pyrexia* (fever), *anorexia* (reduced appetite), and (perhaps) headache but not *anergia* (lethargy), *anhedonia* (reduced pleasure), *malaise* (low mood), and (perhaps) *hyperalgesia* (pain sensitivity/muscle soreness). Compare *sickness*.

ingressive phonation: A voiced wheezing sound produced when inhaling. The sound produced when vocalizing while inhaling (vocalized gasping). Commonly heard as part of weeping behaviors. Compare with *gasping*. See *weeping*. See also *breaking voice, creaky voice, falsetto voice, pharyngealization, sobbing*.

inhibition: A type of executive control in which a displayer endeavors to suppress or thwart an otherwise innate or involuntary behavioral tendency—such as thinking happy thoughts in order to avoid weeping or biting one's lips to avoid smiling. See also *AMIE, amplification, automaticity, emulation, executive control, masking, signal*.

interoception: The process of corporeal self-sensing that helps define one's feeling state. Interoception relies on a combination of conscious perception and unconscious monitoring of various bodily states, including such aspects as temperature, heart rate, hormone levels, muscle tension, posture, immune status, and gastrointestinal activity.

kama muta: A term borrowed from Sanskrit (meaning "moved by love") and adopted by anthropologist Alan Fiske and psychologists Beate Seibt and Thomas Schubert to describe the general amorphous positive feelings of being touched or moved, stirred, smitten, infatuated, entranced, or transported. Commonly associated with positive prosocial feelings. See *being moved, choked up, globus sensation*.

lacrimation: Technical term for the shedding of tears.

laryngealization: See creaky voice.

lethargy: A feeling of low energy; also known as *anergia*. See also *melancholy, sickness*.

limbic: Pertaining to the limbic system, a set of brain structures in proximity to the thalamus and associated with low-level (noncognitive) emotional functions. A now largely obsolete term.

long face: A phrase used in many languages to refer to melancholy or depression. A relaxed facial display symptomatic of low physiological arousal in which the chin tends to drop and the cheeks tend to flatten, giving the face an elongated appearance. As used in this book, a term to denote the relaxed facial display that is commonly observed when a person feels relaxed, fatigued, sleepy, or *melancholy*.

malaise: A glum, forlorn, despondent, morose, or cheerless feeling. See *grief, melancholy, sickness.*

masking: A type of executive control in which a displayer endeavors to hide or conceal an otherwise innate or involuntary behavioral tendency—such as by turning away, obscuring the display by placing a hand in front of one's face, or seeking isolation. See also *AMIE, executive control, amplification, inhibition, emulation.*

melancholic realism: A pattern of prudent thought associated with *melancholy.* A cognitive disposition that typically reduces risk by analyzing problems more thoroughly. Compared with a happy or neutral mood, melancholy promotes more detail-oriented thinking, reduced stereotyping, less judgment bias, greater memory accuracy, reduced gullibility, more patience and greater task perseverance, more social attentiveness and politeness, more accurate assessments of the emotional states of others, a heightened sense of fairness, improved reasoning related to social risks, and a disposition to favor delayed larger rewards over immediate smaller rewards. Also known as "depressive realism." See *melancholy.* Compare *broaden-and-build theory.*

melancholy: A form of sadness characterized by feeling glum, blue, or forlorn. A response to stress involving low physiological arousal; associated with relaxed muscles, reduced activity, and social withdrawal. Relaxed facial muscles cause the chin to be lowered and the cheeks to flatten, resulting in a "long face" appearance. Cognitively, melancholy is associated with sustained *reflection* (as opposed to *rumination*) regarding one's situation. Compare *depression, grief.* See also *melancholic realism, nostalgia, sadness, triadic theory of immunological responses to stress.*

modal voice: The normal voice register or pitch region, as contrasted with *falsetto.* See *breaking voice, weeping.*

mood: An affective state lasting on the order of hours. Commonly used in contrast with *emotion* (shorter durations) and *temperament* (longer durations).

motivation: The impetus for behaviors. Two types of motivation can be distinguished: ultimate and proximal. *Ultimate motivations* include biological goals such as the avoidance of injury, procreation, and metabolic sustenance. Corresponding *proximal motivations* include the feeling of pain, the allure of sexual intercourse, and the pleasure of eating and/or the discomfort of hunger.

mourning cycle: A stress-related behavioral pattern characterized by a recurring alternation between periods of active *weeping* and periods of quiescent *melancholy.*

mutually beneficial: An attribute of *signals*—namely, that they benefit both the *signaler* and the *observer.* See *non-zero-sum game.*

neurotransmitter: An endogenous compound that modulates communication between neurons.

non-zero-sum game: Technical term for any interaction in which both individuals gain or benefit. See *mutually beneficial, signal.* Contrast with *zero-sum game.*

nostalgia: A sentimental reminiscence regarding some past situation with positive personal associations. A state characterized by mixed happy and sad ("bittersweet") feelings. See also *melancholy, reverie, sadness.*

observer: In this book, a person (or nonhuman animal) who observes or is the target recipient of an ethological signal. In ethological signaling theory, the observer is more commonly referred to as the *receiver.* However, in order to avoid the implication that signals are intended to be informative rather than manipulative, our preferred term is observer.

peripheral: Pertaining to the body excluding the brain. Contrasts with *central.* See also *blood–brain barrier.*

pharyngealization: A sort of "pinched" sound quality arising from constricting the muscles of the pharynx. A common feature of all weeping vocalizations. See *weeping.* See also *breaking voice, creaky voice, falsetto voice, gasping, ingressive phonation, sobbing.*

polygenic proclivity of signals: The idea that once a signal is able to induce a stereotypic behavior in an observer, that signal can be recruited to induce the observer behavior in situations other than the one in which the signal originally evolved. As long as the new situation continues to benefit both individuals, the signal will continue to exist due to positive selection pressures. For example, although weeping might have evolved in the context of grief, the pertinent prosocial behavior can be induced in an observer by weeping in other circumstances, such as in tears of joy, tears of loyalty, and tears of laughter.

polygenic signals: The observation that many signals can be evoked by more than one motivating emotion. For example, smiling can arise from feelings of happiness, stress, embarrassment, or deference. Weeping can be motivated by grief, joy, loyalty, or extreme humor. See also *emotion expression model, polygenic proclivity of signals, signal.*

posed display: Any voluntary emulated display, including *social cues* and *pseudo-signals.* Contrasts with *spontaneous display.*

prosocial: Behaviors that benefit other people, or the feelings motivating such behaviors. Prosocial behaviors include responding to pleas for aid or assistance; any act of bonding, solidarity, or affiliation; terminating aggression; or showing a generally favorable attitude or disposition to another person. See *signal.*

prostaglandins: A class of lipid compounds that produce hormone-like effects. Notable for causing widening of blood vessels as part of inflammation, inducing fever (*pyrexia*), and causing muscles to become more pain sensitive (*hyperalgesia*).

protomelancholy: In this book, a conjectured biologically defunct state in human evolution; an archaic stress state representing an intermediate stage or bridge between feeling sick and feeling melancholy. Like sickness and melancholy, protomelancholy is conjectured to have entailed feelings of *anergia* (lethargy) and *anhedonia* (reduced pleasure). Like melancholy, protomelancholy is proposed to have excluded inflammation, hyperalgesia, fever, or headache—common features of sickness. However, protomelancholy differs from melancholy insofar as it does not include enhanced cognition (i.e., melancholic realism). In short, protomelancholy might be regarded as melancholy without melancholic realism. See *hypermelancholy, melancholy, melancholic realism.*

proximal motivation: The immediate or psychological impulse or appeal leading to some behavior. For example, the proximal motivations for consuming food include feelings of hunger and the enjoyment or gustatory pleasure of eating; a proximal motivation encouraging reproduction includes the pleasure of sexual intercourse. See *motivation.* Contrasts with *ultimate motivation.*

pseudo-signal: A voluntary, fake, feigned, or emulated version of a *spontaneous signal.* Examples include feigned "social smiles" (as opposed to spontaneous involuntary smiles) and feigned laughter (as opposed to spontaneous involuntary laughter). A type of *posed display.* Compare *social cues.*

pyrexia: Technical term for fever.

receiver: In conventional ethological signaling theory, a person (or nonhuman animal) who observes or is the target recipient of an ethological signal. In order to avoid the implication that signals are intended to be informative rather than manipulative, in this book, the preferred term is *observer.* See also *signal.*

redundant: The tendency to repeat or sustain a *signal* over time. A property of many, though not all, ethological signals. See also *conspicuousness, signal.*

reflection: A term used to denote helpful or positive forms of cognitive engagement as opposed to brooding or negative forms of thought (i.e., *rumination*). See *melancholic realism.* Compare *reverie, rumination.*

repercussive emotion: Any emotion that arises in response to the emotion of an observed person, where the induced emotion differs from the observed person's emotion. For example, where a person experiences fear in response to an observed person's display of anger. Contrasts with *contagious emotion.*

reuptake: The process by which biochemical compounds like *hormones* and *neurotransmitters* are recycled or broken down in the body. The concentration of any compound in the body depends on the speed of reuptake and the speed with which it is created or synthesized. See also *synthesis.*

reverie: A wistful state of reflection about the past, commonly associated with nostalgia. Compare *reflection, rumination.*

ritualization: 1. In anthropology and sociology, the process by which informal actions become standardized as formal rituals. Rituals can include formal ceremonies but also informal patterned behaviors whose form is stereotypic and socially defined. Ritualization is the process by which such patterned behaviors become established in a given culture. An example of ritualization would be the historical emergence of a practice like shaving one's head when in mourning. Another example would be the use of yawning to convey boredom. 2. In ethology, the process by which a *signal* evolves from an earlier *cue*. Ritualization entails changes such as increasing the conspicuousness of a display and inducing automaticity for both the display behavior and the observer's response. See *automaticity, conspicuousness, cue, signal*.

rumination: A cognitive state characteristic of depression that features brooding negative self-assessments, often linked to repeatedly recalling past situations or failures. A pathological condition that is broadly destructive and unhelpful. See also *depression*. Compare *melancholy, melancholic realism, reflection, reverie*.

sadness: In this book, a general term that refers to the class of sorrowful, despondent, or regretful emotional states, including *melancholy, grief, nostalgia*, and *depression*.

separation distress call: A type of signal commonly produced by infants, observed in many animal species, typically exclusively sonic, intended to attract the attention of a caregiver. Human infant bawling resembles the separation distress calls of many mammal species. Deer mothers have been observed to respond to recordings of human infant bawling in a manner similar to their species-specific separation distress calls. See *bawling*.

serotonin: An endogenous compound that is involved in numerous central and peripheral functions, including digestion, appetite, organ development, growth, tissue repair, memory, learning, and mood. In the brain, low levels of serotonin are associated with social stress, sadness, and depression. High-status male primates have been shown to have twice the level of serotonin in the brain as subordinate males and females. See *central, depression, malaise, melancholy, neurotransmitter*.

sickness: In this book, a distinctive feeling state induced by immune responses to infection or disease, commonly characterized by inflammation, *pyrexia* (fever), *anorexia* (reduced appetite), headache, and feelings of *anergia* (lethargy), *anhedonia* (reduced pleasure), *malaise* (depressed feelings), *hyperalgesia* (pain sensitivity/muscle soreness), and *somnolence* (sleepiness). Compare *illness*. See also *triadic theory of immunological responses to stress*.

signal: In ethology, an evolved innate or instinctive act or display that commonly induces a stereotypic observer behavior and whose consequences benefit both individuals. The mutual benefits provide the selection pressure for the signaling and responding behaviors to coevolve. Examples of human signals include weeping, smiling, laughing, disgust, and threat displays. See also *automaticity, conspicuousness, pseudo-signal, redundancy, signal theory of emotion-associated displays (STEAD), stereotypy*. Compare *cue*.

signaler: In signaling theory, a person (or nonhuman animal) who generates or produces an ethological *signal*. See also *observer*.

signaling theory: A well-known theory in the field of animal behavior (ethology) developed by Konrad Lorenz. The theory distinguishes three types of animal communication: *signals, cues,* and (less commonly) indexes.

signal theory of emotion-associated displays (STEAD): As proposed in this book, a theory whose main claim is that what are commonly called emotional expressions (such as smiling, weeping, laughter, frowning) are not forms of emotional communication but instead are evolved signals in the ethological sense. Signals evolve in non-zero-sum conditions where the signal induces a change of observer behavior that benefits both the displayer and the observer. A given signal can arise from many motivating emotions. Except in mutually beneficial situations, communicating one's emotional state is undesirable since it allows observers to predict one's behavior. Consequently, emotions are typically covert rather than intentionally expressed. Observers *decipher* rather than *recognize* emotional states. The emotion that is most reliably related to a given signal is the emotion induced in the observer (which provides the proximal motivation for the behavior), rather than the emotion motivating the display. See *emotion expression model, signals, polygenic proclivity of signals, polygenic signals.*

sleep deprivation: In sadness research, a simple (but short-lived) way to eliminate depressive symptoms.

size symbolism: The widely observed principle in animal behavior where threat is associated with efforts to look or sound bigger whereas submission or deference is associated with efforts to look or sound smaller. Visual examples include arching one's back or couching low to the ground. Acoustic examples include lowering or raising the frequency of vocalizations.

sleepiness: Also known as *somnolence.* A feeling commonly associated with feeling *sick, depression,* and sometimes *melancholy.*

sobbing: In this book, a characteristic sonic component of *weeping.* The distinctive sound of vocalized coughing or punctuated exhaling (ah-ah-ah-ah-ah . . .). A sound that is closely shared with laughter (usually transcribed as ha-ha-ha-ha . . .). See also *breaking voice, creaky voice, gasping, ingressive phonation, pharyngealization.* Compare *bawling.*

social cue: Any voluntary, fake, feigned, or emulated version of a spontaneous cue that is recruited to communicate to an observer. Examples include yawning as an indicator of boredom or coughing as a way of attracting attention. A type of *posed display.* See *cue.* Compare *pseudo-signals.*

social ostracism: Punishment or chastising by others for behaviors, most commonly failures to act, in situations that call for moral action, such as failing to help someone in need or failing to come to the assistance of a friend.

social smile: A voluntary or posed smile as compared with a spontaneous signal smile. A willful pseudo-signal. See also *Duchenne smile, posed display, pseudo-signal, signal.*

somnolence: Technical term for sleepiness. Commonly associated with feeling sick, *depression*, and sometimes *melancholy*. See *sickness*.

spandrel: A feature of an organism that originates as a by-product or artifact; a trait that has no fitness value. Compare *exaptation*.

spontaneous display: Any involuntary or innate cue or signal display. Examples include such signals as unacted smiling, unprompted frowning, laughing, or screaming and such cues as reflexive sneezing, gasping, scratching, or blinking. Contrasts with *posed display*.

staccato breathing: The punctuated vocalized exhaling characteristic of both laughter (ha-ha-ha) and *sobbing* (ah-ah-ah).

stereotypy: For signals, the tendency for both display behaviors and observer responses to be standardized or formulaic. See also *automaticity*.

STEAD: See *signal theory of emotion-associated displays (STEAD)*.

synthesis: The process by which biochemical compounds like hormones or *neurotransmitters* are created in the body. The concentration of any compound in the body depends on the speed of synthesis and the speed with which it is broken down. See also *reuptake*.

temperament: A long-term affective disposition, such as a generally happy disposition or a generally anxious disposition. Temperaments can change due to life events such as severe trauma. Compare *emotion, mood*.

triadic theory of immunological responses to stress: (TTIRS, pronounced "tears"). In this book, a theory that proposes three general human responses to stress: *sickness* (a corporeal response), *melancholy* (a cognitive response), and *grief* (a social response). Sickness is a state that enhances the immune response to injury or illness by commandeering supplementary metabolic resources primarily by suppressing voluntary movement. Melancholy is a state that enhances our response to stressors, principally by encouraging cognitive reflection and by tailoring cognitive processes so they are more realistic. Grief is a state that enhances our response to stressors by recruiting social assistance through the ethological signal of weeping. The physiological basis for all three states can be traced to the immune system. Each is associated with a distinctive feeling state or emotion. See also *melancholic realism*.

TTIRS: (Pronounced "tears") Initialism for *triadic theory of immunological responses to stress*.

ultimate motivation: The underlying biological purpose for some behavior. For example, the ultimate motivation for eating is the provision of metabolic resources; the ultimate motivation for sexual intercourse is procreation. See *motivation*. Contrasts with *proximal motivation*.

valence: The positive or negative character of an emotion. Positively valenced emotions are states that a person wants to repeat or continue, whereas negatively valenced emotions are states that a person wants to avoid or end. In animal behavior, positive valence is associated with advance or approach, whereas negative valence is associated with avoidance or withdrawal.

vasodilation: The enlargement of blood vessels. The main symptom of the inflammatory response—producing the warm/red/swollen appearance at an injury site.

virtuous: The positive private feeling of having acted in an exemplary, meritorious, or laudable manner. A feeling that commonly arises following (or in anticipation of) prosocial acts of charity or altruism.

vocal fry: See *creaky voice*.

weeping: A type of *crying* behavior, rarely observed in the first six months of life. Characterized by the production of tears, nasal congestion, a choked-up feeling, and distinctive vocalizations including sobbing (cough-like staccato convulsions: *ah-ah-ah-ah-ah* . . .), breaking voice, gasping, and pharyngealized voice. In this book, weeping is claimed to be an ethological *signal* functionally distinct from infant *bawling* (another signal). Weeping commonly evokes a broadly prosocial disposition in observers—feelings that lead to reduced aggression, the offering of altruistic assistance, a feeling of connection or bonding, or simply a favorable attitude toward the weeping individual. See *breaking voice, ingressive phonation, sobbing*. Compare *bawling, choking up*. See also *crying*.

zero-sum game: Technical term for any interaction in which one person's gain is another person's loss. See *signal*. Contrasts with *non-zero-sum game*.

Notes

Preface

1. See Sloman et al. (2003) and Nettle (2004) for possible ways in which depression may be functional.

2. Research on sadness, grief, and nostalgia has a long history. The perspective advocated in this book aside, my presentation necessarily draws on both the empirical and the theoretical work of a number of seminal scholars, including from the research legacies of William Frey, Jeffrey Kottler, Tom Lutz, Ad Vingerhoets, and Constantin Sedikides.

3. Huron (2018).

4. In the strict sense, *fitness* refers solely to the number of offspring produced by an individual. However, our use of the word "fitness" follows the probabilistic definition advocated by Maynard Smith. In this book, fitness is a property not of an individual or an individual behavior but of a class of individuals with a propensity to behave in a particular way that is likely to increase their contribution to the ensuing generation's gene pool.

Preamble

1. Kleinginna & Kleinginna (1981).

2. Gross (2010).

3. Moors (2022).

4. See, e.g., Ortony & Turner (1990); Crivelli & Fridlund (2019).

5. Ben-Ze'ev (2001).

6. Nesse (1991).

7. Nesse (1991).

8. "Nothing's worse than having an itch you can never scratch."—*Blade Runner* (1982).

9. Huron (2006).

10. An example of a widely accepted modern definition of emotion is "intense, conscious and directed affective state with clear cognitive content" (Forgas, 2019, p. 361).

11. As in the chord progression musicians call a "deceptive cadence."

12. Many scholars have voiced similar skepticism regarding definitions of emotion, including Moors (2022). At the same time, Moors has made the strong case that unless one is able to identify what is, and is not, an emotion, the concept of emotion lacks any scientific status. However, that's also the case for virtually all concepts—such as cognition, consciousness, motivation, goals, pleasure, evolution, or mechanism. As Karl Popper observed, many poorly defined concepts remain useful in scientific endeavors despite their nebulous meaning.

Chapter 1

1. Archer (1999), however, has documented a competing research lineage in which the distinction between the "active" and "passive" behaviors described by Darwin has been retained. Research by Shand (1920), Becker (1933), Engel (1962), Klinger (1975, 1977), Archer (1999), and others has developed and clarified this distinction.

2. E.g., Bonanno et al. (2008); Goldie (2011).

3. Frick (1985, p. 420).

4. E.g., Allen & Badcock (2003).

5. E.g., Siegel & Sapru (2006).

6. E.g., Viggiano et al. (2004).

7. Raleigh et al. (1991); Erritzoe et al. (2023).

8. The role of serotonin in depression is controversial. See Moncrieff et al. (2022), although see also Erritzoe et al. (2023).

9. E.g., Frijda (1986).

10. Rucker & Petty (2004).

11. Neth & Martinez (2009).

12. Quieter voice (Banse & Scherer, 1996; Scherer, 1986; Skinner, 1935), more slowly (Breitenstein et al., 2001), lower pitch (Fairbanks & Pronovost, 1939), more monotone (Eldred & Price, 1958), more mumbled articulation (Dalla Bella et al., 2001), and breathier and darker timbre (Ohala, 1980, 2010; Scherer et al., 2003; see also Erickson et al., 2006). Most of these features were already described by Kraepelin (1899).

13. First noted by Kraepelin (1899).

14. Nesse (1991, 2018); Andrews & Thomson (2009).

15. Nesse (1991, 2018).

16. Davidson (2003).

17. Raby (2012).

18. Ventevogel et al. (2013).

19. Levy (1973). Other synonyms for *haumani* include boring, dull, dreary, lonely, and mournful.

20. Vingerhoets (2013).

21. Rosenblatt et al. (1976, p. 15).

22. Frey (1985).

23. In his earliest work, Paul Ekman referred to this as a "sad" face, but in later writings, Ekman refers to this as the "agony" expression.

24. Ekman & Friesen (2003).

25. Frey (1985).

26. Lutz (1999, pp. 67–68).

27. Frey (1985).

28. Provine (2000).

29. Rosenblatt et al. (1976); Vingerhoets (2013).

30. I.e., vasodilation of the blood vessels of the conjunctiva; Provine et al. (2011).

31. E.g., Gertsman (2011); Maguire (1977).

32. Frick (1985, p. 420).

33. Švec & Pešák (1994).

34. E.g., Gertsman (2011).

35. E.g., Ebersole (2000); Hockey et al. (2001); Marsella et al. (1985); Murphy et al. (1964).

36. E.g., Hasson (2009); Montagu (1960); Murube (2009); Roes (1989); Trimble (2012).

37. Panksepp (1998).

38. Provine et al. (2009).

39. Vingerhoets (2013).

40. Bowlby (1961, 1973).

41. Spencer-Booth & Hinde (1971); Hinde & Spencer-Booth (1971); as cited in Archer (1999, p. 56).

42. Hofer (1984).

43. The rate of alternation between melancholy and grief is not fixed. In some cases, a quick alternation may appear to suggest that grief and melancholy are being experienced at the same time or that they share certain features in common. For example, in the transition from grief to melancholy, unevaporated tears lingering on a person's cheeks might lead one to falsely conclude that tears are also a component of melancholy.

Chapter 2

1. Lorenz (1937). It should be noted that terms like *sign, signal, signifier,* and *signified* have a long history in the fields of linguistics and semiotics. Different generations of scholars have interpreted these terms differently, and consequently, they are commonly a source of confusion. For example, the ethological concept of signal differs somewhat from its meaning in semiotics (see Moran, 2009). In this book, we will use the terms *signal* and *cue* as understood in the field of ethology.

2. See, for example, Bradbury & Vehrenkamp (1998); Maynard Smith & Harper (2003).

3. It should be noted that ethologists distinguish different types of signals, including *indices, handicaps,* and *amplifiers.* An *index* is a signal whose magnitude varies over some domain of interest to the observer, such as the health of the signaler (Maynard Smith & Harper, 1995). Indices are inherently reliable due to the close causal relationship between the signal and the quality of interest. A *handicap* is a signal that is costly to the displayer and whose cost ensures the reliability of the signal (Zahavi, 1975). An *amplifier* is a feature that is not in itself informative but facilitates the observer's ability to decipher or assess another trait signal (Gualla et al., 2008).

4. Ward et al. (2008), although see Norman et al. (2015).

5. Partan & Marler (1999).

6. Johnstone (1997).

7. This example comes from Zahavi & Zahavi (1997).

8. The main proponents of the notion of costly signaling are the Israeli husband and wife duo, Amotz and Avishag Zahavi (see Zahavi, 1975; Zahavi & Zahavi, 1997). As formulated in their book, *The Handicap Principle,* this idea has been somewhat controversial among biologists. My presentation here follows a common reformulation and modification of that theory.

9. Explanatory accounts of the peacock's tail remain controversial. See Ryan (2019).

10. See Davis & O'Donald (1976); Eshel (1978); Kirkpatrick (1986); Maynard Smith (1985); Pomiankowski (1987).

11. Caro (1986).

12. This image is licensed under the Creative Commons Attribution-Share Alike 3.0 Unported license. See https://commons.wikimedia.org/wiki/File:Springbokpronk.jpg.

13. FitzGibbon & Fanshawe (1988).

14. Dawkins & Krebs (1978); Krebs & Dawkins (1984).

15. Johnstone & Grafen (1993); Rowell et al. (2006); Searcy & Nowicki (2005).

16. Searcy & Nowicki (2005).

17. I would conjecture that there are perhaps only a dozen signals in the human behavioral repertoire—including smiling, laughing, threat, fear, surprise, disgust, bawling, weeping, and screaming—but perhaps several hundred cues.

18. Based on Maynard Smith & Harper (2003, p. 15); see also Scott-Phillips (2008, p. 388).

19. Based on Hasson (1994).

20. Maynard Smith & Harper (2003).

21. Bradbury & Vehrenkamp (1998).

22. This account comes from native Inuit oral history but has not been independently confirmed by biologists.

23. Tinbergen (1952, 1964); see also Maynard Smith & Harper (2003).

24. Maynard Smith & Harper (2003).

25. The idea that the purpose of all animal communication is to change the behavior of observers was first proposed by Dawkins & Krebs (1978; Krebs & Dawkins, 1984) and has been developed by others, including Rendall & Owren (2013). Despite its broad acceptance among ethologists, dissemination of this idea has nevertheless been slow. For example, only recently have primatologists suggested that primate calls might exist to change the receiver's behavior (see Schamberg et al., 2018). Dawkins and Krebs emphasized deception and cheating in animal communication. Although deception is common in animal interactions, ethologists regard *signals* as a special case in which deception is rare (e.g., Johnstone & Grafen, 1993). Fridlund's (1994) notable work on emotion-related displays was inspired by the earlier conceptions by Dawkins and Krebs. In this book, we follow more recent signaling theory, where signals are recognized as typically "honest."

Chapter 3

1. Siegel & Sapru (2006).

2. Hollien (1960).

3. Sundberg (1987).

4. Tartter (1979, 1980); Tartter & Braun (1994); Schröder et al. (1998).

5. Levy (1973).

6. Johnstone (1997); Partan & Marler (1999); Wiley (1983).

7. Unless the grief-stricken individual begins to speak, in which case the voice quality will be noticeably affected.

8. Nor is choking up a cue, since there is nothing to be observed.

9. Ekman (1982, 2003).

10. Earley, personal communication (2013); Kottler (1996, p. 63).

11. Zahn-Waxler et al. (1992).

12. The photograph is of Nigeria celebrity/actress Hilda Dokubo crying while speaking on how hunger affects poor people at the HungerFREE Campaign of ActionAid. This image is licensed under the Creative Commons Attribution-Share Alike 3.0 Unported license. See https://commons.wikimedia.org/wiki/File:Hilda_Dokubo_crying _1.jpg.

13. Balsters et al. (2013).

14. Russell (1994).

15. Ansfield (2007); LaFrance et al. (2003).

16. Uncomfortable (Ansfield, 2007), embarrassed (Edelmann et al., 1989), shy or polite (Ambadar et al., 2009; Hess et al., 2002), and socially apprehensive (Ickes et al., 1982).

17. Incidentally, the fact that English distinguishes grief from weeping but has no word for a melancholic expression is consistent with the idea that melancholy is a cue rather than a signal.

Chapter 4

1. Becker (1933); Bowlby (1961); Engel (1962); Farberow & Shneidman (1961); Gorer (1965); Henderson (1974); Lewis (1934); and many others.

2. E.g., Cornelius (1997); Frey (1985, 1992); Kottler (1996); Lutz (1999); Vingerhoets (2013); Vingerhoets & Cornelius (2001); Vingerhoets et al. (2000).

3. Kottler (1996, pp. 68–69).

4. Lane (2006).

5. Landreth (1941).

6. Kottler & Montgomery (2001, p. 10).

7. Ellis (1995); Hopcroft (2006).

8. Zerjal et al. (2003). The estimate is based on Y-chromosomal lineage for males. A similar proportion would be expected for females.

9. Khan (2010).

10. Landreth (1941) observed that among children between two and five years of age, boys cry (slightly) more often than girls.

11. Delp & Sackeim (1987).

12. Plato, *Republic* X.605d–606b.

13. Rosenblatt et al. (1976).

14. Vingerhoets (2013).

15. The proximal/ultimate distinction can be understood as addressing two different questions regarding some trait: the "how" and the "why" (Tinbergen, 1963). It should be noted that the proximal/ultimate distinction in evolutionary theory has been the subject of some criticism, especially in the context of development (Lickliter & Berry, 1990). However, see Hochman (2013) for a robust defense.

16. Haldane (1932, 1955); Hamilton (1963, 1964); see also Nowak et al. (2010) with rejoinder by Foster et al. (2006).

17. Trivers (1971); see also Axelrod & Hamilton (1981).

18. Haley (2002); Haley & Fessler (2005); Milinski et al. (2002); Semmann et al. (2005); Silk & Boyd (2010).

19. Silk & Boyd (2010).

20. Latané & Nida (1981).

21. Frijda (1986); Tomkins (1980).

22. As quoted in Ridley (1997, p. 1).

23. Davis (1983). In Davis's IRI model, this component is referred to as "personal distress."

24. Singer et al. (2004).

25. The role of commiseration or observer distress in motivating helping behaviors is described at length in the influential theory of Batson et al. (1987).

26. Evidence consistent with the effect of observer distress can be found in Hendriks et al. (2008) and Hendriks & Vingerhoets (2006). The motivating effect of observer distress is much weaker when observing weeping strangers (Zickfeld et al., 2021), consistent with the experimental results from Singer et al. (2004).

27. Bentham (1789).

28. Satow (1975).

29. Izuma et al. (2008).

30. William Shakespeare, *Othello*, act 2, scene 3.

31. Vermeir et al. (2017).

32. Apart from experimental evidence, there exists a rich game-theoretic literature showing the importance of reputation in models of cooperation. E.g., Axelrod (1984).

33. In Davis's IRI model (1983), this component is referred to as "empathic concern."

34. Zahn-Waxler et al. (1992).

35. Harbaugh et al. (2007).

36. Gospel According to Matthew, chapter 6, verses 2–4.

37. Zickfeld et al. (2021).

38. Berridge & Robinson (1998); Gebauer et al. (2012); Weiss et al. (1993).

39. See Lutz (1999).

40. Breuer & Freud (1895/1968); Labott (2001).

41. Although Cornelius's study was limited to American culture, the assumption that crying is beneficial appears to be widespread throughout Western culture.

42. Borgquist (1906); Cornelius (1997, 2001); Hendriks et al. (2008); Kottler (1996); Mélinand (1902); Nelson (1998); Roes (1990); Stougie et al. (2004).

43. Hendriks et al. (2001).

44. E.g., Cornelius (1997); Stougie et al. (2004).

45. See Vingerhoets et al. (2000); more recent studies include Baker (2019); Sharman et al. (2020).

46. Stougie et al. (2004, p. 13).

47. E.g., Bindra (1972); Frey et al. (1983).

48. Becht & Vingerhoets (2002).

49. Bylsma et al. (2008).

50. Cornelius (1997, 2001).

51. Once again, recall that emotions act as behavioral amplifiers that promote certain action tendencies: we are more likely to act in a particular way in the presence of some emotion than in its absence (Tomkins, 1980).

Chapter 5

1. Apart from motivating immediate action, the idea that emotions facilitate long-term learning has been highlighted by Baumeister et al. (2007).

2. Ginsburg (2007); Wyver & Spence (1999).

3. Fredrickson (2004); Fredrickson & Branigan (2005).

4. More flexible (Isen & Daubman, 1984), more integrative (Isen et al., 1991), more creative (Isen et al., 1987), and wider attention (Derryberry & Tucker, 1994; Basso et al., 1996).

5. Forgas & Moylan (1987).

6. Fredrickson (1998, 2001, 2004).

7. See reviews by Isen (2000) and Fredrickson (2004).

8. Love et al. (2015); Sharot (2011); Wright & Bower (1992); Weinstein (1980).

9. Arkes et al. (1988); Bassi et al. (2013); Gibson & Sanbonmatsu (2004); Gilovich (1983); Isen et al. (1982); Isen & Patrick (1983); Kuhnen & Knutson (2011); Schulreich et al. (2014).

10. Au et al. (2003).

11. Otto et al. (2016).

12. E.g., Sharot (2011).

13. Nesse (1991, 2018).

14. Alloy & Abramson (1979); Dobson & Franche (1989); Ackermann & DeRubeis (1991); Pacini et al. (1998); Moore & Fresco (2012).

15. More detail-oriented thinking (Clore & Huntsinger, 2007), reduced stereotyping (Bless & Fiedler, 2006), less judgment bias (Clore & Huntsinger, 2007; Tan & Forgas, 2010), greater memory accuracy (Forgas et al., 2009; Kraemer et al., 1989; Storbeck & Clore, 2005), reduced gullibility (Forgas, 2019; Forgas & East, 2008), more patience (Zhou et al., 2021), greater task perseverance (Goldenberg & Forgas, 2012), more social attentiveness and politeness (Forgas, 1995, 2002), more accurate assessments of the emotional states of others (Weary & Edwards, 1994; Yost & Weary, 1996), a heightened sense of fairness (Harlé & Sanfey, 2007), improved reasoning related to social

risks (Badcock & Allen, 2003), and a disposition to favor delayed larger rewards over immediate smaller rewards (Zhou et al., 2021). An exception to the cognitive benefits of melancholic realism was found by Bodenhausen et al. (2000), who showed that sad participants are more susceptible to anchoring effects—a type of cognitive bias where an individual's decisions or judgments are unconsciously influenced by an irrelevant reference point or anchor.

16. Huntsinger et al. (2014); Huntsinger & Ray (2016).

17. Huntsinger & Ray (2016).

18. Andrews & Thomson (2009).

19. Nesse (1991, 2018).

20. E.g., Andrews & Thomson (2009); Hagen (2011); Horwitz & Wakefield (2007); Keedwell (2008); Nesse (2000, 2018); Sharot (2011); Wilson (2008).

21. Fiedler (2000); Piaget (1936/1952).

22. Fiedler (2000); see also Koch et al. (2013).

23. Matheson et al. (2008).

24. Matheson et al. (2008). An earlier study by Harding et al. (2004) reported comparable results for mice.

25. As quoted in Wilson (2008, p. 3).

26. Forgas (2007).

27. Fredrickson (2001).

28. Nolen-Hoeksema (1991); Papageorgiou & Wells (2004).

29. Trapnell & Campbell (1999); see also Joireman et al. (2002).

30. Gilbert (2000); Gilbert's original quote refers to depression rather than melancholy.

31. Reviewed by Andrews & Thomson (2009).

32. E.g., Gortner et al. (2006); Graf et al. (2008).

33. Andrews & Thomson (2009).

34. Freed (2009).

35. Nesse (1991, 2018).

36. See also Hagen (2011).

37. As cited in Barrett (2017, p. 147).

38. Schwarz (1990).

39. Strunk et al. (2006).

Chapter 6

1. Pliny the Elder (AD 77/1900, vii, 1.4).

2. Birney & Teevan (1961); Blumberg (2017).

3. Wikipedia article "Instinct." Accessed January 13, 2020.

4. Lorenz (1937).

5. Immelmann & Beer (1989).

6. Gorenstein (1982); Chen et al. (2007); Miller & Cummings (2007).

7. Other contenders include smiling, laughing, fear, and disgust, which can all be observed in newborn infants. However, even these displays are controversial among some scholars. For example, smiling rarely appears before a newborn has first witnessed the smiling of a caregiver.

8. Miller & Cummings (2007); Bornstein et al. (2017).

9. Kraemer & Hastrup (1988).

10. Kraemer & Hastrup (1988).

11. Child (2011).

12. Gračanin et al. (2017), as cited in Baker (2018).

13. Fridlund disagrees with this interpretation and argues that, when alone, displays like weeping arise because we imagine the presence of some audience (Fridlund, 1994).

14. Balsters et al. (2013).

15. Miller & Cummings (2007); Bornstein et al. (2017).

16. Fröhlich et al. (2019).

17. Russon et al. (2009).

18. Russon et al. (2009).

19. Hobolth et al. (2011).

20. Game-theoretic models suggest that the likelihood for cooperation depends on the probability of repeated future interaction. Ahn et al. (2017); Andreoni & Miller (1993); Axelrod & Hamilton (1981).

21. Bloom (2016).

22. Levinson (2006).

23. Bonanno (2019).

24. Bonanno et al. (2015); Galatzer-Levy et al. (2018).

25. Bonanno (2019).

26. Fridlund (1994, p. 132).

27. See Johnstone & Grafen (1993); Rowell et al. (2006); Searcy & Nowicki (2005).

28. Fridlund further argues that displays are not automatic. He gives the following example: "One does not smile upon opening a gift, but rather when catching the gaze of the gift-giver. . . . A child will smile and look excited when expecting candy from an adult, but consume it impassively after obtaining it (though she may smile toward the adult who gave it to her)" (Fridlund, 1994, p. 154). The child's smiling is not correlated with the happy state of consuming the candy but instead directed at the candy-supplying adult. This is consistent with signaling theory that suggests the purpose of a display is to influence the state of the observer rather than communicating the displayer's feelings. But Fridlund also regards such audience effects as evidence against displays as having any instinctive automaticity. Here we part company with Fridlund. While audience effects offer evidence of the role of cognitive assessment, they do not suggest that displays are simply willful socially normative acts. Such displays are "instinctive behaviors" according to the classic definition—namely, complex behaviors that are not learned and that arise spontaneously in predictable circumstances.

29. It should be noted that different displays may exhibit greater or lesser automaticity. For example, smiling appears to be under greater cognitive control than weeping.

Chapter 7

1. iStock photo.

2. One can't discuss "joyful weeping" without mentioning that men appear to be much less likely to exhibit it than women. Nevertheless, I have witnessed at least one male academic reduced to tears when receiving a scholarly society's lifetime achievement award. It may be that this sex-related difference is unique to Western culture, but perhaps not.

3. Provine (2000, p. 40).

4. Provine (2000, p. 45).

5. Vingerhoets (2013).

6. The idea of the unconscious mind preceded Freud's description. Franz Brentano discussed the concept in his *Psychology from an Empirical Standpoint* (1874/2012), and Freud was also familiar with the earlier discussion in Eduard von Hartmann's book *The Philosophy of the Unconscious* (1869/1884).

7. Nisbett & Wilson (1977).

8. Fiske et al. (2019) identify a number of possible synonyms in dozens of languages for this somewhat amorphous feeling. See also Fiske et al. (2017).

9. http://kamamutalab.org/about/kama-muta-in-other-languages/. Accessed October 17, 2020.

Chapter 8

1. Potegal et al. (2010).

2. The idea was first proposed by Alexander (1939).

3. E.g., Siegman (1993); Vandervoort et al. (1996); al'Absi & Bongard (2006).

4. Alan Fridlund offers the same critique in his 1994 book—a model he calls the "Emotions View." The idea originates in the work of Dawkins & Krebs (1978; Krebs & Dawkins, 1984).

5. Russell (1994); see also the review by Barrett (2017).

6. Durán et al. (2017); see also Gendron et al. (2014).

7. Durán et al. (2017).

8. Durán et al. (2017).

9. Fridlund (1994); Crivelli & Fridlund (2019); see also Leys (2017).

10. It should be noted that my use of the word "manipulation" is meant in the sense of influence, maneuver, or encourage, rather than in the sense of deceive or mislead.

11. Scottish cognitive scientist Thom Scott-Phillips has offered a detailed argument noting that while signals are communicative, they are not "informative" ("The informational view [of signals] is . . . conceptually unsound"; 2008, p. 387). I prefer using the words "informative" and "communicative" in the opposite manner to Scott-Phillips. For example, I regard cues as informative but not communicative.

12. It should be noted that in ethological signaling theory, a person (or nonhuman animal) who observes or is the target recipient of the ethological signal is usually referred to as the *receiver*. However, in order to avoid the implication that signals are intended to be informative rather than manipulative, throughout this book, we use a more neutral term, *observer*.

13. Of course, researchers *are* asking the right question from the perspective of an observer. Observers benefit from inferring the affective state of the signaler. The main point is that signals are not designed to reveal the signaler's emotion. See additionally the ensuing discussion.

14. E.g., Calder et al. (2003); Isaacowitz et al. (2007); Lambrecht et al. (2012); Lima et al. (2013); Mill et al. (2009); Orgeta (2010); Orgeta & Phillips (2008); Schlegel et al. (2014); Williams et al. (2009).

15. Ruffman et al. (2008).

16. Lima et al. (2013).

17. E.g., Pollak & Sinha (2002); Elfenbein & Ambady (2002, 2003).

18. See especially Baker (2019).

19. De Sousa (1987); Lench (2018).

Chapter 9

1. Maier et al. (1993).

2. Dantzer & Kelley (2007); Hart (1988).

3. Selye (1974).

4. Kantha (1992).

5. Exton (1997); Murray & Murray (1979).

6. Nesse & Williams (1994, pp. 27–28).

7. Kluger & Rothenburg (1979).

8. Charlton (2000); Dantzer (2009); Exton (1997); Exton et al. (1995); Holmes & Miller (1963); Johnson (2002); Kelley et al. (2003); Konsman et al. (2002); Miller (1964); Murray & Murray (1979); Mullington et al. (2000); Weinberg (1984); Yirmiya (1996).

9. Pruimboom (2020).

10. Nesse (2018).

11. George Engel (1962) was perhaps the first to suggest that sadness-related anergia might serve a resource-conserving function that is an appropriate response to stress. Benjamin Hart (1988) more specifically speculated that sickness behaviors in general permit the reallocation of energy resources from activities like foraging to improved immune responsiveness. Andrew Miller has argued more pointedly that anergia frees metabolic resources for wound healing and fighting infection (Miller et al., 2013).

12. Notably prostaglandin E.

13. Rang (2003, p. 234).

14. Capuron et al. (2002).

15. Capuron et al. (2002); Musselman et al. (2001); Raison et al. (2006).

16. Nesse & Williams (1994).

17. E.g., Nesse (1991, 2018).

18. E.g., Allen & Badcock (2003); Andrews & Thomson (2009); Horwitz & Wakefield (2007); Keedwell (2008); Nesse (2000, 2018); Nesse & Williams (1994); Sharot (2011); Wilson (2008).

19. Chan et al. (2020); Felger & Lotrich (2013); Miller et al. (2013); Raison et al. (2006).

20. Everson (1993); Rogers et al. (2001).

21. Dallaspezia & Benedetti (2011); Pflug & Tölle (1971).

22. Further evidence in support of this interaction between sleep, immune function, and depression is evident in changes to psychological rewards. As already noted, melancholy and depression are associated with anhedonia: when we are sad, normally alluring stimuli lose their appeal. Interestingly, research has shown that sleep deprivation restores or enhances sensitivity to rewards (Gujar et al., 2011; Venkatraman et al., 2007; Venkatraman et al., 2011). For example, sleep deprivation increases activity in the ventral striatum in response to a reward (Mullin et al., 2013). Reduce the effectiveness of the immune system through sleep deprivation, and people become more motivated by reward: anhedonia is supplanted by prohedonia.

23. Berk et al. (2013).

24. Specifically, the dorsal anterior cingulate cortex; Eisenberger & Lieberman (2004).

25. Slavich et al. (2010).

26. Kenis & Maes (2002).

27. Mangino et al. (2017); Fernandez-Pujals et al. (2015).

28. Raison & Miller (2013).

29. Miller et al. (2013).

30. Saul & Bernstein (1941).

31. Brown (1946).

32. Pioro et al. (2010).

33. Nguyen et al. (2016).

34. Alvarez (2009); Schneider et al. (2014).

35. The tuberomammillary nucleus is located in the posterior region of the hypothalamus.

36. Okuda et al. (2009); Nuutinen et al. (2011); see review by Schneider et al. (2014).

37. For example, histamine is a target for studies related to Parkinson's-like motor problems: Inoue et al. (1996); Kubota et al. (2002); Takahashi et al. (2002); Toyota et al. (2002); Nuutinen et al. (2011); see review by Schneider et al. (2014).

38. Alvarez (2009); Blandina et al. (2004); Hancock & Fox (2004); McAfoose & Baune (2009); Schneider et al. (2014).

39. Niver (1948).

40. Ito (2004).

41. Panula et al. (1998); Mazurkiewicz-Kwilecki & Nsonwah (1989); Shan et al. (2012); see review by Schneider et al. (2014).

42. Tashiro et al. (2002).

43. E.g., Eidi et al. (2003); Köhler et al. (2011); Klein et al. (2016).

44. E.g., Anderson (2001).

45. Alvarez (2009); Brioni et al. (2011); Hancock & Fox (2004); McAfoose & Baune (2009); Schneider et al. (2014).

Chapter 10

1. Ehrlich (2000); Lloyd (1999).

2. Futuyma (2017).

3. Bateson (2000).

4. Elster (1989, p. 8).

5. Pardridge (2005).

6. Since fever involves a rise in temperature throughout the body, one might suppose it is a peripheral phenomenon. However, regulation of body temperature is a central process controlled by the hypothalamus.

7. Not on this list is hyperalgesia (pain sensitivity), which is more complicated. Primary (or focal) and secondary hyperalgesia are peripheral processes that are localized to particular (primary) damaged tissue or surrounding (secondary) undamaged tissue. However, there are also learned forms of hyperalgesia that have central origins.

8. Hart (1988).

9. Probably not insects.

10. Lopes (2014).

11. See Bonneaud et al. (2003); Brannelly et al. (2016); Duffield et al. (2017). For review, see Lopes (2014).

12. Lopes (2014).

13. Or grief.

14. Johnson (2002).

15. See review by Johnson (2002).

16. Attributed to Albert Einstein.

Chapter 11

1. Tinbergen (1952, 1964).

2. De Waal (1996).

3. Darley & Latané (1968); Darley & Batson (1973).

4. To be sure, people can be mutually helpful in treating parasites like lice and fleas—although not parasitic diseases like malaria.

5. Gould & Vrba (1982).

Chapter 12

1. Gustafson et al. (2000).

2. E.g., Barr et al. (2000).

3. E.g., Barr (1990).

4. Barry & Paxson (1971).

5. In the West, fears of reinforcing crying behaviors motivated the influential advice of Dr. Benjamin Spock, a famous Western parenting expert from the mid-twentieth century who explicitly cautioned parents against being too responsive to their crying infants (Spock, 1968).

6. Barry & Paxson (1971).

7. Barry & Paxson (1971, p. 487).

8. Soltis (2004, p. 448).

9. Darwin (1872).

10. Zeifman & Brown (2011).

11. Rao et al. (1997).

12. Provine (2000).

13. E.g., Panksepp & Bernatzky (2002).

14. Panksepp & Bernatzky (2002).

15. Wiesenfeld et al. (1981).

16. Vuorenkoski et al. (1969); Mead & Newton (1967).

17. Lenneberg et al. (1965, p. 31).

18. Lenneberg et al. (1965, p. 29).

19. See also Zeifman (2001).

20. E.g., Panksepp & Bernatzky (2002); Gračanin et al. (2017).

21. Soltis (2004).

22. Soltis (2004).

23. Newman (1985); Newman & Symmes (1982); Panksepp (1998); Soltis (2004).

24. van Lawick-Goodall (1968); Fossey (1972).

25. Newman (2004, 2007); Zeskind (2013).

26. Lingle et al. (2012).

27. Lingle & Riede (2014).

28. By contrast, the punctuated vocalized puffs characteristic of weeping (*ah-ah-ah-ah . . .*) better resemble so-called *pant-laughter* in great apes.

29. Potegal (2019, p. 117).

30. Einon & Potegal (2013); Potegal (2019, p. 106).

31. Belden et al. (2008); Gorer (1965); Hill & Hurtado (1996); Maretzki & Maretzki (1963).

32. Green et al. (2011).

33. Potegal & Davidson (2003); see also Potegal et al. (2003).

34. Potegal & Davidson (2003).

35. Potegal & Davidson (2003). Potegal (2019) labels the two factors anger and distress rather than anger and sadness. Potegal notes that "the major emotion in distress is sadness, but the term 'distress' is used to cover behaviors that might not routinely be included in sadness, such as dropping to the floor" (Potegal, 2019, p. 108). Potegal notes that the lowering of the body or a supine posture is a common act of submission or capitulation. Since an act of submission is consistent with a weeping signal, we will refer to this factor here as "sadness" rather than "distress."

36. This dual sadness/anger model of tantrums has been replicated in Eisbach et al. (2014) and Giesbrecht et al. (2010). See Potegal (2019).

37. Morton (1977, 1994).

38. Bolinger (1964); Chuenwattanapranithi et al. (2008).

39. Fitch (1997, 2000, 2010); Fitch & Hauser (1995); Smith & Patterson (2005).

40. Ohala (1980, 1984, 2010).

41. Hinton et al. (1994); Ohala (1980, 2010).

42. Grant (1969, p. 535).

43. Green et al. (2011).

44. Owings & Morton (1998, p. 115).

45. Screaming is produced by funneling a large volume of air through constricted vocal folds. The high tension in the vocal folds commonly results in energy being concentrated in higher vibrational modes—rather than the fundamental—with the result that the scream can exhibit an exceptionally high pitch. The large volume of rapid air passing through the vocal folds creates considerable turbulence, causing the vocal folds to vibrate in a chaotic manner. Typically, this results in amplitude modulation in the region of 30–150 Hz characteristic of the perception of roughness.

46. Anikin et al. (2018); Anikin & Persson (2017).

47. Screaming is thought to serve a number of functions. Driver and Humphries (1969) proposed that screaming originated as a way to foil predator attacks. One way to thwart an attack is simply to startle the predator. Many predators will abandon stalking a prey once their presence is detected. Screaming can also draw the attention of other predators, and the resulting competition between predators can sometimes allow the screaming animal to escape (Chivers et al. 1996; Högstedt, 1983). Screaming can be beneficial when it attracts the assistance of other conspecifics (Rohwer et al., 1976). A specific form of conspecific assistance is predator mobbing, where the screaming call results in conspecifics jointly harassing a predator—commonly seen when birds mob a hawk or owl (Dominey, 1983). Of course, screaming can also draw the attention of a predator to the screaming individual and so increase the likelihood of being caught. Despite the individual danger, when surrounded by related individuals, screaming may have value through kin selection—warning related individuals to avoid the predator (see discussion in Cheney & Seyfarth, 1985).

48. Arnal et al. (2015).

49. From the perspective of biological fitness, notice that acts of actual or threatened self-injury (such as breath holding or head banging) should be alarming to kin observers. Suicide is extremely rare among young children. However, threats of suicide or serious self-injury occur with some regularity in temper tantrums (a kitchen knife offers a convenient appliance for such threats). Among young children, threats of suicide occur most often at the peak of a temper tantrum (e.g., Bush & Pargament, 1994). Although other explanations are possible, such behaviors are most easily understood as a form of kin selection blackmail.

50. Larsen & McGraw (2011); Schimmack (2001).

51. Berrios et al. (2015).

52. Larsen & Green (2013).

53. Hunter et al. (2008, 2010); Larsen & Stastny (2011).

54. Geangu et al. (2010); Hatfield et al. (1993).

55. Potegal (2019, p. 113).

56. Frodi & Lamb (1980); Frodi (1985).

57. Trivers (1974).

58. Wolff (1987).

Chapter 13

1. As quoted in Matt (2011).

2. Summarized from Matt (2011).

3. Batcho (2013).

4. This English-language definition of nostalgia is ubiquitous in online dictionaries. The original source is not known.

5. Ad Vingerhoets, personal communication (2022).

6. Batcho (2007); Sedikides et al. (2006); Sedikides et al. (2008).

7. Batcho (2007); Sedikides et al. (2004); Barrett et al. (2010).

8. For music-induced nostalgia, see Barrett et al. (2010, p. 396).

9. Rubin et al. (1998); Conway & Haque (1999).

10. Nawas & Platt (1965).

11. Barrett et al. (2010).

12. Hepper et al. (2014).

13. Lovallo (1975); Walsh et al. (1989); Lowery et al. (2003).

14. Pert (1997).

15. Zhou et al. (2012).

16. Lench et al. (2015); Lench (2018).

17. See Batcho (2007); Sedikides et al. (2006); Sedikides et al. (2008).

18. For further discussion about the evolutionary roles of different forms of memory, see Huron (2006, chap. 12); see also Nawas & Platt (1965).

19. Batcho (1995).

20. Ritivoi (2002); Matt (2011, p. 145).

21. Damasio (1994); Ariely (2009); Kahneman (2011).

22. Damasio (1994).

23. Vingerhoets (2013).

24. Active research on nostalgia offers evidence suggestive of palliative functions apart from any presumed biological functions. Wildschut et al. (2006) suggest that nostalgia can serve to counteract negative mood. Similarly, Zhou et al. (2012) have proposed that nostalgia is a form of psychological "escape." They note that nostalgia provides a useful psychological tool that counteracts bad feelings, such as feelings of loneliness or in response to threat.

25. Memory also provides an essential basis for forming accurate expectations—another critical adaptive function. See Huron (2006) for an extended exposition regarding the psychology of expectation.

Chapter 14

1. de Léry (1578).

2. Métraux (1928).

3. As, for example, when encountering the Tupinambá.

4. Axelrod & Hamilton (1981).

5. Of course many cultures do have ritualized forms of weeping (e.g., Urban, 1988). One form of ritualized weeping evident in Western culture is fake or mock crying which will be discussed in chapter 15.

6. Blackman (1927, pp. 114, 123), cited in Rosenblatt et al. (1976, p. 43).

7. See, e.g., Brenchley (1873); Ewers (1958); Sumitra (2011); Zimmerman (2011).

8. Favazza (1996).

9. Favazza (1996).

10. Bentley et al. (2014); Gholamrezaei et al. (2017); Nock (2010).

11. Johnstone (1997).

12. Rosenblatt et al. (1976, p. 24).

13. Rosenblatt et al. (1976).

14. Gholamrezaei et al. (2017). Moreover, in Turkish society, self-harm is more accept-able in low-income regions (Toprak et al., 2011).

15. Zahavi & Zahavi (1997).

16. Darwin (1872, p. 155).

17. Forster (1908).

18. Clifton (2012).

19. Mackenzie (1771).

20. Dixon (2015).

21. Dixon notes that the era of the British stiff upper lip occupies a relatively brief period from about 1880 to 1945.

22. Dixon (2013a).

23. Indeed, Dixon himself concludes that "it is impossible to pin tears down." "A tear . . . can mean anything" (Dixon, 2013b).

24. Identified by Dixon (2013a).

25. Williams (1792, Vol. 1, p. 181), as quoted in Dixon (2013a).

26. Zahavi & Zahavi (1997).

Chapter 15

1. Examples of passive cues could include footprints, fingernail clippings, a set of keys, a cardboard box, and a skyscraper.

2. Recall that all signals evolve from preexisting cues that are recruited in situations· where the cue now benefits both the displayer and observer.

3. In classical Chinese, 呃 (è) means hiccup.

4. Precoda, personal communication (2021).

5. The idea that the Duchenne smile represents a genuine spontaneous display is controversial. Refer to the glossary entry for "Duchenne smile." See Duchenne (1862/1990); Ekman et al. (1990); Fridlund (1994, pp. 116–117).

6. Also, "pseudo" has the additional benefit of being more easily recognized by non-English speakers.

7. See Richard Nisbett's excellent 2003 book, *The Geography of Thought: How Asians and Westerners Think Differently . . . And Why*.

8. Laird (2007); Stepper & Strack (1993).

9. Alternatively, "ich" or "yecch."

10. As described in Attardo (2014, p. 45).

11. Mockery is thought to be unique to humans. This makes sense since it appears that only humans have the sophisticated cognitive control that allows us to produce pseudo-signals and social cues.

Chapter 16

1. Ickes et al. (1982); Edelmann et al. (1989); Ochanomizu (1991); Hess et al. (2002); Ambadar et al. (2009).

2. Notwithstanding the use of laughter in mocking behavior.

3. Note that in early *Homo sapiens* environments there would have been little privacy. Consequently, unlike modern environments where one can often find an empty room, most weeping would have been observed. That is, private crying would have been relatively rare in early hominin environments where weeping would have evolved.

4. Stephens et al. (2009).

5. See Austin (1962, p.105)

Chapter 17

1. It is interesting to note that modern research suggests that depression appears to be an autoimmune disorder.

Chapter 18

1. How fear (such as evoked by horror films) might lead to a positively valenced affect is discussed in my earlier book, *Sweet Anticipation: Music and the Psychology of Expectation*. In this book, we will avoid further discussion regarding the paradoxical enjoyment of fear.

2. A useful modern survey can be found in Levinson (2013). See also work by Jerrold Levinson, Carole Talon-Hugon, Aaron Smuts, Elizabeth Belfiore, Noel Carroll, Susan Feagin, and Katherine Allen.

3. Talon-Hugon (2014).

4. Notable exceptions include Juslin (2013); Eerola et al. (2016); Schubert (2016); Garrido (2017); Menninghaus et al. (2017); Vuoskoski & Eerola (2017).

5. Dunbar (1998).

6. Silk & Boyd (2010).

7. E.g., De Clercq (2014).

8. Strange & Leung (1999); Green & Brock (2000).

9. cf. Green et al. (2012); Koopman (2015).

10. Our discussion here avoids any consideration of the mechanism by which stories might engage spectator emotions. A plausible mechanism can be found in the so-called dual process theory (Evans, 1984; Evans & Frankish, 2009; Kahneman, 2003).

11. Plato, *The Republic*, book IV, 439e–440a. Translated by Benjamin Jowett.

12. Livio (2017).

13. Kang et al. (2009).

14. Jepma et al. (2012).

15. Loewenstein (1994); Litman (2005); Litman & Silvia (2006); Gottlieb et al. (2013).

16. Davies (2014); see also Korsmeyer (2014).

17. The enjoyment of a "happily ever after" video is typically reduced if the character is purely fictional: in portrayals of happiness, we tend to prefer factual (documentary) over fictional accounts.

18. E.g., Fehr & Fischbacher (2004).

Chapter 19

1. Gabrielsson (2011).

2. The research literature pertaining to sad music is extensive. Excluding the author's publications, a small sample of pertinent research includes Davies (1997a); Eerola et al. (2021); Garrido (2017); Garrido & Schubert (2011a, 2011b); Kawakami et al. (2013); Levinson (1982, 2013); Matsumoto (2002); Sachs et al. (2015, 2021); Schellenberg et al. (2008); Schubert (1996); Taruffi & Koelsch (2014); van den Tol (2016).

3. Huron (2015, 2016, 2018a, 2018b).

4. Juslin & Laukka (2003); Juslin (2013).

5. Slower overall tempos (Post & Huron, 2009), quieter dynamic levels (Turner & Huron, 2008), lower overall pitch (Bowling et al., 2012; Huron, 2008), small melodic intervals (Huron, 2008), legato articulation (Schutz et al., 2008), and darker timbres (Schutz et al., 2008).

6. Huron et al. (2014).

7. Warrenburg (2020).

8. Davies (1997b).

9. Garrido & Schubert (2011a).

10. Taruffi & Koelsch (2014).

11. Huron & Vuoskoski (2020).

12. Ladinig et al. (2019).

13. The ensuing account is based on Huron & Vuoskoski (2020).

14. The cognitive/affective distinction in empathy harks back at least 200 years (e.g., Smith, 1759; Spencer, 1870). Modern research has helped to delineate the behavioral, neuroanatomical, and neurochemical correlates for both affective empathy and cognitive empathy (Hurlemann et al., 2010; Lackner et al., 2010; Shamay-Tsoory & Aharon-Peretz, 2007; Shamay-Tsoory et al., 2009).

15. Affective empathy is considered to include emotion recognition, emotional contagion, motor empathy, and shared pain. The pertinent neuroanatomical network has been shown to include the inferior frontal gyrus, inferior parietal lobe, anterior cingulate cortex, and anterior insula. Notably, the inferior frontal gyrus is part of the human mirror neuron system (Kilner et al., 2009) and also involved in the internal simulation of emotional facial expressions (Van der Gaag et al., 2007). Affective empathy is thought to develop early in infancy and is linked to oxytocin (Shamay-Tsoory, 2011).

16. Cognitive empathy plays an essential role in perspective-taking and is thought to provide the basis for the ability to attribute mental states (such as beliefs, intentions, and desires) to other people (Baron-Cohen, 1995; Dennett, 1987; Premack & Woodruff, 1978). The pertinent neuroanatomical network includes the ventromedial and dorsomedial prefrontal cortices, the temporoparietal junction, and the medial temporal lobe. Cognitive empathy develops comparatively late, during adolescence, and is linked to dopamine (Shamay-Tsoory, 2011).

17. Heyes (2010).

18. Hatfield et al. (1993).

19. Decety & Jackson (2006).

20. Bryant (2013).

21. For a review, see Seyfarth & Cheney (2013).

22. Davis (1980, 1983). The subscales of the IRI have been associated with individual differences at the neural level, including greater gray matter volume in brain regions belonging to the mirror neuron system (Cheng et al., 2009) and greater activation of the mirror neuron system while observing facial expressions (Pfeifer et al., 2008) or listening to action sounds (Gazzola et al., 2006).

23. IRI, Davis (1980, 1983).

24. Klimecki et al. (2014); Singer & Klimecki (2014).

25. These sex differences are less apparent in the case of psychophysiological and nonverbal behavioral measures and most marked in the case of verbal self-report (Eisenberg & Lennon, 1983).

26. Vuoskoski & Eerola (2017).

27. Vuoskoski et al. (2012).

28. Replicated by later work in Vuoskoski & Eerola (2017).

29. In Finland, Vuoskoski & Eerola (2017), Eerola et al. (2016); in Japan, Kawakami & Katahira (2015); in Austria, Sattmann & Parncutt (2018).

30. Delving deeper into the principles underlying the evocation of sadness through music, Taruffi & Koelsch (2014) discovered that global trait empathy (and its subscales fantasy and empathic concern) correlated most significantly with emotion induction via *social functions* ($r=.40$)–"Sad music makes me feel sad because I am touched by the sadness of others"), further suggesting a link between music-induced sadness and compassionate responding to others' emotions.

31. It should be noted that all five studies cited above employed listening experiments that relied on experimenter-selected (unfamiliar) nominally sad music. It is possible that the use of unfamiliar music, or music deemed by experimenters to be "sad," is biased in some way. Two survey studies provide converging evidence that circumvent this possible confound. Garrido & Schubert (2011a) and Taruffi & Koelsch (2014) conducted survey studies and found similar patterns of correlations between trait empathy and an overall preference for nominally sad music. Since these surveys allowed respondents to interpret in their own way "sad music" or "music which makes me feel sadness or grief" (Garrido & Schubert, 2011a), these studies provide complementary evidence consistent with the broad picture of the association between trait empathy and sad music preferences.

32. Huron & Vuoskoski (2020).

33. Hume (1759/2004, p. 243).

34. Strange & Leung (1999); Green & Brock (2000); Green et al. (2003).

35. For further discussion regarding "plural pleasures," see Huron (2005).

36. E.g., Tversky & Kahneman (1986); Damasio (1994); Kahneman (2011).

37. Heider & Simmel (1944).

38. E.g., Watt & Ash (1998); Broze (2013); Aucouturier & Canonne (2017).

39. Friedman & Taylor (2014).

40. Zentner et al. (2008).

41. In the GEMS model by Zentner et al., terms that load significantly on tenderness include the French equivalents of affectionate, softened up, melancholic, nostalgic, dreamy, and sentimental.

42. Bråten (2007); Cova & Deonna (2014); Hanich et al. (2014); Menninghaus et al. (2015); Konečni (2005, 2015); Kuehnast et al. (2014); Fiske et al. (2017).

43. E.g., Carroll (1993); Konečni (2005).

44. Hanich et al. (2014); Kuehnast et al. (2014); Menninghaus et al. (2015).

45. Menninghaus et al. (2015).

46. Menninghaus et al. (2015, p. 24).

47. Examples abound. The literature on misattribution provides a rich source of examples. See also Nisbett & Wilson (1977).

48. Vuoskoski & Eerola (2017).

49. Huron & Vuoskoski (2020).

50. Huron (2006).

51. In the research literature, "empathic concern" and "personal distress," respectively.

Chapter 20

1. This theme is expanded in the theoretical work of Barrett (2017).

2. Dutton & Aron (1974).

3. Although this particular experimental design was challenged by other researchers, a subsequent flurry of research has established that misattribution is commonplace. Activation of the limbic system can often lead to incorrect attributions of the cause—resulting in indiscriminate or arbitrary associations.

4. Barrett (2017, p. 30).

5. There remains considerable controversy regarding the extent to which context shapes how people interpret interoceptive cues. Classic literature includes Schachter & Singer (1962, 1979); Schachter & Wheeler (1962); Marshall & Zimbardo (1979); Maslach (1979). See also Wilson (2004); Barrett (2017).

6. Chandler (2002).

7. Goleman (1995).

8. Barrett (2017, p. 101).

9. Barrett (2017, p. 102).

10. Wright (1601). As cited in Sullivan (2016, p. 16).

11. Bilson (1599). As cited in Sullivan (2016, p. 16).

12. Barrett (2017, p. 147).

13. This view is hardly new. In his voluminous and insightful ethnography of traditional Tahitian thought and experience, Robert Levy suggested, "'Feeling' becomes associated with cultural understandings which designate the cause of the feeling and what should be done about it. Feelings are halfway stations to action and are amenable to considerable cultural manipulation" (Levy, 1973, pp. 323–324).

14. Rendall et al. (2009).

15. Notable exceptions include mole-rats, meerkats, and dwarf mongooses.

16. My thanks to Federico Lauria for the suggested Latin.

Postscript

1. Leys (2017).

2. Engelking (2017).

3. Griffiths (1997).

Appendix

1. Fridlund's perspective echoes early behaviorist views that disparage unseen mental states. For example, Fridlund has written, "I suggest that facial displays can be understood without recourse to emotions or emotion terms. . . . In the end, emotion may be to the face as Ryle's 'ghost' (Ryle, 1949) was to the machine" (Fridlund, 1994, p. 186; repeated Fridlund, 2002, p. 75).

References

Ackermann, R., & DeRubeis, R. J. (1991). Is depressive realism real? *Clinical Psychology Review*, *11*(5), 565–584.

Ahn, T. K., Janssen, M. A., & Ostrom, E. (2017). Signals, symbols, and human cooperation. In R. W. Sussman & A. R. Chapman (Eds.), *The origins and nature of sociality* (pp. 122–140). Aldine de Gruyter.

al'Absi, M., & Bongard, S. (2006). Neuroendocrine and behavioral mechanisms mediating the relationship between anger expression and cardiovascular risk: Assessment considerations and improvements. *Journal of Behavioral Medicine*, *29*(6), 573–591.

Alexander, F. (1939). Emotional factors in essential hypertension. *Psychosomatic Medicine*, *1*, 173–179.

Allen, N. B., & Badcock, P. B. T. (2003). The social risk hypothesis of depressed mood: Evolutionary, psychosocial, and neurobiological perspectives. *Psychological Bulletin*, *129*(6), 887–913.

Alloy, L. B., & Abramson, L. Y. (1979). Judgment of contingency in depressed and nondepressed students: Sadder but wiser? *Journal of Experimental Psychology: General*, *108*(4), 441–485.

Alvarez, E. O. (2009). The role of histamine on cognition. *Behavioural Brain Research*, *199*(2), 183–189.

Ambadar, Z., Cohn, J. F., & Reed, L. I. (2009). All smiles are not created equal: Morphology and timing of smiles perceived as amused, polite, and embarrassed/nervous. *Journal of Nonverbal Behavior*, *33*, 17–34.

Anderson, M. C. (2001). Active forgetting: Evidence for functional inhibition as a source of memory failure. *Journal of Aggression, Maltreatment & Trauma*, *4*(2), 185–210.

Andreoni, J., & Miller, J. H. (1993). Rational cooperation in the finitely repeated prisoner's dilemma: Experimental evidence. *Economic Journal*, *103*(418), 57–85.

Andrews, P. W., & Thomson Jr., J. A. (2009). The bright side of being blue: Depression as an adaption for analyzing complex problems. *Psychological Review, 116*(3), 620–654.

Anikin, A., Bååth, R., & Persson, T. (2018). Human non-linguistic vocal repertoire: Call types and their meaning. *Journal of Nonverbal Behavior, 42*(1), 53–80.

Anikin, A., & Persson, T. (2017). Nonlinguistic vocalizations from online amateur videos for emotion research: A validated corpus. *Behavior Research Methods, 49*(2), 758–771.

Ansfield, M. E. (2007). Smiling when distressed: When a smile is a frown turned upside down. *Personality and Social Psychology Bulletin, 33*(6), 763–775.

Archer, J. (1999). *The nature of grief: The evolution and psychology of reaction to loss.* Routledge.

Ariely, D. (2009). *Predictably irrational.* Harper Perennial.

Arkes, H. R., Herren, L. T., & Isen, A. M. (1988). The role of potential loss in the influence of affect on risk-taking behavior. *Organizational Behavior and Human Decision Processes, 42*(2), 181–193.

Arnal, L. H., Flinker, A., Kleinschmidt, A., Giraud, A. L., & Poeppel, D. (2015). Human screams occupy a privileged niche in the communication soundscape. *Current Biology, 25*(15), 2051–2056.

Attardo, S. (2014). *Encyclopedia of humor studies.* Sage.

Au, K., Chan, F., Wang, D., & Vertinsky, I. (2003). Mood in foreign exchange trading: Cognitive processes and performance. *Organizational Behavior and Human Decision Processes, 91*(2), 322–338.

Aucouturier, J. J., & Canonne, C. (2017). Musical friends and foes: The social cognition of affiliation and control in improvised interactions. *Cognition, 161*, 94–108.

Austin, J. L. (1962). *How to do things with words.* Oxford University Press.

Axelrod, R. (1984). *The evolution of cooperation.* Basic Books.

Axelrod, R., & Hamilton, W. D. (1981). The evolution of cooperation. *Science, 211*(4489), 1390–1396.

Badcock, P. B. T., & Allen, N. B. (2003). Adaptive social reasoning in depressed mood and depressive vulnerability. *Cognition & Emotion, 17*(4), 647–670.

Baker, M. (2018). Recent advances in the crying literature. *PsyPAG Quarterly, 107*, 15–19.

Baker, M. (2019). *Blood, sweat and tears: The intra- and interindividual function of adult emotional weeping* [PhD dissertation, Department of Psychology, University of Portsmouth].

Balsters, M. J., Krahmer, E. J., Swerts, M. G., & Vingerhoets, A. J. (2013). Emotional tears facilitate the recognition of sadness and the perceived need for social support. *Evolutionary Psychology*, *11*(1), 148–158.

Banse, R., & Scherer, K. R. (1996). Acoustic profiles in vocal emotion expression. *Journal of Personality and Social Psychology*, *70*(3), 614–636.

Baron-Cohen, S. (1995). *Mindblindness: An essay on autism and theory of mind*. MIT Press.

Barr, R. G. (1990). The normal crying curve: What do we really know? *Developmental Medicine & Child Neurology*, *32*(4), 356–362.

Barr, R. G., Hopkins, B., & Green, J. A. (Eds.). (2000). *Crying as a sign, a symptom and a signal: Clinical, emotional and developmental aspects of infant and toddler crying*. Mac Keith Press.

Barrett, F. S., Grimm, K. J. Robins, R. W., Wildschut, T., Sedikides, C., & Janata, P. (2010). Music-evoked nostalgia: Affect, memory, and personality. *Emotion*, *10*(3), 390–403.

Barrett, L. F. (2017). *How emotions are made: The secret life of the brain*. Houghton Mifflin Harcourt.

Barry, H., & Paxson, L. M. (1971). Infancy and early childhood: Cross-cultural codes 2. *Ethnology*, *10*(4), 466–508.

Bassi, A., Colacito, R., & Fulghieri, P. (2013). 'O sole mio: An experimental analysis of weather and risk attitudes in financial decisions. *Review of Financial Studies*, *26*(7), 1824–1852.

Basso, M. R., Schefft, B. K., Ris, M. D., & Dember, W. N. (1996). Mood and global-local visual processing. *Journal of the International Neuropsychological Society*, *2*(3), 249–255.

Batcho, K. I. (1995). Nostalgia: A psychological perspective. *Perceptual and Motor Skills*, *80*(1), 131–143.

Batcho, K. I. (2007). Nostalgia and the emotional tone and content of song lyrics. *American Journal of Psychology*, *120*(3), 361–381.

Batcho, K. I. (2013). Nostalgia: The bittersweet history of a psychological concept. *History of Psychology*, *16*(3), 165–176.

Bateson, P. (2000). Taking the stink out of instinct. In H. Rose & S. Rose (Eds.), *Alas poor Darwin: Arguments against evolutionary psychology* (pp. 157–173). Harmony Books.

Batson, C. D., Fultz, J., & Schoenrade, P. A. (1987). Distress and empathy: Two qualitatively distinct vicarious emotions with different motivational consequences. *Journal of Personality*, *55*(1), 19–39.

Baumeister, R. F., Vohs, K. D., DeWall, N., & Zhang, L. (2007). How emotion shapes behavior: Feedback, anticipation, and reflection, rather than direct causation. *Personality and Social Psychology Review*, *11*(2), 167–203.

Becht, M. C., & Vingerhoets, A. J. (2002). Crying and mood change: A cross-cultural study. *Cognition & Emotion*, *16*(1), 87–101.

Becker, H. (1933). The sorrow of bereavement. *Journal of Abnormal and Social Psychology*, *27*(4), 391–410.

Belden, A. C., Thomson, N. R., & Luby, J. L. (2008). Temper tantrums in healthy versus depressed and disruptive preschoolers: Defining tantrum behaviors associated with clinical problems. *Journal of Pediatrics*, *152*(1), 117–122.

Bentham, J. (1789). *Principles of morals and legislation*. Clarendon.

Bentley, K. H., Nock, M. K., & Barlow, D. H. (2014). The four-function model of nonsuicidal self-injury: Key directions for future research. *Clinical Psychological Science*, *2*(5), 638–656.

Ben-Ze'ev, A. (2001). *The subtlety of emotions*. MIT Press.

Berk, M., Williams, L. J., Jacka, F. N., O'Neil, A., Pasco, J. A., Moylan, S., Allen, N. B., Stuart, A. L., Hayley, A. C., Byrne, M. L., & Maes, M. (2013). So depression is an inflammatory disease, but where does the inflammation come from? *BMC Medicine*, *11*(1), 1–16.

Berridge, K. C., & Robinson, T. E. (1998). What is the role of dopamine in reward: Hedonic impact, reward learning, or incentive salience? *Brain Research Reviews*, *28*(3), 309–369.

Berrios, R., Totterdell, P., & Kellett, S. (2015). Eliciting mixed emotions: a meta-analysis comparing models, types, and measures. *Frontiers in Psychology*, *6*, Article 428.

Bilson, T. (1599). *The effect of certain sermons touching the full redemption of mankind by the death and blood of Christ Jesus*. Peter Short.

Bindra, D. (1972). Weeping: A problem of many facets. *Bulletin of the British Psychological Society*, *25*(89), 281–284.

Birney, R. C., & Teevan, R. C. (Eds.). (1961). *Instinct: An enduring problem in psychology*. Van Nostrand.

Blackman, W. S. (1927). *The Fellahin of Upper Egypt*. Harrap.

Blandina, P., Efoudebe, M., Cenni, G., Mannaioni, P., & Passani, M. B. (2004). Acetylcholine, histamine, and cognition: Two sides of the same coin. *Learning and Memory*, *11*(1), 1–8.

Bless, H., & Fiedler, K. (2006). Mood and the regulation of information processing and behavior. In J. P. Forgas (Ed.), *Affect in social thinking and behavior* (pp. 65–84). Psychology Press.

Bloom, P. (2016). *Against empathy: The case for rational compassion.* HarperCollins.

Blumberg, M. S. (2017). Development evolving: The origins and meanings of instinct. *Wiley Interdisciplinary Reviews: Cognitive Science, 8*(1–2), e1371.

Bodenhausen, G. V., Gabriel, S., & Lineberger, M. (2000). Sadness and susceptibility to judgmental bias: The case of anchoring. *Psychological Science, 11*(4), 320–323.

Bolinger, D. L. (1964). Intonation across languages. In J. Greenberg, C. Ferguson, & E. Moravcsik (Eds.), *Universals of human language: Vol. 2. Phonology* (pp. 417–524). Stanford University Press.

Bonanno, G. A. (2019). *The other side of sadness: What the new science of bereavement tells us about life after loss* (Rev. ed.). Basic Books.

Bonanno, G. A., Goorin, L., & Coifman, K. G. (2008). Sadness and grief. In M. lewis, J. Haviland-Jones, & L. F. Barrett (Eds.), *Handbook of emotions* (3rd ed., pp. 797–810). Guilford Press.

Bonanno, G. A., Romero, S. A., & Klein, S. I. (2015). The temporal elements of psychological resilience: An integrative framework for the study of individuals, families, and communities. *Psychological Inquiry, 26*(2), 139–169.

Bonneaud, C., Mazuc, J., Gonzalez, G., Haussy, C., Chastel, O., Faivre, B., & Sorci, G. (2003). Assessing the cost of mounting an immune response. *American Naturalist, 161*(3), 367–379.

Borgquist, A. (1906). Crying. *American Journal of Psychology, 17*(2), 149–205.

Bornstein, M. H., Putnick, D. L., Rigo, P., Esposito, G., Swain, J. E., Suwalsky, J. T. D., Su, X., Du, X., Zhang, K., Cote, L. R., De Pisapia, N., & Venuti, P. (2017). Neurobiology of culturally common maternal responses to infant cry. *Proceedings of the National Academy of Sciences, 114*(45), E9465–E9473.

Bowlby, J. (1961). Process of mourning. *International Journal of Psycho-Analysis, 42*(4–5), 317–340.

Bowlby, J. (1973). *Attachment and loss: Vol. 2. Separation.* Hogarth.

Bowling, D. L., Sundararajan, J., Han, S. E., & Purves, D. (2012). Expression of emotion in Eastern and Western music mirrors vocalization. *PLoS One, 7*(3), e31942.

Bradbury, J. W., & Vehrenkamp, S. L. (1998). *Principles of animal communication.* Sinauer.

Brannelly, L. A., Webb, R., Skerratt, L. F., & Berger, L. (2016). Amphibians with infectious disease increase their reproductive effort: Evidence for the terminal investment hypothesis. *Open Biology, 6*(6), 150251.

Bråten, S. (2007). *On being moved: From mirror neurons to empathy.* John Benjamins.

Breitenstein, C., van Lancker, D., & Daum, I. (2001). The contribution of speech rate and pitch variation to the perception of vocal emotions in a German and an American sample. *Cognition & Emotion, 15*(1), 57–79.

Brenchley, J. L. (1873). *Jottings during the Cruise of HMS 'Curacoa' among the South Seas Islands in 1865.* Longmans, Green, and Company.

Brentano, F. (1874/2012). *Psychologie vom empirischen Standpunkt.* Dunker & Humblot. Translated as *Psychology from an empirical standpoint.* Routledge.

Breuer, J., & Freud, S. (1895/1968). *Studien über Hysterie.* In Gesamtwerk Band 1. Frankfurt: S. Fischer. Trans. as *Studies on hysteria.* Hogarth Press.

Brioni, J. D., Esbenshade, T. A., Garrison, T. R., Bitner, S. R., & Cowart, M. D. (2011). Discovery of histamine H3 antagonists for the treatment of cognitive disorders and Alzheimer's disease. *Journal of Pharmacology and Experimental Therapeutics, 336*(1), 38–46.

Brown, W. T. (1946). The probable role of histamine in some emotionally precipitated allergic conditions. *Yale Journal of Biology and Medicine, 19*(1), 63–65.

Broze, G. J. (2013). Animacy, anthropomimesis, and musical line. [PhD dissertation, School of Music, Ohio State University.]

Bryant, G. A. (2013). Animal signals and emotion in music: Coordinating affect across groups. *Frontiers in Psychology, 4,* Article 990.

Bush, E. G., & Pargament, K. I. (1994). A quantitative and qualitative analysis of suicidal preadolescent children and their families. *Child Psychiatry and Human Development, 25*(4), 241–252.

Bylsma, L. M., Vingerhoets, A. J., & Rottenberg, J. (2008). When is crying cathartic? An international study. *Journal of Social and Clinical Psychology, 27*(10), 1165–1187.

Calder, A. J., Keane, J., Manly, T., Sprengelmeyer, R., Scott, S., Nimmo-Smith, I., & Young, A. W. (2003). Facial expression recognition across the adult life span. *Neuropsychologia, 41*(2), 195–202.

Capuron, L., Gumnick, J. F., Musselman, D. L., Lawson, D. H., Reemsnyder, A., Nemeroff, C. B., & Miller, A. H. (2002). Neurobehavioral effects of interferon-alpha in cancer patients: Phenomenology and paroxetine responsiveness of symptom dimensions. *Neuropsychopharmacology, 26*(5), 643–652.Caro, T. M. (1986). The functions of stotting in Thomson's gazelles: Some tests of the predictions. *Animal Behaviour, 34*(3), 663–684.

Carroll, N. (1993). On being moved by nature: Between religion and natural history. In S. Kemal & I. Gaskell (Eds.), *Landscape, natural beauty and the arts* (pp. 244–266). Cambridge University Press.

Chan, R. F., Turecki, G., Shabalin, A. A., Guintivano, J., Zhao, M., Xie, L. Y., van Grootheest, G., Kaminsky, Z., Dean, B., Penninx, B., & Aberg, K. A. (2020). Cell type–specific methylome-wide association studies implicate neurotrophin and innate immune signaling in major depressive disorder. *Biological Psychiatry, 87*(5), 431–442.

Chandler, D. (2002). *Semiotics: The basics*. Routledge.

Charlton, B. G. (2000). The malaise theory of depression: Major depressive disorder is sickness behavior and antidepressants are analgesic. *Medical Hypotheses, 54*(1), 126–130.

Chen, A. C., Porjesz, B., Rangaswamy, M., Kamarajan, C., Tang, Y., Jones, K. A., Chorlian, D. B., Stimus, A. G., & Begleiter, H. (2007). Reduced frontal lobe activity in subjects with high impulsivity and alcoholism. *Alcoholism: Clinical and Experimental Research, 31*(1), 156–165.

Cheney, D. L., & Seyfarth, R. M. (1985). Vervet monkey alarm calls: Manipulation through shared information? *Behaviour, 94*(1–2), 150–166.

Cheng, Y., Chou, K. H., Decety, J., Chen, I. Y., Hung, D., Tzeng, O. L., & Lin, C. P. (2009). Sex differences in the neuroanatomy of human mirror-neuron system: A voxel-based morphometric investigation. *Neuroscience, 158*, 713–720.

Child, B. (2011, August 18). Virgin Atlantic in-flight films to carry 'weepy warnings.' *The Guardian*. Retrieved April 14, 2022, from https://www.theguardian.com/film/2011/aug/18/virgin-atlantic-films-warning

Chivers, D. P., Brown, G. E., & Smith, R. J. F. (1996). The evolution of chemical alarm signals: Attracting predators benefits alarm signal senders. *American Naturalist, 148*(4), 649–659.

Chuenwattanapranithi, S., Xu, Y., Thipakorn, B., & Maneewongvatana, S. (2008). Encoding emotions in speech with the size code. *Phonetica, 65*(4), 210–230.

De Clercq, R. (2014). A simple solution to the paradox of negative emotion. In J. Levinson (Ed.), *Suffering art gladly: The paradox of negative emotion in art* (pp. 111–122). Springer.

Clifton, J. (2012, November 21). Singapore ranks as least emotional country in the world residents living in the Philippines are the most emotional. *Gallup News*. Retrieved August 16, 2021, from https://news.gallup.com/poll/158882/singapore-ranks-least-emotional-country-world.aspx#1

Clore, G. L., & Huntsinger, J. R. (2007). How emotions inform judgment and regulate thought. *Trends in Cognitive Science, 11*(9), 393–399.

Conway, M. A., & Haque, S. (1999). Overshadowing the reminiscence bump: Memories of a struggle for independence. *Journal of Adult Development, 6*(1), 35–44.

Cornelius, R. R. (1997). Toward a new understanding of weeping and catharsis? In A. Vingerhoets, F. Van Bussel, & A. Boelhouwer (Eds.), *The (non)expression of emotions in health and disease* (pp. 303–321). Tilburg University Press.

Cornelius, R. R. (2001). Crying and catharsis. In A. J. Vingerhoets & R. R. Cornelius (Eds.), *Adult crying: A biopsychosocial approach* (pp. 199–211). Brunner-Routledge.

Cova, F., & Deonna, J. A. (2014). Being moved. *Philosophical Studies, 169*, 447–466.

Crivelli, C., & Fridlund, A. J. (2019). Inside-out: From basic emotions theory to the behavioral ecology view. *Journal of Nonverbal Behavior, 43*(2), 161–194.

Dalla Bella, S., Peretz, I., Rousseau, L., & Gosselin, N. (2001). A developmental study of the affective value of tempo and mode in music. *Cognition, 80*(3), B1–B10.

Dallaspezia, S., & Benedetti, F. (2011). Chronobiological therapy for mood disorders. *Expert Review of Neurotherapeutics, 11*(7), 961–970.

Damasio, A. R. (1994). *Descartes' error: Emotion, reason and the human brain.* G. P. Putnam's Sons.

Dantzer, R. (2009). Cytokine, sickness behavior, and depression. *Immunology and Allergy Clinics, 29*(2), 247–264.

Dantzer, R., & Kelley, K. W. (2007). Twenty years of research on cytokine-induced sickness behavior. *Brain, Behavior, and Immunity, 21*(2), 153–160.

Darley, J. M., & Batson, C. D. (1973). "From Jerusalem to Jericho": A study of situational and dispositional variables in helping behavior. *Journal of Personality and Social Psychology, 27*(1), 100–108.

Darley, J. M., & Latané, B. (1968). Bystander intervention in emergencies: Diffusion of responsibility. *Journal of Personality and Social Psychology, 8*(4), 377–383.

Darwin, C. (1872). *The expression of emotions in man and animals.* John Murray.

Davidson, R. J. (2003). Affective neuroscience and psychophysiology: Toward a synthesis. *Psychophysiology, 40*(5), 655–665.

Davies, D. (2014). Watching the unwatchable: Irréversible, Empire, and the paradox of intentionally inaccessible art. In J. Levinson (Ed.), *Suffering art gladly: The paradox of negative emotion in art* (pp. 246–266). Macmillan Palgrave.

Davies, S. (1997a). Contra the hypothetical persona in music. In M. Hjort & S. Laver (Eds.), *Emotion and the arts* (pp. 95–109). Oxford University Press.

Davies, S. (1997b). Why listen to sad music if it makes one feel sad? In J. Robinson (Ed.), *Music and meaning* (pp. 242–253). Cornell University Press.

Davis, J. W. F., & O'Donald, P. (1976). Sexual selection for a handicap: A critical analysis of Zahavi's model. *Journal of Theoretical Biology, 57*(2), 345–354.

Davis, M. H. (1980). A multidimensional approach to individual differences in empathy. *JSAS Catalog of Selected Documents in Psychology, 10*, 85.

Davis, M. H. (1983). Measuring individual differences in empathy: Evidence for a multidimensional approach. *Journal of Personality and Social Psychology, 44*(1), 113–126.

Dawkins, R., & Krebs, J. R. (1978). Animal signals: Information or manipulation. In J. R. Krebs & N. B. Davies (Eds.), *Behavioural ecology: An evolutionary approach* (pp. 282–309). Blackwell.

Decety, J., & Jackson, P. L. (2006). A social-neuroscience perspective on empathy. *Current Directions in Psychological Science, 15*(2), 54–58.

Delp, M. J., & Sackeim, H. A (1987). Effects of mood on lacrimal flow: Sex differences and asymmetry. *Psychophysiology, 24*, 550–556.

Dennett, D. C. (1987). *The intentional stance*. MIT Press.

Derryberry, D., & Tucker, D. M. (1994). Motivating the focus of attention. In P. M. Neidenthal & S. Kitayama (Eds.), *The heart's eye: Emotional influences in perception and attention* (pp. 167–196). Academic Press.

De Sousa, R. (1987). *The rationality of emotion*. MIT Press.

Dixon, T. (2013a). Enthusiasm delineated: Weeping as a religious activity in eighteenth-century Britain. *Litteraria Pragensia, 22*(43), 59–81.

Dixon, T. (2013b, February 22). The Waterworks: Tears of sorrow, tears of joy, tears of incontinence or of ecstasy. Crying must mean something—but what? *Aeon*. Retrieved August 21, 2021, from https://aeon.co/essays/read-it-and-weep-what-it-means-when-we-cry

Dixon, T. (2015). *Weeping Britannia: Portrait of a nation in tears*. Oxford University Press.

Dobson, K., & Franche, R. L. (1989). A conceptual and empirical review of the depressive realism hypothesis. *Canadian Journal of Behavioural Science/Revue canadienne des sciences du comportement, 21*(4), 419–433.

Dominey, W. J. (1983). Mobbing in colonially nesting fishes, especially the bluegill, Lepomis macrochirus. *Copeia, 76*(4), 1086–1088.

Driver, P. M., & Humphries, D. A. (1969). The significance of the high-intensity alarm call in captured passerines. *Ibis, 111*(2), 243–244.

Duchenne, G. (1862/1990). *Mecanisme de la physionomie humaine*. Translated by R. A. Cuthbertson as *The mechanism of human facial expression*. Cambridge University Press.

Duffield, K. R., Bowers, E. K., Sakaluk, S. K., & Sadd, B. M. (2017). A dynamic threshold model for terminal investment. *Behavioral Ecology and Sociobiology*, *71*(12), 1–17.

Dunbar, R. I. M. (1998). *Grooming, gossip, and the evolution of language*. Harvard University Press.

Durán, J. I., Reisenzein, R., & Fernández-Dols, J. M. (2017). Coherence between emotions and facial expressions. In J. Fernández-Dols & J. Russell (Eds.), *The science of facial expression* (pp. 107–129). Oxford University Press.

Dutton, D. G., & Aron, A. P. (1974). Some evidence for heightened sexual attraction under conditions of high anxiety. *Journal of Personality and Social Psychology*, *30*(4), 510–517.

Ebersole, G. L. (2000). The function of ritual weeping revisited: Affective expression and moral discourse. *History of Religions*, *39*(3), 211–246.

Edelmann, R. J., Asendorpf, J., Contarello, A., Zammuner, V., Georgas, J., & Villanueva, C. (1989). Self-reported expression of embarrassment in five European cultures. *Journal of Cross-Cultural Psychology*, *20*(4), 357–371.

Eerola, T., Vuoskoski, J. K., Kautiainen, H. (2016). Being moved by unfamiliar sad music is associated with high empathy. *Frontiers in Psychology*, *7*, Article 1176.

Eerola, T., Vuoskoski, J. K., Kautiainen, H., Peltola, H. R., Putkinen, V., & Schäfer, K. (2021). Being moved by listening to unfamiliar sad music induces reward-related hormonal changes in empathic listeners. *Annals of the New York Academy of Sciences*, *1502*(1), 121–131.

Ehrlich, P. R. (2000). *Human natures: Genes, cultures, and the human prospect*. Island Press.

Eidi, M., Zarrindast, M. R., Eidi, A., Oryan, S., & Parivar, K. (2003). Effects of histamine and cholinergic systems on memory retention of passive avoidance learning in rats. *European Journal of Pharmacology*, *465*(1–2), 91–96.

Einon, D., & Potegal, M. (2013). Temper tantrums in young children. In M. Potegal & J. F. Knutson (Eds.), *The dynamics of aggression: Biological and social processes in dyads and groups* (pp. 175–212). Psychology Press.

Eisbach, S. S., Cluxton-Keller, F., Harrison, J., Krall, J. R., Hayat, M., & Gross, D. (2014). Characteristics of temper tantrums in preschoolers with disruptive behavior in a clinical setting. *Journal of Psychosocial Nursing and Mental Health Services*, *52*(5), 32–40.

Eisenberg, N., & Lennon, R. (1983). Sex differences in empathy and related capacities. *Psychological Bulletin*, *94*(1), 100–131.

Eisenberger, N. I., & Lieberman, M. D. (2004). Why rejection hurts: A common neural alarm system for physical and social pain. *Trends in Cognitive Science*, *8*(7), 294–300.

Ekman, P. (1982). *Emotion in the human face*. Cambridge University Press.

Ekman, P. (2003). *Emotions revealed: Understanding faces and feelings*. Times Books.

Ekman, P., Davidson, R. J., & Friesen, W. V. (1990). The Duchenne smile: Emotional expression and brain physiology II. *Journal of Personality and Social Psychology, 58*(2), 342–353.

Ekman, P., & Friesen, W. V. (2003). *Unmasking the face: A guide to recognizing emotions from facial clues*. Malor Books.

Elfenbein, H. A., & Ambady, N. (2002). On the universality and cultural specificity of emotion recognition: A meta-analysis. *Psychological Bulletin, 128*(2), 203–235.

Elfenbein, H. A., & Ambady, N. (2003). When familiarity breeds accuracy: Cultural exposure and facial emotion recognition. *Journal of Personality and Social Psychology, 85*(2), 276–290.

Ellis, L. (1995). Dominance and reproductive success among nonhuman animals: A cross-species comparison. *Ethology and Sociobiology, 16*(4), 257–333.

Elster, J. (1989). *Nuts and bolts for the social sciences*. Cambridge University Press.

Engel, G. L. (1962). Anxiety and depression-withdrawal. The primary affects of unpleasure. *International Journal of Psychoanalysis, 43*(2–3), 89–97.

Engelking, C. (2017). What is an organ? The definition of an organ in our bodies is more slippery than you think. *Discover Magazine*. Retrieved September 9, 2022, from https://www.discovermagazine.com/health/the-mesentery-isnt-the-organ-you-think -it-is

Erickson, D., Yoshida, K., Menezes, C., Fujino, A., Mochida, T., & Shibuya, Y. (2006). Exploratory study of some acoustic and articulatory characteristics of sad speech. *Phonetica, 61*(1), 1–25.

Erritzoe, D., Godlewska, B. R., Rizzo, G., Searle, G. E., Agnorelli, C., Lewis, Y., Ashok, A. H., Colasanti, A., Boura, I., Farrell, C., Parfit, H., Howes, O., Passchier, J., Gunn, R. N., Nutt, D. J., Cowen, P. J., Knudsen, G., & Rabiner, E. A. (2023). Brain serotonin release is reduced in patients with depression: A [11C] Cimbi-36 positron emission tomography study with a d-amphetamine challenge. *Biological Psychiatry, 93*(12), 1089–1098.

Eshel, I. (1978). On the handicap principle—A critical defense. *Journal of Theoretical Biology, 70*(2), 245–250.

Evans, J. S. B. (1984). Heuristic and analytic processes in reasoning. *British Journal of Psychology, 75*(4), 451–468.

Evans, J. S. B., & Frankish, K. E. (2009). *In two minds: Dual processes and beyond*. Oxford University Press.

Everson, C. A. (1993). Sustained sleep deprivation impairs host defense. *American Journal of Physiology, 265*(5), R1148–R1154.

Ewers, J. C. (1958). *The Blackfeet: Raiders of the northwestern plains.* University of Oklahoma Press.

Exton, M. S. (1997). Infection-induced anorexia: Active host defence strategy. *Appetite, 29*(3), 369–383.

Exton, M. S., Bull, D. F., King, M. G., & Husband, A. J. (1995). Modification of body temperature and sleep state using behavioral conditioning. *Physiology & Behavior, 57*(4), 723–729.

Fairbanks, G., & Pronovost, W. (1939). An experimental study of the pitch characteristics of the voice during the expression of emotion. *Speech Monographs, 6*(1), 87–104.

Farberow, N. L., & Shneidman, E. S. (1961). *The cry for help.* Blakiston Division, McGraw-Hill.

Favazza, A. R. (1996). *Bodies under siege: Self-mutilation and body modification in culture and psychiatry* (2nd ed.). Johns Hopkins University Press.

Fehr, E., & Fischbacher, U. (2004). Third-party punishment and social norms. *Evolution and Human Behavior, 25*(2), 63–87.

Felger, J. C., & Lotrich, F. E. (2013). Inflammatory cytokines in depression: Neurobiological mechanisms and therapeutic implications. *Neuroscience, 246*, 199–229.

Fernandez-Pujals, A. M., Adams, M. J., Thomson, P., McKechanie, A. G., Blackwood, D. H., Smith, B. H., Dominiczak, A. F., Morris, A. D., Matthews, K., Campbell, A., Linksted, P., Haley, C. S., Deary, I. J., Porteous, D. J., MacIntyre, D. J., & McIntosh, A. M. (2015). Epidemiology and heritability of major depressive disorder, stratified by age of onset, sex, and illness course in generation Scotland: Scottish Family Health Study (GS:SFHS). *PLoS One, 10*(11), e0142197.

Fiedler, K. (2000). Affective states trigger processes of assimilation and accommodation. In G. L. Clore & L. L. Martin (Eds.), *Theories of mood and cognition: A user's guidebook* (pp. 85–99). Psychology Press.

Fiske, A. P., Schubert, T., & Seibt, B. (2017). "Kama muta" or "being moved by love": A bootstrapping approach to the ontology and epistemology of an emotion. In J. Cassaniti & U. Menon (Eds.), *Universalism without uniformity: Explorations in mind and culture* (pp. 79–100). University of Chicago Press.

Fiske, A. P., Seibt, B., & Schubert, T. (2019). The sudden devotion emotion: Kama Muta and the cultural practices whose function is to evoke it. *Emotion Review, 11*(1), 74–86.

Fitch, W. T. (1997). Vocal tract length and formant frequency dispersion correlate with body size in rhesus macaques. *Journal of the Acoustical Society of America, 102*(2), 1213–1222.

Fitch, W. T. (2000). The evolution of speech: A comparative review. *Trends in Cognitive Sciences, 4*(7), 258–267.

Fitch, W. T. (2010). *The evolution of language.* Cambridge University Press.

Fitch, W. T., & Hauser, M. D. (1995). Vocal production in nonhuman primates: Acoustics, physiology, and functional constraints on 'honest' advertisement. *American Journal of Primatology, 37*(3), 191–219.

FitzGibbon, C. D., & Fanshawe, J. H. (1988). Stotting in Thomson's gazelles: An honest signal of condition. *Behavioral Ecology and Sociobiology, 23*(2), 69–74.

Forgas, J. P. (1995). Mood and judgment: The affect infusion model (AIM). *Psychological Bulletin, 117*(1), 39–66.

Forgas, J. P. (2002). Feeling and doing: Affective influences on interpersonal behavior. *Psychological Inquiry, 13*(1), 1–28.

Forgas, J. P. (2007). When sad is better than happy: Negative affect can improve the quality and effectiveness of persuasive messages and social influence strategies. *Journal of Experimental Social Psychology, 43*(4), 513–528.

Forgas, J. P. (2019). Happy believers and sad skeptics? Affective influences on gullibility. *Current Directions in Psychological Science, 28*(3), 306–313.

Forgas, J. P., & East, R. (2008). How real is that smile? Mood effects on accepting or rejecting the veracity of emotional facial expressions. *Journal of Nonverbal Behavior, 32*(3), 157–170.

Forgas, J. P., Goldenberg, L., & Unkelbach, C. (2009). Can bad weather improve your memory? A field study of mood effects on memory in a real-life setting. *Journal of Experimental Social Psychology, 45*(1), 254–257.

Forgas, J. P., & Moylan, S. J. (1987). After the movies: The effects of transient mood states on social judgments. *Personality and Social Psychology Bulletin, 3*(4), 478–489.

Forster, E. M. (1908). *A room with a view.* Edward Arnold.

Fossey, D. (1972). Vocalizations of the mountain gorilla (*Gorilla gorilla beringei*). *Animal Behaviour, 20*(1), 36–53.

Foster, K. R., Wenseleers, T., & Ratnieks, F. L. (2006). Kin selection is the key to altruism. *Trends in Ecology & Evolution, 21*(2), 57–60.

Fredrickson, B. L. (1998). What good are positive emotions? *Review of General Psychology, 2*(3), 300–319.

Fredrickson, B. L. (2001). The role of positive emotions in positive psychology. The broaden-and-build theory. *American Psychologist, 56*(3), 218–226.

Fredrickson, B. L. (2004). The broaden-and-build theory of positive emotions. *Philosophical Transactions of the Royal Society of London. Series B: Biological Sciences, 359*(1449), 1367–1377.

Fredrickson, B. L., & Branigan, C. (2005). Positive emotions broaden the scope of attention and thought-action repertoires. *Cognition & Emotion, 19*(3), 313–332.

Freed, P. (2009). Is sadness an evolutionarily conserved brain mechanism to dampen reward seeking? Depression may be a "sadness disorder." *Neuropsychoanalysis, 11*(1), 61–66.

Freud, S. (1917). Trauer und Melancholie. *Internationale Zeitschrift für Psychoanalyse, 4*(6), 288–301.

Frey, W. H. (1985). *Crying: The mystery of tears*. Winston Press.

Frey, W. H. (1992). Tears: Medical research helps explain why you cry. *Mayo Clinic Health Letter, 10*, 4–5.

Frey, W. H., Hoffman-Ahern, C., Johnson, R. A., Lykken, D. T., & Tuason, V. B. (1983). Crying behavior in the human adult. *Integrative Psychiatry, 1*(3), 94–100.

Frick, R. W. (1985). Communicating emotion: The role of prosodic features. *Psychological Bulletin, 97*(3), 412–429.

Fridlund, A. J. (1994). *Human facial expression: An evolutionary view*. Academic Press.

Fridlund, A. J. (2002). The behavioral ecological view of smiling and other facial expressions. In M. H. Abel (Ed.), *An empirical reflection on the smile* (pp. 45–82). Edwin Mellen Press.

Friedman, R. S., & Taylor, C. L. (2014). Exploring emotional responses to computationally-created music. *Psychology of Aesthetics, Creativity, and the Arts, 8*(1), 87–95.

Frijda, N. H. (1986). *The emotions: studies in emotion and social interaction*. Cambridge University Press.

Frodi, A. (1985). When empathy fails: Aversive infant crying and child abuse. In B. Lester & C. F. Boukydis (Eds.), *Infant crying* (pp. 263–277). Plenum Press.

Frodi, A., & Lamb, M. E. (1980). Child abusers' responses to infant smiles and cries. *Child Development, 51*, 238–241.

Fröhlich, M., Lee, K., Mitra Setia, T., Schuppli, C., & van Schaik, C. P. (2019). The loud scratch: A newly identified gesture of Sumatran orangutan mothers in the wild. *Biology Letters, 15*(7), 20190209.

Futuyma, D. J. (2017). *Evolution* (4th ed.). Sinauer Associates.

van der Gaag, C., Minderaa, R. B., & Keysers, C. (2007). Facial expressions: What the mirror neuron system can and cannot tell us. *Social Neuroscience, 2*(3–4), 179–222.

Gabrielsson, A. (2011). *Strong experiences with music.* Oxford University Press.

Galatzer-Levy, I. R., Huang, S. H., & Bonanno, G. A. (2018). Trajectories of resilience and dysfunction following potential trauma: A review and statistical evaluation. *Clinical Psychology Review, 63*, 41–55.

Garrido, S. (2017). *Why are we attracted to sad music?* Palgrave Macmillan.

Garrido, S., & Schubert, E. (2011a). Individual differences in the enjoyment of negative emotion in music: A literature review and experiment. *Music Perception, 28*(3), 279–296.

Garrido, S., & Schubert, E. (2011b). Negative emotion in music: What is the attraction? A qualitative study. *Empirical Musicology Review, 6*(4), 214–230.

Gazzola, V., Aziz-Zadeh, L., & Keysers, C. (2006). Empathy and the somatotopic auditory mirror system in humans. *Current Biology, 16*, 1824–1829.

Geangu, E., Benga, O., Stahl, D., & Striano, T. (2010). Contagious crying beyond the first days of life. *Infant Behavior and Development, 33*(3), 279–288.

Gebauer, L., Kringelbach, M. L., & Vuust, P. (2012). Ever-changing cycles of musical pleasure: The role of dopamine and anticipation. *Psychomusicology: Music, Mind, and Brain, 22*(2), 152–167.

Gendron, M., Roberson, D., van der Vyver, J. M., & Barrett, L. F. (2014). Perceptions of emotion from facial expressions are not culturally universal: Evidence from a remote culture. *Emotion, 14*(2), 251–262.

Gertsman, E. (2011). *Crying in the middle ages: Tears of history.* Routledge.

Gholamrezaei, M., De Stefano, J., & Heath, N. L. (2017). Nonsuicidal self-injury across cultures and ethnic and racial minorities: A review. *International Journal of Psychology, 52*(4), 316–326.

Gibson, B., & Sanbonmatsu, D. M. (2004). Optimism, pessimism, and gambling: The downside of optimism. *Personality and Social Psychology Bulletin, 30*(2), 149–160.

Giesbrecht, G. F., Miller, M. R., & Müller, U. (2010). The anger-distress model of temper tantrums: Associations with emotional reactivity and emotional competence. *Infant and Child Development, 19*(5), 478–497.

Gilbert, P. R. (2000). *Counselling for depression.* Sage.

Gilovich, T. (1983). Biased evaluation and persistence in gambling. *Journal of Personality and Social Psychology, 44*(6), 1110–1126.

Ginsburg, K. R. (2007). The importance of play in promoting healthy child development and maintaining strong parent-child bonds. *Pediatrics, 119*(1), 182–191.

Goldenberg, L., & Forgas, J. P. (2012, January). *Can happiness make us lazy? Hedonistic discounting can reduce perseverance and the motivation to perform.* Poster presented at the Thirteenth Annual Meeting of the Society for Personality and Social Psychology, San Diego, CA.

Goldie, P. (2011). Grief: A narrative account. *Ratio, 24*(2), 119–137.

Goleman, D. (1995). *Emotional intelligence: Why it can matter more than IQ.* Bantam Books.

Gorenstein, E. E. (1982). Frontal lobe functions in psychopaths. *Journal of Abnormal Psychology, 91*(5), 368–379.

Gorer, G. (1965). *Death, grief, and mourning in contemporary Britain.* Doubleday.

Gortner, E. M., Rude, S. S., & Pennebaker, J. W. (2006). Benefits of expressive writing in lowering rumination and depressive symptoms. *Behavior Therapy, 37*(3), 292–303.

Gottlieb, J., Oudeyer, P. Y., Lopes, M., & Baranes, A. (2013). Information-seeking, curiosity, and attention: computational and neural mechanisms. *Trends in Cognitive Sciences, 17*(11), 585–593.

Gould, S. J., & Vrba, E. S. (1982). Exaptation—a missing term in the science of form. *Paleobiology, 8*(1), 4–15.

Gračanin, A., Bylsma, L. M., & Vingerhoets, A. J. (2017). The communicative and social functions of human crying. In J.-M. Fernández-Dols & J. A. Russell (Eds.), *The science of facial expression* (pp. 217–233). Oxford University Press.

Gračanin, A., Krahmer, E., Balsters, M., Küster, D., & Vingerhoets, A. J. (2017, July). *Interactive effects of tears and muscular facial expressions.* Paper presented at the International Society for Research in Emotion, St. Louis, MO.

Graf, M. C., Gaudiano, B. A., & Geller, P. A. (2008). Written emotional disclosure: A controlled study of the benefits of expressive writing homework in outpatient psychotherapy. *Psychotherapy Research, 18*(4), 389–399.

Grant, E. C. (1969). Human facial expression. *Man, 4*(4), 525–536.

Green, J. A., Whitney, P. G., & Potegal, M. (2011). Screaming, yelling, whining, and crying: Categorical and intensity differences in vocal expressions of anger and sadness in children's tantrums. *Emotion, 11*(5), 1124–1133.

Green, M. C., & Brock, T. C. (2000). The role of transportation in the persuasiveness of public narratives. *Journal of Personality and Social Psychology, 79*(5), 701–721.

Green, M. C., Chatham, C., & Sestir, M. A. (2012). Emotion and transportation into fact and fiction. *Scientific Study of Literature, 2*(1), 37–59.

Green, M. C., Strange, J. J., & Brock, T. C. (2003). *Narrative impact: Social and cognitive foundations*. Psychology Press.

Griffiths, P. (1997). *What emotions really are: The problem of psychological categories*. University of Chicago Press.

Gross, J. J. (2010). The future's so bright, I gotta wear shades. *Emotion Review, 2*(3), 212–216.

Gualla, F., Cermelli, P., & Castellano, S. (2008). Is there a role for amplifiers in sexual selection? *Journal of Theoretical Biology, 252*(2), 255–271.

Gujar, N., Yoo, S. S., Hu, P., & Walker, M. P. (2011). Sleep deprivation amplifies reactivity of brain reward networks, biasing the appraisal of positive emotional experiences. *Journal of Neuroscience, 31*(12), 4466–4474.

Gustafson, G. E., Wood, R. M., & Green, J. A. (2000). Can parents "tell" what their infants' cries mean. In R. Barr, B. Hopkins, & J. Green (Eds.), *Crying as a sign, a symptom and a signal: Clinical, emotional and developmental aspects of infant and toddler crying* (pp. 8–22). Mac Keith Press.

Hagen, E. H. (2011). Evolutionary theories of depression: A critical review. *Canadian Journal of Psychiatry, 56*(12), 716–726.

Haldane, J. B. S. (1932). *The causes of evolution*. Princeton University Press.

Haldane, J. B. (1955). Population genetics. *New Biology, 18*(1), 34–51.

Haley, K. J. (2002). *With us or against us? Reputational psychology and the punishment of norm violations* [Master of Arts thesis, University of California, Los Angeles].

Haley, K. J., & Fessler, D. M. T. (2005). Nobody's watching? Subtle cues affect generosity in an anonymous economic game. *Evolution and Human Behavior, 26*(3), 245–256.

Hamilton, W. D. (1963). The evolution of altruistic behavior. *American Naturalist, 97*(896), 354–356.

Hamilton, W. D. (1964). The genetical evolution of social behaviour. *Journal of Theoretical Biolology, 7*(1), 1–16.

Hancock, A. A., & Fox, G. B. (2004). Cognitive enhancing effects of drugs that target histamine receptors. In J. J. Buccafusco (Ed.), *Cognitive enhancing drugs* (pp. 97–114). Birkhäuser Verlag.

Hanich, J., Wagner, V., Shah, M., Jacobsen, T., & Menninghaus, W. (2014). Why we like to watch sad films: The pleasure of being moved in aesthetic experiences. *Psychology of Aesthetics, Creativity, and the Arts, 8*(2), 130–143.

Harbaugh, W. T., Mayr, U., & Burghart, D. R. (2007). Neural responses to taxation and voluntary giving reveal motives for charitable donations. *Science, 316*(5831), 1622–1625.

Harding, E. J., Paul, E. S., & Mendl, M. (2004). Cognitive bias and affective state. *Nature, 427*(6972), 311–312.

Harlé, K. M., & Sanfey, A. G. (2007). Incidental sadness biases social economic decisions in the ultimatum game. *Emotion, 7*(4), 876–881.

Hart, B. L. (1988). Biological basis of the behavior of sick animals. *Neuroscience and Biobehavioral Reviews, 12*(2), 123–137.

von Hartmann, E. (1869/1884). *Philosophie des Unbewussten*. Duncker. Trans. by W. C. Coupland as *Philosophy of the unconscious: Speculative results according to the induction method of the physical sciences*. Trübner.

Hasson, O. (1994). Cheating signals. *Journal of Theoretical Biology, 167*(3), 223–238.

Hasson, O. (2009). Emotional tears as biological signals. *Evolutionary Psychology, 7*(3), 363–370.

Hatfield, E., Cacioppo, J. T., & Rapson, R. L. (1993). *Emotional contagion: Studies in emotion and social interaction*. Cambridge University Press.

Heider, F., & Simmel, M. (1944). An experimental study of apparent behavior. *American Journal of Psychology, 57*(2), 243–259.

Henderson, S. (1974). Care-eliciting behavior in man. *Journal of Nervous and Mental Disease, 159*(3), 172–181.

Hendriks, M. C. P., Becht, M. C., Vingerhoets, A. J., Van heck, G. L. (2001, June). *Crying and health*. Poster presented at the 2001 Amsterdam Emotion Symposium, Amsterdam, Netherlands.

Hendriks, M. C., Nelson, J. K., Cornelius, R. R., & Vingerhoets, A. J. (2008). Why crying improves our well-being: An attachment-theory perspective on the functions of adult crying. In A. J. Vingerhoets, I. Nyklíček, & J. Denollet (Eds.), *Emotion regulation: conceptual and clinical issues* (pp. 87–96). Springer.

Hendriks, M. C., & Vingerhoets, A. J. (2006). Social messages of crying faces: Their influence on anticipated person perception, emotions and behavioural responses. *Cognition and Emotion, 20*(6), 878–886.

Hepper, E. G., Wildschut, T., Sedikides, C., Ritchie, T., Yung, Y.-F., Hansen, N., Abakoumkin, G., Arikan, G., Cisek, S., Demassosso, B. D., Gebauer, J., Gerber, J. P., Gonzales, R., Kusumi, T., Misra, G., Rusu, M., Ryan, O., Stephan, E., Vingerhoets, A., & Zhou, X. (2014). Pancultural nostalgia: Prototypical conceptions across cultures. *Emotion, 14*(4), 733–747.

Hess, U., Beaupré, M. G., & Cheung, N. (2002). Who to whom and why—cultural differences and similarities in the function of smiles. In M. Abel (Ed.), *An empirical reflection on the smile* (pp. 187–216). Edwin Mellen Press.

Heyes, C. (2010). Where do mirror neurons come from? *Neuroscience & Biobehavioral Reviews, 34*(4), 575–583.

Hill, K., & Hurtado, A. M. (1996). *Ache life history: The ecology and demography of a foraging people.* Aldine Transaction.

Hinde, R. A., & Spencer-Booth, Y. (1971). Effects of brief separation from mother on rhesus monkeys. *Science, 173*(3992), 111–118.

Hinton, L., Nichols, J., & Ohala, J. J. (Eds.). (1994). *Sound symbolism.* Cambridge University Press.

Hobolth, A., Dutheil, J. Y., Hawks, J., Schierup, M. H., & Mailund, T. (2011). Incomplete lineage sorting patterns among human, chimpanzee, and orangutan suggest recent orangutan speciation and widespread selection. *Genome Research, 21*(3), 349–356.

Hochman, A. (2013). The phylogeny fallacy and the ontogeny fallacy. *Biology & Philosophy, 28*(4), 593–612.

Hockey, J. L., Katz, J., & Small, N. (2001). *Grief, mourning, and death ritual.* Open University Press.

Hofer, M. A. (1984). Relationships as regulators: A psychobiologic perspective on bereavement. *Psychosomatic Medicine, 46*(3), 183–197.

Högstedt, G. (1983). Adaptation unto death: Function of fear screams. *American Naturalist, 121*(4), 562–570.

Hollien, H. (1960). Some laryngeal correlates of vocal pitch. *Journal of Speech and Hearing Research, 3*(1), 52–58.

Holmes, J. E., & Miller, N. E. (1963). Effects of bacterial endotoxin on water intake, food intake, and body temperature in the albino rat. *Journal of Experimental Medicine, 118*(4), 649–658.

Hopcroft, R. L. (2006). Sex, status, and reproductive success in the contemporary United States. *Evolution and Human Behavior, 27*(2), 104–120.

Horwitz, A., & Wakefield, J. (2007). *The loss of sadness: How psychiatry transformed normal sadness into depressive disorder.* Oxford University Press.

Hume, D. (1759/2004). Letter to Adam Smith, 18 July 1759. In J. R. Otteson (Ed.), *Adam Smith: Selected philosophical writings* (p. 243). Imprint Academic.

Hunter, P. G., Schellenberg, E. G., & Schimmack, U. (2008). Mixed affective responses to music with conflicting cues. *Cognition & Emotion, 22*(2), 327–352.

Hunter, P. G., Schellenberg, E. G., & Schimmack, U. (2010). Feelings and perceptions of happiness and sadness induced by music: Similarities, differences and mixed emotions. *Psychology of Aesthetics, Creativity, and the Arts, 4*(1), 47–56.

Huntsinger, J. R., Isbell, L. M., & Clore, G. L. (2014). The affective control of thought: Malleable, not fixed. *Psychological Review, 121*(4), 600–618.

Huntsinger, J. R., & Ray, C. (2016). A flexible influence of affective feelings on creative and analytic performance. *Emotion, 16*(6), 826–837.

Hurlemann, R., Patin, A., Onur, O. A., Cohen, M. X., Baumgartner, T., Metzler, S., Dziobeck, I., Gallinat, J., Wagner, M., Maier, W., & Kendrick, K. M. (2010). Oxytocin enhances amygdala-dependent, socially reinforced learning and emotional empathy in humans. *Journal of Neuroscience, 30*(14), 4999–5007.

Huron, D. (2005). The plural pleasures of music. In W. Brunson & J. Sundberg (Eds.), *Proceedings of the 2004 Music and Science Conference* (pp. 65–78). Kungliga Musikhöskolan Förlaget.

Huron, D. (2006). *Sweet anticipation: Music and the psychology of expectation.* MIT Press.

Huron, D. (2008). A comparison of average pitch height and interval size in major- and minor-key themes: Evidence consistent with affect-related pitch prosody. *Empirical Musicology Review, 3*(2), 59–63.

Huron, D. (2015). Affect induction through musical sounds: An ethological perspective. *Philosophical Transactions of the Royal Society of London, Series B, Biological Sciences, 370*(1664), 1–7.

Huron, D. (2016). Cues and signals: An ethological approach to music-related emotion. In J. R. do Carmo Jr. & P. A. Brandt (Eds.), *Sémiotique de la Musique—Music and Meaning, Signata No. 6, Annales des Sémiotiques / Annals of Semiotics* (pp. 333–353). Presses Universitaires de Liège.

Huron, D. (2018a). On the functions of sadness and grief. In H. C. Lench (Ed.), *Functions of emotion: When and why emotions help us* (pp. 59–92). Springer.

Huron, D. (2018b). Affect induction through musical sounds: An ethological perspective. In H. Honing (Ed.), *The origins of musicality* (pp. 309–322). MIT Press.

Huron, D., Anderson, N., & Shanahan, D. (2014). You can't play sad music on a banjo: Acoustic factors in the judgment of instrument capacity to convey sadness. *Empirical Musicology Review, 9*(1), 29–41.

Huron, D., & Vuoskoski, J.K. (2020). On the enjoyment of sad music: Pleasurable compassion theory and the role of trait empathy. *Frontiers in Psychology, 11*, 1–16.

Ickes, W., Patterson, M. L., Rajecki, D. W., & Tanford, S. (1982). Behavioral and cognitive consequences of reciprocal versus compensatory responses to preinteraction expectancies. *Social Cognition, 1*(2), 160–190.

Immelmann, K., & Beer, C. (1989). *A dictionary of ethology.* Harvard University Press.

Inoue, I., Yanai, K., Kitamura, D., Taniuchi, I., Kobayashi, T., Niimura, K., Watanabe, T., & Watanabe, T. (1996). Impaired locomotor activity and exploratory behavior in mice lacking histamine H1 receptors. *Proceedings of the National Academy of Sciences of the U.S.A., 93*(23), 13316–13320.

Isaacowitz, D. M., Löckenhoff, C., Lane, R., Wright, R., Sechrest, L., Riedel, R., & Costa, P. T. (2007). Age differences in recognition of emotion in lexical stimuli and facial expressions. *Psychology and Aging, 22*(1), 147–159.

Isen, A. M. (2000). Positive affect and decision making. In M. Lewis & J. M. Haviland-Jones (Eds.). *Handbook of emotions* (2nd ed., pp. 417–435). Guilford.

Isen, A. M. & Daubman, K. A. (1984). The influence of affect on categorization. *Journal of Personality and Social Psychology, 47*(6), 1206–1217.

Isen, A. M., Daubman, K. A., & Nowicki, G. P. (1987). Positive affect facilitates creative problem solving. *Journal of Personality and Social Psychology, 52*(6), 1122–1131.

Isen, A. M., Means, B., Patrick, R., & Nowicki, G. (1982). Some factors influencing decision making strategy and risk taking. In M. S. Clark & S. T. Fiske (Eds.), *Affect and cognition: The 17th Annual Carnegie Symposium on Cognition* (pp. 243–261). Erlbaum.

Isen, A. M., & Patrick, R. (1983). The effect of positive feelings on risk taking: When the chips are down. *Organizational Behavior and Human Performance, 31*(2), 194–202.

Isen, A. M., Rosenzweig, A. S., & Young, M. J. (1991). The influence of positive affect in clinical problem solving. *Medical Decision Making, 11*(3), 221–227.

Ito, C. (2004). The role of the central histaminergic system on schizophrenia. *Drug News & Perspectives, 17*(6), 383–387.

Izuma, K., Saito D., & Sadato, N. (2008). Processing of social and monetary rewards in the human striatum. *Neuron, 58*(2), 284–294.

Jepma, M., Verdonschot, R. G., Van Steenbergen, H., Rombouts, S. A., & Nieuwenhuis, S. (2012). Neural mechanisms underlying the induction and relief of perceptual curiosity. *Frontiers in Behavioral Neuroscience, 6*, 5.

Johnson, R. W. (2002). The concept of sickness behavior: A brief chronological account of four key discoveries. *Veterinary Immunology and Immunopathology, 87*(3–4), 443–450.

Johnstone, R. A. (1997). The evolution of animal signals. In J. R. Krebs & N. B. Davies (Eds.), *Behavioural ecology* (pp. 155–178). Oxford University Press.

Johnstone, R. A., & Grafen, A. (1993). Dishonesty and the handicap principle. *Animal Behaviour, 46*(4), 759–764.

Joireman, J., Parrott, L., & Hammersla, J. (2002). Empathy and the self-absorption paradox: Support for the distinction between self-rumination and self-reflection. *Self and Identity, 1*(1), 53–65.

Juslin, P. N. (2013). From everyday emotions to aesthetic emotions: Towards a unified theory of musical emotions. *Physics of Life Review, 10,* 235–266.

Juslin, P. N., & Laukka, P. (2003). Communication of emotions in vocal expression and music performance: Different channels, same code? *Psychological Bulletin, 129,* 770–814.

Kahneman, D. (2003). A perspective on judgment and choice: Mapping bounded rationality. *American Psychologist, 58*(9), 697–720.

Kahneman, D. (2011). *Thinking, fast and slow.* Macmillan.

Kang, M. J., Hsu, M., Krajbich, I. M., Loewenstein, G., McClure, S. M., Wang, J. T. Y., & Camerer, C. F. (2009). The wick in the candle of learning: Epistemic curiosity activates reward circuitry and enhances memory. *Psychological Science, 20*(8), 963–973.

Kantha, S. S. (1992). Clues to prolific productivity among prominent scientists. *Medical Hypotheses, 39*(2), 159–163.

Kawakami, A., Furukawa, K., Katahira, K., & Okanoya, K. (2013). Sad music induces pleasant emotion. *Frontiers in Psychology, 4,* Article 311, 1–15.

Kawakami, A., & Katahira, K. (2015). Influence of trait empathy on the emotion evoked by sad music and on the preference for it. *Frontiers in Psychology, 6,* Article 1541.

Keedwell, P. (2008). *How sadness survived: The evolutionary basis of depression.* Radcliffe Publishing.

Kelley, K. W., Bluthé, R. M., Dantzer, R., Zhou, J. H., Shen, W. H., Johnson, R. W., & Broussard, S. R. (2003). Cytokine-induced sickness behavior. *Brain, Behavior, and Immunity, 17*(Suppl. 1), S112–S118.

Kenis, G., & Maes, M. (2002). Effects of antidepressants on the production of cytokines. *International Journal of Neuropsychopharmacology, 5*(4), 401–412.

Khan, R. (2010, August 5). Gene Expression: 1 in 200 men direct descendants of Genghis Khan. *Discover Magazine.* Retrieved March 16, 2018, from http://blogs.disco vermagazine.com/gnxp/2010/08/1-in-200-men-direct-descendants-of-genghis-khan /#.Wqy53GbMx24

Kilner, J. M., Neal, A., Weiskopf, N., Friston, K. J., & Frith, C. D. (2009). Evidence of mirror neurons in human inferior frontal gyrus. *Journal of Neuroscience, 29,* 10153–10159.

Kirkpatrick, M. (1986). The handicap mechanism of sexual selection does not work. *American Naturalist, 127*(2), 222–240.

Klein, B., Mrowetz, H., Thalhamer, J., Scheiblhofer, S., Weiss, R., & Aigner, L. (2016). Allergy enhances neurogenesis and modulates microglial activation in the hippocampus. *Frontiers in Cellular Neuroscience, 10,* Article 169.

Kleinginna, P. R., & Kleinginna, A. M. (1981). A categorized list of emotion definitions, with suggestions for a consensual definition. *Motivation and Emotion, 5*(4), 345–379.

Klimecki, O. M., Leiberg, S., Ricard, M., & Singer, T. (2014). Differential pattern of functional brain plasticity after compassion and empathy training. *Social Cognitive and Affective Neuroscience, 9*(6), 873–879.

Klinger, E. (1975). Consequences of commitment to and disengagement from incentives. *Psychological Review, 82*(1), 1–25.

Klinger, E. (1977). *Meaning and void: Inner experience and the incentives in people's lives.* University of Minnesota Press.

Kluger, M. J., & Rothenburg, B. A. (1979). Fever and reduced iron: Their interaction as a host defense response to bacterial infection. *Science, 203*(4378), 374–376.

Koch, A. S., Forgas, J. P., & Matovic, D. (2013). Can negative mood improve your conversation? Affective influences on conforming to Grice's communication norms. *European Journal of Social Psychology, 43*(5), 326–334.

Köhler, C. A., da Silva, W. C., Benetti, F., & Bonini, J. S. (2011). Histaminergic mechanisms for modulation of memory systems. *Neural Plasticity, 2011*, 328602, 1–16.

Konečni, V. J. (2005). The aesthetic trinity: Awe, being moved, thrills. *Bulletin of Psychology and the Arts, 5*(2), 27–44.

Konečni, V. J. (2015). Being moved as one of the major aesthetic emotional states: A commentary on "Being moved: linguistic representation and conceptual structure." *Frontiers in Psychology, 6*, Article 343.

Konsman, J. P., Parnet, P., & Dantzer, R. (2002). Cytokine-induced sickness behaviour: Mechanisms and implications. *Trends in Neurosciences, 25*(3), 154–159.

Koopman, E. M. E. (2015). Empathic reactions after reading: The role of genre, personal factors and affective responses. *Poetics, 50*, 62–79.

Korsmeyer, C. (2014). A lust of the mind: curiosity and aversion in eighteenth-century British aesthetics. In J. Levinson (Ed.), *Suffering art gladly: The paradox of negative emotion in art* (pp. 45–67). Springer.

Kottler, J. A. (1996). *The language of tears.* Jossey-Bass Publishers.

Kottler, J. A., & Montgomery, M. J. (2001). Theories of crying. In A. J. Vingerhoets & R. R. Cornelius (Eds.), *Adult crying: A biopsychosocial approach* (pp. 1–7). Brunner-Routledge.

Kraemer, D. L., & Hastrup, J. L. (1988). Crying in adults: Self-control and autonomic correlates. *Journal of Social and Clinical Psychology, 6*(1), 53–68.

Kraemer, D. L., Avants, S. K., Wilkins, W. P., & Melnick, S. (1989, April). *A laboratory study of crying in adults: Memory processes and concomitant psychological adjustment.* Paper presented at the 75th Annual Meeting of the Eastern Psychological Association, Boston, MA.

Kraepelin, E. (1899). *Psychiatrie. Ein Lehrbuch für Studierende und Ärzte, ed. 2. Klinische Psychiatrie. II.* Johann Ambrosius Barth.

Krebs, J. R., & Dawkins, R. (1984). Animal signals: Mind-reading and manipulation. In J. R. Krebs & N. B. Davies (Eds.), *Behavioural ecology: An evolutionary approach* (2nd ed., pp. 380–402). Oxford University Press.

Kubota, Y., Ito, C., Sakurai, E., Sakurai, E., Watanabe, T., & Ohtsu, H. (2002). Increased methamphetamine-induced locomotor activity and behavioral sensitization in histamine-deficient mice. *Journal of Neurochemistry, 83*(4), 837–845.

Kuehnast, M., Wagner, V., Wassiliwizky, E., Jacobsen, T., & Menninghaus, W. (2014). Being moved: Linguistic representation and conceptual structure. *Frontiers in Psychology, 5*, Article 1242.

Kuhnen, C. M., & Knutson, B. (2011). The influence of affect on beliefs, preferences, and financial decisions. *Journal of Financial and Quantitative Analysis, 46*(3), 605–626.

Labott, S. M. (2001). Crying in psychotherapy. In A. J. Vingerhoets & R. R. Cornelius (Eds.), *Adult crying: A biopsychosocial approach* (pp. 213–226). Brunner-Routledge.

Lackner, C. L., Bowman, L. C., & Sabbagh, M. A. (2010). Dopaminergic functioning and preschoolers' theory of mind. *Neuropsychologia, 48*, 1767–1774.

Ladinig, O., Brooks, C., Hansen, N. C., Horn, K., & Huron, D. (2019). Enjoying sad music: A test of the prolactin theory. *Musicae Scientiae, 23*(4), 1–20.

LaFrance, M., Hecht, M. A., & Paluck, E. L. (2003). The contingent smile: A meta-analysis of sex differences in smiling. *Psychological Bulletin, 129*(2), 305–334.

Laird, J. D. (2007). *Feelings: The perception of self.* Oxford University Press.

Lambrecht, L., Kreifelts, B., & Wildgruber, D. (2012). Age-related decrease in recognition of emotional facial and prosodic expressions. *Emotion, 12*(3), 529–539.

Landreth, C. (1941). Factors associated with crying in young children in the nursery school and the home. *Child Development, 12*(2), 81–97.

Lane, C. J. (2006). *Evolution of gender differences in adult crying* [PhD dissertation, University of Texas at Arlington].

Larsen, J. T., & Green, J. D. (2013). Evidence for mixed feelings of happiness and sadness from brief moments in time. *Cognition & Emotion, 27*(8), 1469–1477.

Larsen, J. T., & McGraw, A. P. (2011). Further evidence for mixed emotions. *Journal of Personality and Social Psychology, 100*(6), 1095–1110.

Larsen, J. T., & Stastny, B. J. (2011). It's a bittersweet symphony: Simultaneously mixed emotional responses to music with conflicting cues. *Emotion, 11*(6), 1469–1473.

Latané, B., & Nida, S. (1981). Ten years of research on group size and helping. *Psychological Bulletin, 89*(2), 308–324.

van Lawick-Goodall, J. (1968). The behaviour of free-living chimpanzees in the Gombe Stream Reserve. *Animal Behaviour Monographs, 1*(3), 161–311, IN1–IN12.

Lench, H. C. (Ed.). (2018). *Functions of emotion: When and why emotions help us.* Springer.

Lench, H. C., Bench, S. W., Darbor K. E., & Moore, M. (2015). Functionalist manifesto: Goal-related emotions from an evolutionary perspective. *Emotion Review, 7*(1), 90–98.

Lenneberg, E. H., Rebelsky, F. G., & Nichols, I. A. (1965). The vocalizations of infants born to deaf and to hearing parents. *Human Development, 8*(1), 23–37.

de Léry, J. (1578). *Histoire d'un voyage fait en la terre du Bresil, autrement dite Amérique.* Antoine Chuppin.

Levinson, D. F. (2006). The genetics of depression: A review. *Biological Psychiatry, 60*(2), 84–92.

Levinson, J. (1982). Music and negative emotion. *Pacific Philosophical Quarterly, 63*(4), 327–346.

Levinson, J. (2013). *Suffering art gladly: The paradox of negative emotion in art.* Springer.

Levy, R. I. (1973). *Tahitians: Mind and experience in the Society Islands.* University of Chicago Press.

Lewis, A. J. (1934). Melancholia: A clinical survey of depressive states. *British Journal of Psychiatry, 80*(329), 277–378.

Leys, R. (2017). *The ascent of affect: Geneology and critique.* University of Chicago Press.

Lickliter, R., & Berry, T. D. (1990). The phylogeny fallacy: Developmental psychology's misapplication of evolutionary theory. *Developmental Review, 10*(4), 348–364.

Lima, C. F., Alves, T., Scott, S. K., & Castro, S. L. (2013). In the ear of the beholder: How age shapes emotion processing in nonverbal vocalizations. *Emotion, 14*(1), 145–160.

Lingle, S., & Riede, T. (2014). Deer mothers are sensitive to infant distress vocalizations of diverse mammalian species. *American Naturalist, 184*(4), 510–522.

Lingle, S., Wyman, M. T., Kotrba, R., Teichroeb, L. J., & Romanow, C. A. (2012). What makes a cry a cry? A review of infant distress vocalizations. *Current Zoology, 58*(5), 698–726.

Litman, J. (2005). Curiosity and the pleasures of learning: Wanting and liking new information. *Cognition & Emotion, 19*(6), 793–814.

Litman, J. A., & Silvia, P. J. (2006). The latent structure of trait curiosity: Evidence for interest and deprivation curiosity dimensions. *Journal of Personality Assessment*, *86*(3), 318–328.

Livio, M. (2017). *Why? What makes us curious.* Simon and Schuster.

Lloyd, E. A. (1999). Evolutionary psychology: The burdens of proof. *Biology and Philosophy*, *14*(2), 211–233.

Loewenstein, G. (1994). The psychology of curiosity: A review and reinterpretation. *Psychological Bulletin*, *116*(1), 75–98.

Lopes, P. C. (2014). When is it socially acceptable to feel sick? *Proceedings of the Royal Society B: Biological Sciences*, *281*(1788), 20140218.

Lorenz, K. (1937). Über die Bildung des Instinktbegriffes. *Naturwissenschaften*, *25*(19), 289–300, 307–318, 324–331.

Lovallo, W. (1975). The cold pressor test and autonomic function: A review and integration. *Psychophysiology*, *12*(3), 268–282.

Love, B. C., Kopeć, Ł., & Guest, O. (2015). Optimism bias in fans and sports reporters. *PLoS One*, *10*(9), e0137685.

Lowery, D., Fillingim, R. B., & Wright, R. A. (2003). Sex differences and incentive effects on perceptual and cardiovascular responses to cold pressor pain. *Psychosomatic Medicine*, *65*(2), 284–291.

Lutz, T. (1999). *Crying: A natural and cultural history of tears.* W. W. Norton.

Mackenzie, H. (1771). *The man of feeling.* T. Cadell.

Maguire, H. (1977). The depiction of sorrow in Middle Byzantine art. *Dumbarton Oaks Papers*, *31*, 123–74.

Maier, S. F., Wiertelak, E. P., Martin, D., & Watkins, L. R. (1993). Interleukin-1 mediates the behavioral hyperalgesia produced by lithium chloride and endotoxin. *Brain Research*, *623*(2), 321–324.

Mangino, M., Roederer, M., Beddall, M. H., Nestle, F. O., & Spector, T. D. (2017). Innate and adaptive immune traits are differentially affected by genetic and environmental factors. *Nature Communications*, *8*(1), 1–7.

Maretzki, T. W., & Maretzki, H. W. (1963). Taira: An Okinawan village. In B. Whiting (Ed.), *Six cultures: Studies of child rearing* (pp. 363–539). John Wiley and Sons.

Marsella, A. J., Sartorius, N., Jablensky, A., & Fenton, F. R. (1985). Cross-cultural studies of depressive disorders: An overview. In A. Kleinman & B. Good (Eds.), *Culture and depression: Studies in the anthropology and cross-cultural psychiatry of affect and disorder* (pp. 299–324). University of California Press.

Marshall, G. D., & Zimbardo, P. G. (1979). Affective consequences of inadequately explained physiological arousal. *Journal of Personality and Social Psychology, 37*(6), 970–988.

Maslach, C. (1979). Negative emotional biasing of unexplained arousal. *Journal of Personality and Social Psychology, 37*(6), 953–969.

Matheson, S. M., Asher, L., & Bateson, M. (2008). Larger enriched cages are associated with 'optimistic' response biases in captive European starlings (Sturnus vulgaris). *Applied Animal Behaviour Science, 109*(2–4), 374–383.

Matsumoto, J. (2002). Why people listen to sad music: Effects of music on sad moods. *Japanese Journal of Educational Psychology, 50*(1), 23–32.

Matt, S. J. (2011). *Homesickness: An American history.* Oxford University Press.

Maynard Smith, J. (1985). Mini review: Sexual selection, handicaps and true fitness. *Journal of Theoretical Biology, 115*(1), 1–8.

Maynard Smith, J., & Harper, D. G. (1995). Animal signals: Models and terminology. *Journal of Theoretical Biology, 177*(3), 305–311.

Maynard Smith, J. M., & Harper, D. (2003). *Animal signals.* Oxford University Press.

Mazurkiewicz-Kwilecki, I. M., & Nsonwah, S., (1989). Changes in the regional brain histamine and histidine levels in postmortem brains of Alzheimer patients. *Canadian Journal of Physiology and Pharmacology, 67*(1), 75–78.

McAfoose, J., & Baune, B. T. (2009). Evidence for a cytokine model of cognitive function. *Neuroscience and Biobehavioral Reviews, 33*(3), 355–366.

Mead, M., & Newton, N. (1967). Cultural patterning of perinatal behavior. In S. Richardson & A. Guttmacher (Eds.), *Childbearing: Its social and psychological aspects* (pp. 142–244). Williams and Wilkins.

Mélinand, C. (1902). Why do we cry? The psychology of tears. *Current Literature, 32*, 696–699.

Menninghaus, W., Wagner, V., Hanich, J., Wassiliwizky, E., Kuehnast, M., & Jacobsen, T. (2015). Towards a psychological construct of being moved. *PLoS One, 10*(6), e0128451.

Menninghaus, W., Wagner, V., Hanich, J., Wassiliwizky, E., Jacobsen, T., & Koelsch, S. (2017). The distancing-embracing model of the enjoyment of negative emotions in art reception. *Behavioral and Brain Sciences, 40*, e347.

Métraux, A. (1928). *La civilisation matérielle des Tupi-Guarani.* Paul Geuthner.

Milinski, M., Semmann, D., & Krambeck, H. J. (2002). Reputation helps solve the 'tragedy of the commons'. *Nature, 415*(6870), 424–426.

Mill, A., Allik, J., Realo, A., & Valk, R. (2009). Age-related differences in emotion recognition ability: A cross-sectional study. *Emotion, 9*(5), 619–630.

Miller, A. H., Haroon, E., Raison, C. L., & Felger, J. C. (2013). Cytokine targets in the brain: Impact on neurotransmitters and neurocircuits. *Depression and Anxiety, 30*(4), 297–306.

Miller, B. L., & Cummings, J. L. (Eds.). (2007). *The human frontal lobes: Functions and disorders*. Guilford Press.

Miller, N. (1964). Some psychophysiological studies of motivation and of the behavioral effects of illness. *Bulletin of the British Psychological Society, 17*(55), 1–20.

Moncrieff, J., Cooper, R. E., Stockmann, T., Amendola, S., Hengartner, M. P., & Horowitz, M. A. (2022). The serotonin theory of depression: a systematic umbrella review of the evidence. *Molecular Psychiatry*, 1–14.

Montagu, A. (1960). Natural selection and the origin and evolution of weeping in man. *Journal of the American Medical Association, 174*(4), 392–397.

Moore, M. T., & Fresco, D. (2012). Depressive realism: A meta-analytic review. *Clinical Psychology Review, 32*(6), 496–509.

Moors, A. (2022). *Demystifying emotions: A typology of theories in psychology and philosophy*. Cambridge University Press.

Moran, T. (2009). John Maynard Smith's typology of animal signals: A view from semiotics. *Sign Systems Studies, 37*(3–4), 477–95.

Morton, E. S. (1977). On the occurrence and significance of motivation-structural rules in some bird and mammal sounds. *American Naturalist, 111*(981), 855–869.

Morton, E. S. (1994). Sound symbolism and its role in non-human vertebrate communication. In L. Hinton, J. Nichols, & J. Ohala (Eds.), *Sound symbolism* (pp. 348–365). Cambridge University Press.

Mullin, B. C., Phillips, M. L., Siegle, G. J., Buysse, D. J., Forbes, E. E., & Franzen, P. L. (2013). Sleep deprivation amplifies striatal activation to monetary reward. *Psychological Medicine, 43*(10), 2215–2225.

Mullington, J., Korth, C., Hermann, D. M., Orth, A., Galanos, C., Holsboer, F., & Pollmächer, T. (2000). Dose-dependent effects of endotoxin on human sleep. *American Journal of Physiology. Regulatory, Integrative and Comparative Physiology, 278*(4), R947–R955.

Murphy, H. B. M., Wittkower, E., & Chance, N. (1964). Cross-cultural inquiry into the symptomatology of depression. *Transcultural Psychiatric Research Review, 1*(1), 5–21.

Murray, M. J., & Murray, A. B. (1979). Anorexia of infection as a mechanism of host defense. *American Journal of Clinical Nutrition, 32*(3), 593–596.

Murube, J. (2009). Hypotheses on the development of psychoemotional tearing. *The Ocular Surface, 7*(4), 171–175.

Musselman, D. L., Lawson, D. H., Gumnick, J. F., Manatunga, A. K., Penna, S., Goodkin, R. S., Greiner, K., Nemeroff, C. B., & Miller, A. H. (2001). Paroxetine for the prevention of depression induced by high-dose interferon alfa. *New England Journal of Medicine, 344*(13), 961–966.

Nawas, M. M., & Platt, J. J. (1965). A future-oriented theory of nostalgia. *Journal of Individual Psycholology, 21*(1), 51–57.

Nelson, J. K. (1998). The meaning of crying based on attachment theory. *Clinical Social Work Journal, 26*(1), 9–22.

Nesse, R. M. (1991). What good is feeling bad? The evolutionary benefits of psychic pain. *The Sciences, 31*, 30–37.

Nesse, R. M. (2000). Is depression an adaptation? *Archives of General Psychiatry, 57*(1), 14–20.

Nesse, R. M. (2018). *Good reasons for bad feelings: Insights from the frontier of evolutionary psychiatry*. Dutton.

Nesse, R. M., & Williams, G. C. (1994). *Why we get sick: The new science of Darwinian medicine*. Times Books.

Neth, D., & Martinez, A. M. (2009). Emotion perception in emotionless face images suggests a norm-based representation. *Journal of Vision, 9*(1), 1–11.

Nettle, D. (2004). Evolutionary origins of depression: A review and reformulation. *Journal of Affective Disorders, 81*(2), 91–102.

Newman, J. D. (1985). The infant cry of primates. In B. Lester & Z. Boukydis (Eds.), *Infant crying: Theoretical and research perspectives* (pp. 307–323). Plenum Press.

Newman, J. D. (2004). The primate isolation call: A comparison with precocial birds and non-primate mammals. In L. J. Rogers & G. Kaplan (Eds.), *Comparative vertebrate cognition: Are primates superior to non-primates?* (pp. 171–187). Kluwer Academic/Plenum.

Newman, J. D. (2007). Neural circuits underlying crying and cry responding in mammals. *Behavioural Brain Research, 182*(2), 155–165.

Newman, J. D., & Symmes, D. (1982). Inheritance and experience in the acquisition of primate acoustic behavior. In C. Snowdon, C. Brown, & M. Petersen (Eds.), *Primate communication* (pp. 259–278). Cambridge University Press.

Nguyen, L., Thomas, K. L., Lucke-Wold, B. P., Cavendish, J. Z., Crowe, M. S., & Matsumoto, R. R. (2016). Dextromethorphan: An update on its utility for neurological and neuropsychiatric disorders. *Pharmacology & Therapeutics, 159*, 1–22.

Nisbett, R. (2003). *The geography of thought: How Asians and Westerners think differently . . . and why*. Free Press.

Nisbett, R., & Wilson, T. (1977). Telling more than we can know: Verbal reports on mental processes. *Psychological Review, 84*(3), 231–259.

Niver, E. O. (1948). The use of histamine in preparing patients for psychotherapy. *Psychiatric Quarterly, 22*(1–4), 729–736.

Nock, M. K. (2010). Self-injury. *Annual Review of Clinical Psychology, 6*, 339–363.

Nolen-Hoeksema, S. (1991). Responses to depression and their effects on the duration of depressive episodes. *Journal of Abnormal Psychology, 100*(4), 569–582.

Norman, K., Pellis, S., Barrett, L., & Henzi, S. P. (2015). Down but not out: Supine postures as facilitators of play in domestic dogs. *Behavioural Processes, 110*, 88–95.

Nowak, M. A., Tarnita, C. E., & Wilson, E. O. (2010). The evolution of eusociality. *Nature, 466*(7310), 1057–1062.

Nuutinen, S., Lintunen, M., Vanhanen, J., Ojala, T., Rozov, S., & Panula, P. (2011). Evidence for the role of histamine H3 receptor in alcohol consumption and alcohol reward in mice. *Neuropsychopharmacology, 36*(10), 2030–2040.

Ohala, J. (1980). The acoustic origin of the smile. *Journal of the Acoustical Society of America, 68*, S33.

Ohala, J. (1984). An ethological perspective on common cross-language utilization of F0 in voice. *Phonetica, 41*(1), 1–16.

Ohala, J. J. (2010). What's behind the smile? *Behavioral and Brain Sciences, 33*(6), 456–457.

Okuda, T., Zhang, D., Shao, H., Okamura, N., Takino, N., Iwamura, T., Sakurai, E., Yoshikawa, T., & Yanai, K. (2009). Methamphetamine- and 3,4-methylenedioxymethamphetamine-induced behavioral changes in histamine H3-receptor knockout mice. *Journal of Pharmacological Sciences, 111*(2), 167–174.

Orgeta, V. (2010). Effects of age and task difficulty on recognition of facial affect. *The Journals of Gerontology: Series B: Psychological Sciences and Social Sciences, 65*(3), 323–327.

Orgeta, V., & Phillips, L. H. (2008). Effects of age and emotional intensity on the recognition of facial emotion. *Experimental Aging Research, 34*(1), 63–79.

Ortony, A., & Turner, T. J. (1990). What's basic about basic emotions? *Psychological Review, 97*(3), 315–331.

Otto, A. R., Fleming, S. M., & Glimcher, P. W. (2016). Unexpected but incidental positive outcomes predict real-world gambling. *Psychological Science, 27*(3), 299–311.

Owings, D. H., & Morton, E. S. (1998). *Animal vocal communication: A new approach.* Cambridge University Press.

Pacini, R., Muir, F., & Epstein, S. (1998). Depressive realism from the perspective of cognitive-experiential self-theory. *Journal of Personality and Social Psychology, 74*(4), 1056–1068.

Panksepp, J. (1998). *Affective neuroscience: The foundations of human and animal emotions.* Oxford University Press.

Panksepp, J., & Bernatzky, G. (2002). Emotional sounds and the brain: The neuro-affective foundations of musical appreciation. *Behavioural Processes, 60*(2), 133–155.

Panula, P., Rinne, J., Kuokkanen, K., Eriksson, K. S., Sallmen, T., Kalimo, H., & Relja, M. (1998). Neuronal histamine deficit in Alzheimer's disease. *Neuroscience, 82*(4), 993–997.

Papageorgiou, C., & Wells, A. (2004). *Depressive rumination: Nature, theory and treatment.* Wiley.

Pardridge, W. M. (2005). The blood-brain barrier: Bottleneck in brain drug development. *NeuroRx, 2*(1), 3–14.

Partan, S., & Marler, P. (1999). Communication goes multimodal. *Science, 283*(5406), 1272–1273.

Pert, C. B. (1997). *Molecules of emotion.* Scribner.

Pfeifer, J. H., Iacoboni, M., Mazziotta, J. C., & Dapretto, M. (2008). Mirroring others' emotions relates to empathy and interpersonal competence in children. *Neuroimage, 39,* 2076–2085.

Pflug, B., & Tölle, R. (1971). Disturbance of the 24-hour rhythm in endogenous depression and the treatment of endogenous depression by sleep deprivation. *International Pharmacopsychiatry, 6,* 187–196.

Piaget, J. (1936/1952). *La naissance de l'intelligence chez l'enfant.* Delachaux et Niestlé. Trans. by M. Cook as *The origins of intelligence in children.* W. W. Norton.

Pioro, E. P., Brooks, B. R., Cummings, J., Schiffer, R., Thisted, R. A., Wynn, D., Hepner, A., & Kaye, R. (2010). Dextromethorphan plus ultra low-dose quinidine reduces pseudobulbar affect. *Annals of Neurology, 68*(5), 693–702.

Pliny the Elder. (circa 77CE /1900). *Naturalis historia.* Translated by J. Bostock & H. Riley as *The natural history of Pliny.* George Bell & Sons.

Pollak, S. D., & Sinha, P. (2002). Effects of early experience on children's recognition of facial displays of emotion. *Developmental Psychology, 38*(5), 784–791.

Pomiankowski, A. (1987). Sexual selection: The handicap principle does work sometimes. *Proceedings of the Royal Society B, 231*(1262), 123–145.

Post, O., & Huron, D. (2009). Western classical music in minor modes is slower (except in the Romantic Period). *Empirical Musicology Review, 4*(1), 1–9.

Potegal, M. (2019). On being mad, sad and very young. In A. K. Roy, M. A. Brotman, & E. Leibenluft (Eds.), *Irritability in pediatric psychopathology* (pp. 105–145). Oxford University Press.

Potegal, M., & Davidson, R. J. (2003). Temper tantrums in young children: 1. Behavioral composition. *Journal of Developmental & Behavioral Pediatrics, 24*(3), 140–147.

Potegal, M., Kosorok, M. R., & Davidson, R. J. (2003). Temper tantrums in young children: 2. Tantrum duration and temporal organization. *Journal of Developmental & Behavioral Pediatrics, 24*(3), 148–154.

Potegal, M., Stemmler, G., & Spielberger, C. (Eds.). (2010). *International handbook of anger*. Springer.

Premack, D., & Woodruff, G. (1978). Does the chimpanzee have a theory of mind? *Behavioral and Brain Sciences, 1*, 515–526.

Provine, R. R. (2000). *Laughter: A scientific investigation*. Penguin Books.

Provine, R. R., Cabrera, M. O., Brocato, N. W., & Krosnowski, K. A. (2011). When the whites of the eyes are red: A uniquely human cue. *Ethology, 117*(5), 395–399.

Provine, R. R., Krosnowski, K. A., & Brocato, N. W. (2009). Tearing: Breakthrough in human emotional signaling. *Evolutionary Psychology, 7*(1), 78–81.

Pruimboom, L. (2020). The selfish immune system: When the immune system overrides the 'selfish' brain. *Journal of Immunology and Clinical Microbiology, 5*(1), 1–34.

Raby, D. (2012). "My heart can't do what it wants": Sadness and learned empathy in a Nahua community. *Anthropology and Humanism, 37*(2), 201–213.

Raison, C. L., Capuron, L., & Miller, A. (2006). Cytokines sing the blues: Inflammation and the pathogenesis of depression. *Trends in Immunology, 27*(1), 24–31.

Raison, C. L., & Miller, A. H. (2013). The evolutionary significance of depression in Pathogen Host Defense (PATHOS-D). *Molecular Psychiatry, 18*(1), 15–37.

Raleigh, M. J., McGuire, M. T., Brammer, G. L., Pollack, D. B., & Yuwiler, A. (1991). Serotonergic mechanisms promote dominance acquisition in adult male vervet monkeys. *Brain Research, 559*(2), 181–190.

Rang, H. P. (2003). *Pharmacology* (5th ed.). Churchill Livingstone.

Rao, M., Blass, E. M., Brignol, M. M., Marino, L., & Glass, L. (1997). Reduced heat loss following sucrose ingestion in premature and normal human newborns. *Early Human Development, 48*(1–2), 109–116.

Rendall, D., & Owren, M. J. (2013). Communication without meaning or information: Abandoning language-based and informational constructs in animal communication theory. In U. Stegmann (Ed.), *Animal communication theory: Information and influence* (pp. 151–188). Cambridge University Press.

Rendall, D., Owren, M. J., & Ryan, M. J. (2009). What do animal signals mean? *Animal Behaviour, 78*(2), 233–240.

Ridley, M. (1997). *The origins of virtue.* Penguin UK.

Ritivoi, A. D. (2002). *Yesterday's self: Nostalgia and the immigrant identity.* Rowman & Littlefield.

Roes, F. L. (1989). On the origin of crying and tears. *Human Ethology Newsletter, 5,* 5–6.

Roes, F. L. (1990). Waarom mensen huilen. *Psychologie, 10,* 44–45.

Rogers, N. L., Szuba, M. P., Staab, J. P., Evans, D. L., & Dinges, D. F. (2001). Neuroimmunologic aspects of sleep and sleep loss. *Seminars in Clinical Neuropsychiatry, 6*(4), 295–307.

Rohwer, S., Fretwell, S. D., & Tuckfield, R. C. (1976). Distress screams as a measure of kinship in birds. *American Midland Naturalist, 96*(2), 418–430.

Rosenblatt, P. C., Walsh, R. P., & Jackson, D. A. (1976). *Grief and Mourning in Cross-cultural Perspective.* HRAF Press.

Rowell, J. T., Ellner, S. P., & Reeve, H. K. (2006). Why animals lie: How dishonesty and belief can coexist in a signaling system. *American Naturalist, 168*(16), E180–E204.

Rubin, D. C., Rahhal, T. A., & Poon, L. W. (1998). Things learned in early adulthood are remembered best. *Memory & Cognition, 26*(1), 3–19.

Rucker, D. D., & Petty, R. E. (2004). Emotion specificity and consumer behavior: Anger, sadness, and preference for activity. *Motivation and Emotion, 28*(1), 3–21.

Ruffman, T., Henry, J. D., Livingstone, V., & Phillips, L. H. (2008). A meta-analytic review of emotion recognition and aging: Implications for neuropsychological models of aging. *Neuroscience and Biobehavioral Reviews, 32*(4), 863–881.

Russell, J. A. (1994). Is there universal recognition of emotion from facial expression? A review of the cross-cultural studies. *Psychological Bulletin, 115*(1), 102–141.

Russon, A. E., van Schaik, C. P., Kuncoro, P., Ferisa, A., Handayani, D. P., & van Noordwijk, M. A. (2009). Innovation and intelligence in orangutans. In S. Wich, S. Utami-Atmoko, T. Mitra Setia, & C. van Schaik (Eds.), *Orangutans: Geographic variation in behavioral ecology and conservation* (pp. 279–298). Oxford University Press.

Ryan, M. (2019). *A taste for the beautiful: The evolution of attraction.* Princeton University Press.

Sachs, M. E., Damasio, A., & Habibi, A. (2015). The pleasures of sad music: A systematic review. *Frontiers in Human Neuroscience, 9,* Article 404.

Sachs, M. E., Damasio, A., & Habibi, A. (2021). Unique personality profiles predict when and why sad music is enjoyed. *Psychology of Music, 49*(5), 1145–1164.

Sattmann, S., & Parncutt, R. (2018, July). *The role of empathy in musical chills.* Paper presented at the International Conference on Music Perception and Cognition, Graz, Austria.

Satow, K. L. (1975). Social approval and helping. *Journal of Experimental Social Psychology, 11*(6), 501–509.

Saul, L. J., & Bernstein, C. (1941). The emotional settings of some attacks of urticaria. *Psychosomatic Medicine, 3,* 349–369.

Schachter, S., & Singer, J. (1962). Cognitive, social, and physiological determinants of emotional state. *Psychological Review, 69*(5), 379–399.

Schachter, S., & Singer, J. E. (1979). Comments on the Maslach and Marshall-Zimbardo experiments. *Journal of Personality and Social Psychology, 37*(6), 989–995.

Schachter, S., & Wheeler, L. (1962). Epinephrine, chlorpromazine, and amusement. *Journal of Abnormal and Social Psychology, 65*(2), 121–128.

Schamberg, I., Wittig, R.M., & Crockford, C. (2018). Call type signals caller goal: A new take on ultimate and proximate influences in vocal production. *Biological Reviews, 93*(4), 2071–2082.

Schellenberg, E. G., Peretz, I., & Vieillard, S. (2008). Liking for happy- and sad-sounding music: Effects of exposure. *Cognition & Emotion, 22*(2), 218–237.

Scherer, K. R. (1986). Vocal affect expression: A review and a model for future research. *Psychological Bulletin, 99*(2), 143–165.

Scherer, K. R., Johnstone, T., & Klasmeyer, G. (2003). Vocal expression of emotion. In R. J. Davidson, K. R. Scherer, & H. Goldsmith (Eds.), *Handbook of the affective sciences* (pp. 433–456). Oxford University Press.

Schimmack, U. (2001). Pleasure, displeasure, and mixed feelings? Are semantic opposites mutually exclusive? *Cognition and Emotion, 15*(1), 81–97.

Schlegel, K., Grandjean, D., & Scherer, K. R. (2014). Introducing the Geneva emotion recognition test: An example of Rasch-based test development. *Psychological Assessment, 26*(2), 666–672.

Schneider, E. H., Neumann, D., & Seifert, R. (2014). Modulation of behavior by the histaminergic system: Lessons from HDC-, H3R- and H4R-deficient mice. *Neuroscience and Biobehavioral Reviews, 47,* 101–121.

Schröder, M., Aubergé, V., & Cathiard, M.-A. (1998). Can we hear smile? *Proceedings of the Conference on Spoken Language Processing 3*, 559–562.

Schubert, E. (1996). Enjoyment of negative emotions in music: An associative network explanation. *Psychology of Music, 24*, 18–28.

Schubert, E. (2016). Enjoying sad music: Paradox or parallel processes? *Frontiers in Human Neuroscience, 10*, Article 312.

Schulreich, S., Heussen, Y. G., Gerhardt, H., Mohr, P. N., Binkofski, F. C., Koelsch, S., & Heekeren, H. R. (2014). Music-evoked incidental happiness modulates probability weighting during risky lottery choices. *Frontiers in Psychology, 4*, Article 981.

Schutz, M., Huron, D., Keeton, K., & Loewer, G. (2008). The happy xylophone: Acoustic affordances restrict an emotional pallete. *Empirical Musicology Review, 3*(3), 126–135.

Schwarz, N. (1990). Feelings as information: Information and motivational functions of affective states. In E. T. Higgins & R. M. Sorrentino (Eds.), *Handbook of motivation and cognition: Foundations of social behavior* (Vol. 2, pp. 527–561). Guilford Press.

Scott-Phillips, T. C. (2008). Defining biological communication. *Journal of Evolutionary Biology, 21*(2), 387–395.

Searcy, W. A., & Nowicki, S. (2005). *The evolution of animal communication: Reliability and deception in signaling systems.* Princeton University Press.

Sedikides, C., Wildschut, T., Arndt, J., & Routledge, C. (2006). Affect and the self. In J. P. Forgas (Ed.), *Affect in social thinking and behavior: Frontiers in social psychology* (pp. 197–215). Psychology Press.

Sedikides, C., Wildschut, T., Arndt, J., & Routledge, C. (2008). Nostalgia past, present, and future. *Current Directions in Psychological Science, 17*(5), 304–307.

Sedikides, C., Wildschut, T., & Baden, D. (2004). Nostalgia: Conceptual issues and existential functions. In J. Greenberg, S. Koole, & T. Pyszczynski (Eds.), *Handbook of experimental existential psychology* (pp. 200–214). Guilford Press.

Selye, H. (1974). *Stress without distress.* J. B. Lippincott.

Semmann, D., Krambeck, H. J., & Milinski, M. (2005). Reputation is valuable within and outside one's own social group. *Behavioral Ecology and Sociobiology, 57*(6), 611–616.

Seyfarth, R. M., & Cheney, D. L. (2013). Affiliation, empathy, and the origins of theory of mind. *Proceedings of the National Academy of Sciences, 110*(2), 10349–10356.

Shamay-Tsoory, S. G. (2011). The neural bases for empathy. *Neuroscientist, 17*, 18–24.

Shamay-Tsoory, S. G., & Aharon-Peretz, J. (2007). Dissociable prefrontal networks for cognitive and affective theory of mind: A lesion study. *Neuropsychologia, 45*, 3054–3067.

Shamay-Tsoory, S. G., Aharon-Peretz, J., & Perry, D. (2009). Two systems for empathy: A double dissociation between emotional and cognitive empathy in inferior frontal gyrus versus ventromedial prefrontal lesions. *Brain, 132*, 617–627.

Shan, L., Bossers, K., Unmehopa, U., Bao, A. M., & Swaab, D. F. (2012). Alterations in the histaminergic system in Alzheimer's disease: A postmortem study. *Neurobiology of Aging, 33*(11), 2585–2598.

Shand, A. F. (1920). *The foundations of character: Being a study of the tendencies of the emotions and sentiments*. Macmillan and Co.

Sharman, L. S., Dingle, G. A., Vingerhoets, A. J., & Vanman, E. J. (2020). Using crying to cope: Physiological responses to stress following tears of sadness. *Emotion, 20*(7), 1279–1291.

Sharot, T. (2011). *The optimism bias: A tour of the irrationally positive brain*. Pantheon Books.

Siegel, A., & Sapru, H. N. (2006). *Essential neuroscience*. Lippincott Williams & Wilkins.

Siegman, A. W. (1993). Cardiovascular consequences of expressing, experiencing, and repressing anger. *Journal of Behavioral Medicine, 16*(6), 539–569.

Silk, J. B., & Boyd, R. (2010). From grooming to giving blood: The origins of human altruism. In P. M. Kappeler & J. Silk (Eds.), *Mind the gap* (pp. 223–244). Springer.

Singer, T., & Klimecki, O. M. (2014). Empathy and compassion. *Current Biology, 24*(18), R875–R878.

Singer, T., Seymour, B., O'Doherty, S. K., Kaube, H., Dolan, R. J., & Frith, C. D. (2004). Empathy for pain involves the affective but not sensory components of pain. *Science, 303*(5661), 1157–1162.

Skinner, E. R. (1935). A calibrated recording and analysis of the pitch, force and quality of vocal tones expressing happiness and sadness. *Speech Monographs, 2*(1), 81–137.

Slavich, G. M., Way, B. M., Eisenberger, N. I., & Taylor, S. E. (2010). Neural sensitivity to social rejection is associated with inflammatory responses to social stress. *Proceedings of the National Academy of Sciences of the U.S.A., 107*(33), 14817–14822.

Sloman, L., Gilbert, P., & Hasey, G. (2003). Evolved mechanisms in depression: The role and interaction of attachment and social rank in depression. *Journal of Affective Disorders, 74*(2), 107–121.

Smith, A. (1759). *The theory of moral sentiments*. Andrew Millar, Alexander Kincaid & J. Bell.

Smith, D. R., & Patterson, R. D. (2005). The interaction of glottal-pulse rate and vocal-tract length in judgements of speaker size, sex, and age. *Journal of the Acoustical Society of America, 118*(5), 3177–3186.

Soltis, J. (2004). The signal functions of early infant crying. *Behavioral and Brain Sciences, 27*(4), 443–458.

Spencer, H. (1870). *Principles of psychology* (2 vols.). Williams and Norgate.

Spencer-Booth, Y., & Hinde, R. A. (1971). Effects of brief separations from mothers during infancy on behavior of rhesus monkeys 6–24 months later. *Journal of Child Psychology and Psychiatry, 12*, 157–172.

Spock, B. (1968). *Baby and child care* (3rd ed.). Bodley Head.

Stanley H, E., & Price, D. B. (1958). A linguistic evaluation of feeling states in psychotherapy. *Psychiatry, 21*(2), 115–121.

Stephens, R., Atkins, J., & Kingston, A. (2009). Swearing as a response to pain. *NeuroReport, 20*, 1056–1060.

Stepper, S., & Strack, F. (1993). Proprioceptive determinants of emotional and nonemotional feelings. *Journal of Personality and Social Psychology, 64*(2), 211–220.

Storbeck, J., & Clore, G. L. (2005). With sadness comes accuracy; with happiness, false memory: Mood and the false memory effect. *Psychological Science, 16*(10), 785–791.

Stougie, S., Vingerhoets, A. J., & Cornelius, R. R. (2004). Crying, catharsis, and health. In I. Nyklicek, L. Temoshok, & A. J. Vingerhoets (Eds.), *Emotional expression and health* (pp. 225–288). Brunner-Routledge.

Strange, J. J., & Leung, C. C. (1999). How anecdotal accounts in news and in fiction can influence judgments of a social problem's urgency, causes, and cures. *Personality and Social Psychology Bulletin, 25*(4), 436–449.

Strunk, D. R., Lopez, H., & DeRubeis, R. J. (2006). Depressive symptoms are associated with unrealistic negative predictions of future life events. *Behaviour Research and Therapy, 44*(6), 861–882.

Sullivan, E. (2016). *Beyond melancholy: Sadness and selfhood in renaissance England.* Oxford University Press.

Sumitra. (2011, December 16). Tribe practices finger cutting as a means of grieving. *Oddity Central.* Retrieved August 20, 2021, from http://www.odditycentral.com/pics/tribe-practices-finger-cutting-as-a-means-of-grieving.html

Sundberg, J. (1987). *The science of the singing voice.* Northern Illinois University Press.

Švec, J. G., & Pešák, J. (1994). Vocal breaks from the modal to the falsetto register. *Folia Phoniatrica et Logopaedica, 46*(2), 97–103.

Takahashi, K., Suwa, H., Ishikawa, T., & Kotani, H. (2002). Targeted disruption of H3 receptors results in changes in brain histamine tone leading to an obese phenotype. *Journal of Clinical Investigation, 110*(12), 1791–1799.

Talon-Hugon, C. (2014). The resolution and dissolution of the paradox of negative emotions in the aesthetics of the eighteenth century. In J. Levinson (Ed.), *Suffering art gladly: The paradox of negative emotion in art* (pp. 28–44). Springer.

Tan, H. B., & Forgas, J. P. (2010). When happiness makes us selfish, but sadness makes us fair: Affective influences on interpersonal strategies in the dictator game. *Journal of Experimental Social Psychology, 46*(3), 571–576.

Tartter, V. C. (1979). Grinning from ear-to-ear: The perception of smiled speech. *Journal of the Acoustical Society of America, 65*(Suppl. 1), S112–S113.

Tartter, V. C. (1980). Happy talk: Perceptual and acoustic effects of smiling on speech. *Perception & Psychophysics, 27*(1), 24–27.

Tartter, V. C., & Braun, D. (1994). Hearing smiles and frowns in normal and whisper registers. *Journal of the Acoustical Society of America, 96*(4), 2101–2107.

Taruffi, L., & Koelsch, S. (2014). The paradox of music-evoked sadness: An online survey. *PLoS One, 9*, Article e110490.

Tashiro, M., Mochizuki, H., Iwabuchi, K., Sakurada, Y., Itoh, M., Watanabe, T., & Yanai, K. (2002). Roles of histamine in regulation of arousal and cognition: Functional neuroimaging of histamine H1 receptors in human brain. *Life Sciences, 72*(4–5), 409–414.

Tinbergen, N. (1952). "Derived" activities; their causation, biological significance, origin, and emancipation during evolution. *Quarterly Review of Biology, 27*(1), 1–32.

Tinbergen, N. (1963). On the aims and methods of ethology. *Zietschrift für Tierpsychologie, 20*(4), 410–433.

Tinbergen, N. (1964). The evolution of signaling devices. In W. Etkin (Ed.), *Social behavior and organization among vertebrates* (pp. 206–230). University of Chicago Press.

van den Tol, A. J. (2016). The appeal of sad music: A brief overview of current directions in research on motivations for listening to sad music. *The Arts in Psychotherapy, 49*, 44–49.

Tomkins, S. S. (1980). Affect as amplification: Some modifications in theory. In R. Plutchik & H. Kellerman (Eds.), *Theories of emotion* (pp. 141–164). Academic Press.

Toprak, S., Cetin, I., Guven, T., Can, G., & Demircan, C. (2011). Self-harm, suicidal ideation and suicide attempts among college students. *Psychiatry Research, 187*(1), 140–144.

Toyota, H., Dugovic, C., Koehl, M., Laposky, A. D., Weber, C., Ngo, K., Wu, Y., Lee, D. H., Yanai, K., Sakurai, E., Watanabe, T., Liu, C., Chen, J., Barbier, A. J., Turek, F. W., Fung-Leung, W. P., & Lovenberg, T. W. (2002). Behavioral characterization of mice lacking histamine H3 receptors. *Molecular Pharmacology, 62*(2), 389–397.

Trapnell, P. D., & Campbell, J. D. (1999). Private self-consciousness and the five-factor model of personality: Distinguishing rumination from reflection. *Journal of Personality and Social Psychology, 76*(2), 284–304.

Trimble, M. (2012). *Why humans like to cry: Tragedy, evolution, and the brain.* Oxford University Press.

Trivers, R. L. (1971). The evolution of reciprocal altruism. *Quarterly Review of Biology, 46*(1), 35–57.

Trivers, R. L. (1974). Parent-offspring conflict. *American Zoologist, 14*(1), 249–264.

Turner, B., & Huron, D. (2008). A comparison of dynamics in major- and minor-key works. *Empirical Musicology Review, 3*(2), 64–68.

Tversky, A., & Kahneman, D. (1986). The framing of decisions and the evaluation of prospects. *Studies in Logic and the Foundations of Mathematics, 114*, 503–520.

Urban, G. (1988). Ritual wailing in Amerindian Brazil. *American Anthropologist, 90*(2), 385–400.

Vandervoort, D. J., Ragland, D. R., & Syme, S. L. (1996). Expressed and suppressed anger and health problems among transit workers. *Current Psychology, 15*(2), 179–193.

Venkatraman, V., Chuah, L. Y. M., Huettel, S. A., & Chee, M. W. L. (2007). Sleep deprivation elevates expectation of gains and attenuates response to losses following risky decisions. *Sleep, 30*(5), 603–609.

Venkatraman, V., Huettel, S. A., Chuah, L. Y. M., Payne, J. W., & Chee, M. W. L. (2011). Sleep deprivation biases the neural mechanisms underlying economic preferences. *Journal of Neuroscience, 31*(10), 3712–3718.

Ventevogel, P., Jordans, M., Reis, R., & de Jong, J. (2013). Madness or sadness? Local concepts of mental illness in four conflict-affected African communities. *Conflict and Health, 7*(1), Article 3.

Vermeir, I., De Bock, T., & Van Kenhove, P. (2017). The effectiveness of fear appeals featuring fines versus social disapproval in preventing shoplifting among adolescents. *Psychology & Marketing, 34*(3), 264–274.

Viggiano, D., Ruocco, L. A., Arcieri, S., & Sadile, A. G. (2004). Involvement of norepinephrine in the control of activity and attentive processes in animal models of attention deficit hyperactivity disorder. *Neural Plasticity, 11*(1–2), 133–149.

Vingerhoets, A. J. J. M. (2013). *Why only humans weep: Unravelling the mysteries of tears.* Oxford University Press.

Vingerhoets, A. J. J. M., & Cornelius, R. R. (Eds.). (2001). *Adult crying: A biopsychosocial approach.* Brunner-Routledge.

Vingerhoets, A. J. J. M., Cornelius, R. R., Van Heck, G. L., & Becht, M. C. (2000). Adult crying: A model and review of the literature. *Review of General Psychology, 4*(4), 354–377.

Vuorenkoski, V., Wasz-Höckert, O., Koivisto, E., & Lind, J. (1969). The effect of cry stimulus on the temperature of the lactating breast of primipara. A thermographic study. *Experientia, 25*(12), 1286–1287.

Vuoskoski, J. K., & Eerola, T. (2017). The pleasure evoked by sad music is mediated by feelings of being moved. *Frontiers in Psychology, 8*, Article 439.

Vuoskoski, J. K., Thompson, W. F., McIlwain, D., & Eerola, T. (2012). Who enjoys listening to sad music and why? *Music Perception, 29*, 311–317.

de Waal, F. B. (1996). *Good natured: The origins of right and wrong in humans and other animals*. Harvard University Press.

Walsh, N. E, Schoenfeld, L., Ramamurthy, S., & Hoffman, J. (1989). Normative model for cold pressor test. *American Journal of Physical Medicine and Rehabilitation, 68*(1), 6–11.

Ward, C., Bauer, E. B., Smuts, B. B. (2008). Partner preferences and asymmetries in social play among domestic dog, *Canis lupus familiaris*, littermates. *Animal Behavior, 76*, 1187–1199.

Warrenburg, L. A. (2020). Choosing the right tune: A review of music stimuli used in emotion research. *Music Perception, 37*(3), 240–258.

Watt, R. J., & Ash, R. L. (1998). A psychological investigation of meaning in music. *Musicae Scientiae, 2*(1), 33–53.

Weary, G., & Edwards, J. A. (1994). Social cognition and clinical psychology: Anxiety, depression, and the processing of social information. In R. S. Wyer Jr. & T. K. Srull (Eds.), *Handbook of social cognition* (pp. 289–338). Erlbaum.

Weinberg, E. D. (1984). Iron withholding: A defense against infection and neoplasia. *Physiological Reviews, 64*(1), 65–102.

Weinstein, N. D. (1980). Unrealistic optimism about future life events. *Journal of Personality and Social Psychology, 39*(5), 806–820.

Weiss, F., Lorang, M. T., Bloom, F. E., & Koob, G. F. (1993). Oral alcohol self-administration stimulates dopamine release in the rat nucleus accumbens: Genetic and motivational determinants. *Journal of Pharmacology and Experimental Therapeutics, 267*(1), 250–258.

Wiesenfeld, A. R., Malatesta, C. Z., & Deloach, L. L. (1981). Differential parental response to familiar and unfamiliar infant distress signals. *Infant Behavior and Development, 4*, 281–295.

Wildschut, T., Sedikides, C., Arndt, J., & Routledge, C. (2006). Nostalgia: Content, triggers, functions. *Journal of Personality and Social Psychology, 91*(5), 975–993.

Wiley, R. H. (1983). The evolution of communication: Information and manipulation. In T. R. Halliday & P. J. Slater (Eds.), *Communication* (pp. 82–113). Blackwell.

Williams, H. M. (1792). *Letters from France: Containing a great variety of original information concerning the most important events that have occurred in that country* (2 vols.). J. Chambers.

Williams, L. M., Mathersul, D., Palmer, D. M., Gur, R. C., Gur, R. E., & Gordon, E. (2009). Explicit identification and implicit recognition of facial emotions: I. Age effects in males and females across 10 decades. *Journal of Clinical and Experimental Neuropsychology, 31*(3), 257–277.

Wilson, E. (2008). *Against happiness: In praise of melancholy.* Farrar, Straus and Giroux.

Wilson, E. O. (2019). *Genesis: The deep origin of societies.* W. W. Norton.

Wilson, T. D. (2004). *Strangers to ourselves.* Harvard University Press.

Wolff, P. H. (1987). *The development of behavioral states and the expression of emotions in early infancy: New proposals for investigation.* University of Chicago Press.

Wright, T. (1601). *The passions of the minde in generall.* V. Sims.

Wright, W. F., & Bower, G. H. (1992). Mood effects on subjective probability assessment. *Organizational Behavior and Human Decision Processes, 52*(2), 276–291.

Wyver, S. R., & Spence, S. H. (1999). Play and divergent problem solving: Evidence supporting a reciprocal relationship. *Early Education and Development, 10*(4), 419–444.

Yirmiya, R. (1996). Endotoxin produces a depressive-like episode in rats. *Brain Research, 711*(1–2), 163–174.

Yost, J. H., & Weary, G. (1996). Depression and the correspondent inference bias: Evidence for more effortful cognitive processing. *Personality and Social Psychology Bulletin, 22*(2), 192–200.

Zahavi, A. (1975). Mate selection—a selection for a handicap. *Journal of Theoretical Biology, 53*(1), 205–214.

Zahavi, A., & Zahavi, A. (1997). *The handicap principle: A missing piece of Darwin's puzzle.* Oxford University Press.

Zahn-Waxler, C., Radke-Yarrow, M., Wagner, E., & Chapman, M. (1992). Development of concern for others. *Developmental Psychology, 28*(1), 126–136.

Zeifman, D. M. (2001). An ethological analysis of human infant crying: Answering Tinbergen's four questions. *Developmental Psychobiology, 39*(4), 265–285.

Zeifman, D. M., & Brown, S. A. (2011). Age-related changes in the signal value of tears. *Evolutionary Psychology*, *9*(3), 313–324.

Zentner, M., Grandjean, D., & Scherer, K. R. (2008). Emotions evoked by the sound of music: characterization, classification, and measurement. *Emotion*, *8*(4), 494–521.

Zerjal, T., Xue, Y., Bertorelle, G., Wells, R. S., Bao, W., Zhu, S., Qamar, R., Ayub, Q., Mohyuddin, A., Fu, S., Li, P., Yuldasheva, N., Ruzibakiev, R., Xu, J., Shu, Q., Du, R., Yang, H., Hurles, M., Robinson, E., Gerelsaikhan, T., Dashnyam, B., Mehdi, S. Q., & Tyler-Smith, C. (2003). The genetic legacy of the Mongols. *American Journal of Human Genetics*, *72*(3), 717–721.

Zeskind, P. S. (2013). Infant crying and the synchrony of arousal. In E. Altenmül-ler, S. Schmidt & E. Zimmermann (Eds.), *Evolution of emotional communication: From sounds in nonhuman mammals to speech and music in man* (pp. 155–172). Oxford University Press.

Zhou, L., Yang, Y., & Li, S. (2021). Music-induced emotions influence intertemporal decision making. *Cognition and Emotion*, *36*(2), 211–229.

Zhou, X., Wildschut, T., Sedikides, C., Chen, X., & Vingerhoets, A. (2012). Heartwarming memories: Nostalgia maintains physiological comfort. *Emotion*, *12*(4), 678–684.

Zickfeld, J. H., van de Ven, N., Pich, O., Schubert, T. W., Berkessel, J. B., Pizarro, J. J., Bhushan, B., Mateo, N. J., Barbosa, S., Sharman, L., Kökönyei, G., Schrover, E., Kardum, I., Aruta, J. J. B., Lazarevic, L. B., Escobar, M. J., Stadel, M., Arriaga, P., Dodaj, A., Shankland, R., . . . Vingerhoets, A. (2021). Tears evoke the intention to offer social support: A systematic investigation of the interpersonal effects of emotional crying across 41 countries. *Journal of Experimental Social Psychology*, *95*, 104137.

Zimmerman, F. (2011, October 18). Sioux mourning ritual. *American Indian History Blog Spot*. 2011. Retrieved August 20, 2021, from https://americanindianshistory.blogspot.com/2011/10/sioux-mourning-ritual.html

Index